Simple
Statistical Tests
for Geography

Simple Statistical Tests for Geography

Danny McCarroll

Swansea University, UK

CRC Press
Taylor & Francis Group
Boca Raton London New York

CRC Press is an imprint of the
Taylor & Francis Group, an **informa** business

A CHAPMAN & HALL BOOK

CRC Press
Taylor & Francis Group
6000 Broken Sound Parkway NW, Suite 300
Boca Raton, FL 33487-2742

Printed on acid-free paper
Version Date: 20160826

International Standard Book Number-13: 978-1-4987-5881-9 (Paperback)

Library of Congress Cataloging-in-Publication Data

Names: McCarroll, Danny, author.
Title: Simple statistical tests for geography / Danny McCarroll.
Description: Boca Raton, Flordia : CRC Press, 2016. | Includes
bibliographical references.
Identifiers: LCCN 2016022635| ISBN 9781498758819 (pbk : alk. paper) | ISBN
9781498758932 (e-book) | ISBN 9781498758949 (e-book) | ISBN 9781498758895
(e-book)
Subjects: LCSH: Geography--Statistical methods.
Classification: LCC G70.3 .M39 2016 | DDC 519.5--dc23
LC record available at https://lccn.loc.gov/2016022635

Visit the Taylor & Francis Web site at
http://www.taylorandfrancis.com

and the CRC Press Web site at
http://www.crcpress.com

Printed and bound in the United States of America by
Edwards Brothers Malloy on sustainably sourced paper

To my friend and teacher

Professor John A. Matthews

The Laird of Leirdalen

Contents

Preface

I never expected to write a book on statistics. I hated mathematics at school, was terrified of statistics as a student, and still quail at the sight of equations with Greek letters. However, as a postgraduate student I realised that I needed statistical methods and also that I did not have to be good at mathematics to be able to use them. As a researcher and teacher I have learned that statistical methods are just a useful tool, like a mass spectrometer or a computer, and you do not have to be able to take them apart and put them back together again in order to use them effectively. You do not need to fully understand the mathematics behind a particular statistical test, but you do need to understand the logic that it is based on, so that you can decide when it is appropriate, how to do it and how to interpret the results.

This book is not aimed at teachers or lecturers, it is aimed directly at students. One of the greatest pleasures of my academic career has been helping individual students to analyse the data that they have collected themselves for their dissertation. They often come to me brimming with enthusiasm but lost because they cannot work out what statistical methods they should use. When they find out which test is appropriate they are not particularly interested in learning the mathematics behind it, they just want to be able to do it, preferably quickly, get the correct result and make sense of the results. Although there are many excellent text books on statistical methods, for use in geography and more generally, I have struggled to find an up-to-date book that gives them what they need. This book is my attempt to fill the gap.

Many students of geography will, like me, have endured a course on statistical methods. Students are generally taught either by someone who is very good at mathematics, in which case they struggle to pitch the material at the right level for many students, or they are taught by the most recently appointed lecturer, who is not in a position to refuse. This book is not really designed as a course text, but if you are currently being tortured, in time-honoured fashion, I hope it will help. For those methods where it is likely that you may be forced to make calculations by hand I have presented the relevant equations in the simplest way I can think of and then gone through the calculations one step at a time.

There is really no need any more for making calculations by hand, all of the methods in this book can be conducted very quickly and easily using a range of tools including free online calculators, free add-ins to spreadsheet programmes or using specialist software. The real challenge now is to recognise the kind of data that you have and find the most appropriate statistical tests. My approach is to focus on the logic behind the tests, so that you can see if they are appropriate for your data, and then to briefly describe the options for performing them so that you have a choice. I do not see the point of forcing students to use a single package to perform all statistical tests. I accept that being able to use a package like SPSS, and particularly developing some skills in using R are a great benefit to students and I would encourage students to develop those skills. However, the reality is that many students find the complexity of these packages a barrier to learning. They are not interested in statistics, they are interested in geography, and just want to apply the right methods quickly. I see no reason for not helping them to do that.

Companion Site Calculators

I have noticed in recent years that many students who claim to have little mathematical ability, and who lack the confidence to use statistical packages, are nevertheless very good at manipulating numbers in a spreadsheet and producing nice graphs. Where it is feasible, therefore, I try to explain the easiest way to conduct most of the tests in this book using a spreadsheet. To help, I have produced a series of very simple 'calculators' that cover almost all of the tests covered in this book. They are just spreadsheets that allow you to enter your data and get the results that you need very quickly without the bother of using any fancy software. They are not intended to act as a substitute for a real statistical package, they are just provided for convenience. The companion site also has some tables of critical values that would not fit into this book. The free online calculators that are mentioned in this book are also listed on the companion website with hyperlinks, so that I can keep the lists up to date and you can access the sites at the click of a mouse. No doubt I will think of other things to add to the site in due course.

Companion site: geography.swan.ac.uk/danny-mccarroll-sstg

Acknowledgements

Acknowledge who? I wrote the whole thing myself!

But of course I did have some help along the way. I have a full time job, as a teacher and researcher, so this book had to fit around that, and was written mostly in my 'spare' time. First I want to thank my two lovely daughters, Bethan and Rhiannon. The first drafts of the early chapters were written on a fantastic climbing trip to the wonderful island of Kalymnos with Bethan. We climbed every day, but in the mornings and evenings, as she dozed or revised, I wrote about statistics while gazing across the Aegean. If this book makes any money I intend to spend it all on similar climbing trips. Several chapters were also written on a wonderful trip to the island of Gotland in Sweden with Rhiannon. They have both stoically endured my inane drivel about statistical methods, checked my text for political correctness, and they even did a bit of proof reading. They are both students now, and use statistical methods themselves, so this book may even come in handy.

Several close colleagues have helped and supported me over the years, and in particular I would like to thank Colin Ballantyne, Stefan Doerr, Mary Gagen, Neil Loader, Sietse Los, Adrian Luckman, John Matthews, Geraint Owen, Iain Robertson, Rick Shakesby, Giles Young and Rory Walsh. I would also like to warmly thank my many friends from the EU-funded 'Millennium' project. The aim of that project was to bring together researchers from many different disciplines and from all over Europe and beyond to try to reconstruct the climate of Europe over the last one thousand years. One of the guiding principles was that we should share our expertise, and one of the most challenging aspects of the work was finding appropriate methods for dealing with very disparate types of data, ranging from stable isotope measurements to interpretation of historical documents. It was in that context that I learned many of the methods and skills that were needed to write this book. There are too many to thank individually but I am particularly indebted to Sheila Hicks, Risto Jalkanen, Eduardo Zorita, Anders Moberg, Fidel Gonzalez-Rouco, Rike Wagner-Cremer, Atle Nesje, James Scourse, Bill Austin, Dennis Wheeler, Andrea Kiss, Tom Levanic, Juerg Luterbacher, Rob Wilson, Rudolf Brazdil and Keith Briffa.

One of the dangers of allowing a naturally reclusive academic to write a book is that they inevitably slip even closer to complete insanity. The beautiful harmonies of Mandolin Orange helped to keep me sane, eased the drudgery of formatting and soothed the savage beast of first year statistics lectures. Adrian and Saira Luckman are thanked for forcing me to leave my desk occasionally to eat nice food and drink lots of red wine. Mary Gagen, Neil Loader and Giles Young are thanked for making me discuss other interesting things, even though most of those did involve a bit of data analysis. Special thanks go to John Bullock, who still hasn't completely given up on me, and continues to drag me out climbing at every opportunity. His inextinguishable enthusiasm is a great inspiration.

Writing a book and getting it published is a rather daunting prospect, but it was greatly eased by a chance contact with Stan Wakefield, who gave excellent advice, guided me through the early stages of forming a proposal and then immediately found me an enthusiastic publisher. My editors, Rob Calver and Alex Edwards found me some very helpful reviewers, who helped me to strengthen the second draft, and they have generally made the whole process very easy. It is a strange reflection on this weird electronic world that we inhabit that I have never actually met any of these people!

My final and greatest thanks go to my wife Louise. She has taught me many things, not least that you cannot always rely on statistical probabilities. She was with me in a small tent in the mountains of Norway, when I first learned to use statistics during the fieldwork for my doctoral thesis, and she has supported me in everything I have tried to do since. Against all the odds she is still here now. Actually, as I write, that is not true, she is currently in Australia playing hockey for Wales, but she will be back soon and I can't wait.

Author

Danny McCarroll earned a geography degree at the University of Sheffield in 1983 and a PhD at Swansea University in 1986. He worked for a few years in the Universities of Cardiff and Southampton before returning to Swansea, which he loves. His main research interests are in reconstructing the climate of the past and he currently specialises in using stable isotopes in tree rings. He lives in an old house with a big garden, beside a sub-glacial meltwater channel, on the edge of the beautiful Gower peninsula with his wife Louise, two daughters Bethan and Rhiannon (occasionally), three chickens and a dog. His office is in a tree-house. His great passion is rock climbing, which he doesn't do often enough. He likes poetry, folk music and plays the tin whistle.

1

Introduction

1.1 Is This the Book for You?

If you are interested in statistics and good at mathematics, this is probably not the right book for you. This book is aimed at geography students who are not particularly interested in statistics, perhaps lack a bit of confidence in their mathematical abilities, but want to be able to analyse data properly in order to make sense of projects, including dissertations, and get better marks.

This is very firmly where I stood as a student. I had good reason to lack confidence in my mathematical abilities and I was not just disinterested in statistics, I was terrified. I was, however, very interested in geography and soon realised that I needed to be able to understand and use statistical methods in order to make sense of much of the literature and to analyse my own data. I struggled with the 'methods' courses as an undergraduate, partly because they were taught largely by people who were very good at mathematics and statistics and they could not bring the material down to my level. In the end, I gave up on the lectures, bought a book and worked my way through it at my own very slow pace. Eventually, the 'penny dropped' and I understood enough. My aim was never to be good at statistics, I just wanted to be able to choose the simplest test that was suitable for the problem at hand, apply it properly, and be able to explain the results. My aim in this book is to get you to that same level.

If you know a little bit about statistical methods, you may spot that this book is rather unusual. Most books on statistical methods start with long explanations of abstract concepts like the 'Gaussian' or 'normal distribution' and there are pages and pages of warnings about assumptions before you ever get to any actual tests. That is because most books focus on tests that are called 'parametric' and which require a lot of assumptions. The alternative 'nonparametric' tests are usually reserved for those cases where the assumptions are not met. Here, I take a different approach and focus mainly on the nonparametric tests. This is not because they are simpler to understand or easier to apply, it is because I think they are the most useful and appropriate tests for most of the projects that geography students will tackle. They are suitable for quite small data sets, can be used when you are collecting data that is not in the form of individual measurements, for example, where you just have counts in categories, and they can cope with data that includes the odd extreme value. Because there are very few assumptions, once you have grasped a few basic concepts, like forming hypotheses and understanding probabilities, you can usually apply them without having to use complicated equations or fancy software. Students sometimes worry that the nonparametric method may not be 'as good' as the parametric equivalent but actually, for the kinds of data they are likely to be using, this is rarely the case and if the data seem to fit the assumptions, then a simple solution is to apply both the parametric and nonparametric tests and check that they give similar results. Some common parametric tests, such as 'Student's t-test', are covered as well.

The book includes sections on correlation, using parametric and nonparametric methods, and on regression. I have included them because they are very useful methods and although they used to be regarded as complicated, they are actually very easy to apply using a spreadsheet. When you use these methods, it is important to be aware of the assumptions and to make some simple checks. It is not really necessary to use complicated techniques to do this, just plotting appropriate graphs and applying a bit of common sense will suffice.

1.2 How to Use This Book

If you are confident that you have formulated clear and testable hypotheses, can understand the difference between a one and a two-tail test, and can interpret probabilities, then you can jump straight to 'choosing the right test' (Section 3.4) and get on with it. Otherwise, it is worth reading the introductory sections to make sure that you really understand the basics. There is no point in doing the test properly but then losing marks by demonstrating that you do not really understand what it means. The best way to use this book is to read it before you collect any data, so that you can design your sampling scheme to fit your desired test. In reality, of course, many readers will already have collected the data and are looking for some guidance on how to analyse it. I have tried to lay things out so that you can find the right test for the data you have, even if it has not been collected with any particular test in mind.

Students are often taught to perform all statistical calculations using some specialist software. I really do not see the point in being so prescriptive and have noticed that in many cases, the software is so complicated to use that it acts as a barrier rather than an aid to learning. For me, statistical methods are just tools, it is the research that matters, so I encourage students to use the most convenient method to apply an appropriate test and get the correct result. The most convenient method varies by test, and may include using a calculator, a spreadsheet, or most simply of all a free online calculator. A few tests are easier to perform using specialist software, and I have chosen to use SPSS, because it is commonly used by geographers and will be available to many students and *R* Commander because it is powerful and free. Where I use a spreadsheet I will give instructions for Excel, which is widely used, but most of the instructions will apply also to the free spreadsheet available via Libre Office. Details of how to access the software are given in the section on 'tools of the trade' (Chapter 4). For each of the tests, I have provided some worked examples, so that you can see how the test works, but the focus is on explaining the logic of the test, rather than the mathematics. I realise that students are sometimes required to make calculations, and show the workings, so for most tests, I have added worked examples using equations. These are generally placed into boxes kept separate from the main text and I have tried to show every step in the analysis, so that students who lack confidence can take the same small steps with their own data. I have also produced a set of very simple spreadsheet-based calculators that will perform most of the tests in this book. They use the normal Excel functions and all of the workings are left open to view and editable, with no hidden code, so that you can see how they work.

1.3 Why Bother with Statistics?

The great value of statistical techniques is that they allow you to go beyond simple description of the data that you have collected and say something about the wider world. That step, from simply describing what you can see to making more general statements is called inference. Most of the tests that we use in geography are called inferential statistics. They allow you to make clear statements about your data, and they effectively tell you whether it is safe to go beyond simply describing your data and make inferences about the wider implications of the results. This is perhaps most clearly illustrated by a couple of examples.

Imagine that you have been told to conduct a questionnaire survey for which the answers are either yes or no. The question does not matter for the moment, it might be something like 'Does being stopped in the street and asked questions by geography students annoy you?' We are interested in whether there is some gender difference in the replies and so the results can be plotted as some kind of graph. You could choose a bar graph that shows how many people replied or you could use a pie chart that just shows the proportions or percentages replying yes and no (Figure 1.1).

Without access to statistical tests, all we can do is describe these graphs. For example, one student might find that eight out of the 10 male respondents objected, whereas only three of the 10 female respondents objected. However, with such small samples, it is difficult to say whether the gender difference really reflects the wider population that we are interested in. A reasonable interpretation would be that there is no real difference between

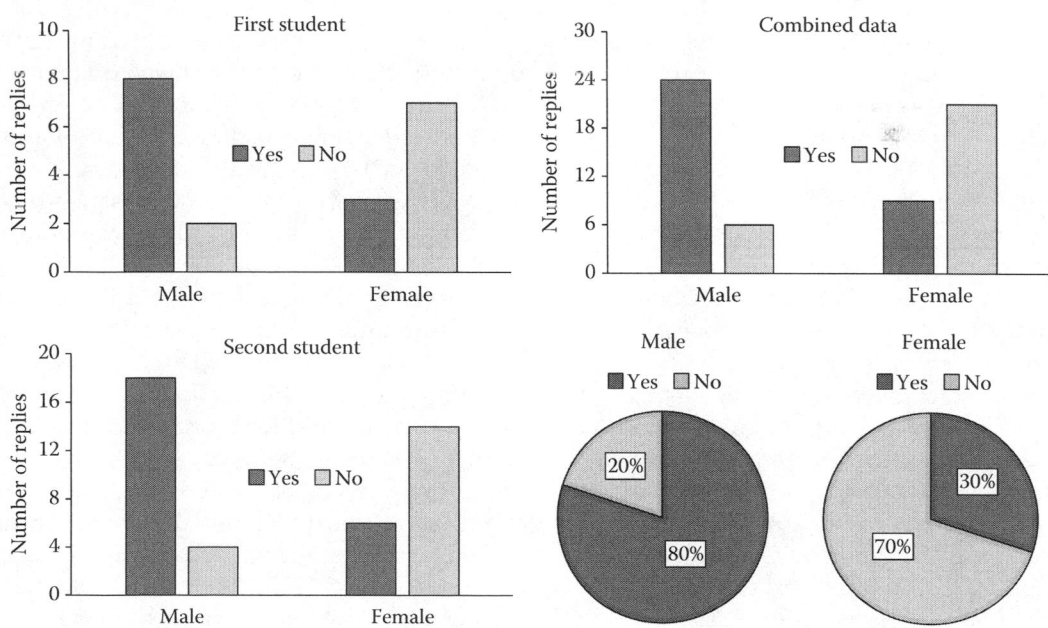

FIGURE 1.1
Replies to a simple yes or no question arranged according to gender. Changing the sample size does not change the appearance of the graphs but it does change the probability that a gender difference of this magnitude could occur just by chance if there was no real difference in the population. Fisher's exact test can be used on this kind of data and is very easy to perform using free online calculators.

males and females and the difference that we see in the small sample is just due to the luck of the draw. Asking a different set of 10 males and 10 females might well result in a very different result.

A second and more industrious student may have conducted the same survey but included more people, finding that 16 out of 20 males objected whereas only 6 from 20 female respondents objected. When this larger data set is plotted on graphs, it does not look much different from those of the first student. The second student might justifiably come to the same conclusion, that there is a difference between males and females in the sample but it is difficult to say whether it is safe to make any inference beyond the sample. Both students are effectively trapped because all they can do is describe the sample that they collected; they cannot say anything about how likely it is that the results reflect similar differences in the population. The second student might justifiably wonder if it was really worth the effort of doubling the size of the sample.

By using a simple statistical test, in this case called 'Fisher's exact test', we are able to come to much clearer conclusions. It is a tricky test to calculate by hand, but luckily there are some free online calculators and all you have to do is enter four numbers to get the result. The only assumption is that the samples are unbiased. The first student, with only 10 respondents from each gender, could conclude that there is a 7% chance, even if there was no real difference between the genders, that such a small sample could yield a difference of the size that was observed 'just by luck'. We generally (just by convention, there is no strict rule) assume that if the chances of something occurring 'just by luck' are more than 5% then it is too risky to accept that the observed difference is real. There would be little point in rambling on about the gender difference in this case, because the sample is just too small to be confident that the difference is real.

The second student, with the larger sample, could apply the same test and would find that the chances of obtaining this split in response just by luck, if there was no gender difference in the wider population, is about 0.4%, which is four in a thousand or one in 250. A 250 to one chance is pretty long odds, so this student can conclude that the sample is large enough to be confident that there is a gender difference and rambling on about the reasons for it is fully justified. If the two students were to combine their data, giving samples of 30 for each gender, they would find that the probability of a difference of this magnitude occurring just by luck is about two in 10,000, or 5,000 to one.

As an alternative example, consider a physical geographer who has trekked into the mountains to look at moraine ridges formed by small glaciers in response to climate changes in the past. She might want to use the degree of rock weathering to see whether two moraines are very similar in age or whether they represent glacial advances that are well separated in time. The Schmidt hammer is a simple hand-held tool used by engineers to measure the hardness of concrete, but since weathering tends to result in a reduction in rock surface hardness, it is also used by geomorphologists (Goudie 2006; Darvill et al. 2015). A reasonable sampling strategy might be to record one 'rebound value' from each of 30 boulders on each moraine ridge crest and compare the results using histograms (Figure 1.2). Without recourse to statistical methods, it is quite difficult to decide whether the two sets of results are different enough to conclude that there is probably a difference in degree of rock weathering and therefore age of moraine.

In fact, there is a very nice little statistical test that can be used to compare two sets of data like this, even when they are clearly a bit skewed to one side and have some extreme values. It goes by a variety of names, but in this book I have called it the 'Mann–Whitney *U*-test'. It is not difficult to calculate using a spreadsheet. It involves putting all 60 values in rank order, ignoring which sample they come from, and then separating the two samples

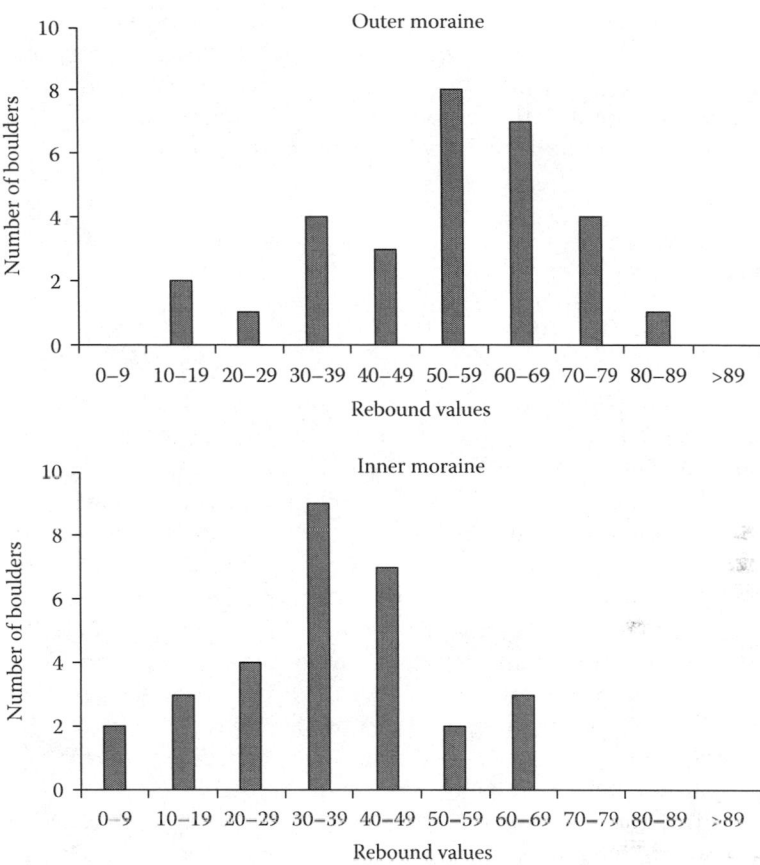

FIGURE 1.2
Schmidt hammer rebound values obtained from two moraine ridges. The Mann–Whitney *U*-test reveals that there is less than a one in a thousand chance that such a large difference would be found between two unbiased samples if the moraines were about the same age and there was no real difference in boulder surface hardness.

and comparing the sums of the ranks. If that sounds like too much work, you can just cut and paste the two sets of data into another online calculator and click on 'calculate'. In this case, the student can conclude that there is less than a one in a thousand chance that the two samples of Schmidt hammer rebound values would be as different as they are if there was no difference in rock hardness, and therefore degree of rock weathering. Of course, the student needs to be careful to ensure that it is sensible to compare the two moraines in this way, making sure, for example, that the geology is the same, but in that case, it would be reasonable to conclude that there is a substantial difference in the age of the two moraines. If she has Schmidt hammer values from moraines that have been independently dated, perhaps using a mixture of old photographs and dendrochronology of trees uprooted by the advancing glacier and incorporated into moraines, she might even be able to use other statistical methods (correlation and regression) to estimate the age of the undated moraines (Shakesby et al. 2006, 2011).

As a geography student, you should not be afraid of 'statistics', even if you are not good at mathematics. This is the twenty-first century and we are surrounded by computers and access to the internet. To apply most statistical techniques, you do not need any mathematical skills at all, you just need to be able to type the numbers into a spreadsheet and then cut and paste them into either an online calculator or a software package. Applying statistical methods has become really quick and easy. However, being able to perform a particular test is not much use unless you know how it works, when to use it and can make sense of the results. I hope that this book will give you those skills and the confidence to use a range of statistical methods and get better marks as a result.

1.4 A Note for Lecturers and Teachers

Teaching statistical techniques and other numerical methods to geography students is not an easy task. Of course, some geography students have excellent numerical skills, but many of those who are drawn to study the subject lack confidence in their mathematical abilities. Do not be fooled into thinking that this means they are stupid; confidence in mathematical ability is often much more to do with the quality of teaching at school than with the raw ability of the student. I have found that many geography students, including some extremely intelligent ones, are simply terrified of anything that involves numbers. The simple option, of course, is to remove numerical techniques from the curriculum, but personally, I think that is an abrogation of responsibility on our part. Our students need to leave university with a skill set that prepares them for the world of employment. It is one of the great benefits of a geography degree that students obtain many transferable skills and although I would not argue that the ability to use statistical tests is the most important, I do think that improving their numerical skills, and especially their confidence in their numerical skills, is one of the best things we can do for them.

Traditionally, numerical methods courses in geography have focussed on parametric statistics and students have been taught using specialist software. This is unfortunate, because to use parametric statistics properly, it is important to understand the concept of probability distributions, and that forms an immediate barrier to students who lack confidence with mathematics. At the same time, they are faced with specialist software that is so complicated that, rather than facilitating the easy application of statistical methods, actually becomes another barrier to learning.

In my own teaching, and in this book, I promote a different approach to teaching statistical methods to geography students, and that is to focus mainly on nonparametric, or distribution-free methods. I have taken this approach mainly because I find that for the scale of study, and type of data that students tend to work with these are the most appropriate techniques. In terms of teaching an introductory course on statistical methods, which I do as soon as the students arrive, there is the enormous advantage that you can get almost straight to using statistical tests on geographical problems without too much emphasis on statistical theory or assumptions, which is where the less confident students tend to get lost. At this level, I completely avoid any specialist software, such as SPSS or R, and encourage the students to focus on the geographical problem and treat the statistical tests as tools, to be applied in the most convenient way possible. I start with tests that can be conducted without any mathematics at all, simply by entering the numbers into a free online calculator.

I begin the course with probability, using a coin tossing example, and then go straight to using the sign test, which is based on the same probabilities (binomial distribution), and involves just entering two numbers into an online calculator or spreadsheet. That leads on to Fisher's exact test, again using an online calculator, and then to the Wilcoxon matched pairs signed-ranks test, followed by the Mann–Whitney *U*-test (instead of the *t*-test) and then the Kruskal–Wallis *H*-test (instead of analysis of variance [ANOVA]). I end the course with correlation and regression, conducted in a spreadsheet, but by this stage, the students are familiar with the concepts of hypothesis formulation, one and two-tail tests and with probabilities, so the added complexity of checking for a normal distribution does not distress them. This approach has worked extremely well in Swansea, and our students have become much more confident in using a range of numerical methods and we have seen significant improvements in the data analysis components of project work throughout the degree and especially in the dissertation.

I have not written this book as a course text, though I hope that it will sometimes be used in that way. It is aimed specifically at students, mainly to help them with data analysis for their dissertations. There are several tests included here that I have rarely, if ever seen, used in published work by academic geographers, but they fit the types of problem that our students have come to me for help with and I hope they will help your students too. A lot of our students collect data as 'counts in categories' and in many statistical text books, the only option presented for such data is the chi square test. Here, I present a range of other options, including Fisher's exact test, McNemar's test and the Kolmogorov–Smirnov two-sample test, and also show how chi square results can be expanded to provide a lot of extra information (risk ratio, odds ratio, Cramer's *V*, percentage deviations, normalised residuals).

Where students have collected individual measurements, and wish to compare two samples, I have promoted the use of the Mann–Whitney *U*-test and Student's t-test simultaneously, which allows them to confirm whether any deviations from normality need to be considered. For comparing more than two independent samples, which in the past often resulted in defaulting back to chi square, I promote the use of the Kruskal–Wallis H-test for nominal categories but of the rarely used Jonckheere–Terpstra trend test where there is a logical order to the categories, which is often the case in geography. In both cases, post hoc tests are available to allow the relationship between individual pairs of categories to be interrogated. For linked data, the equivalent tests are Freidman's two-way analysis of variance by ranks and Page's trend test. For many of the tests, I encourage the students to consider effect size as well as statistical significance, which is a concept that has been rather neglected in geography.

On the companion site, I have supplied a set of spreadsheet-based calculators that will perform most of the tests in this book. They are no substitute for a real statistical package, of course, they are just intended to make it easy for students who lack the confidence to use a package like SPSS to apply their chosen test. My experience is that this does not make the students lazy; on the contrary, when they see that they can apply the methods that they need very easily, they are often inspired to go further and explore a range of other methods. From a teaching perspective, it is all about breaking the fear of numbers and encouraging the students to focus on the geographical problems that interest them, and to use whatever statistical tools they need.

'Teaching the stats course' has traditionally been regarded as one of the less popular jobs in geography departments, often reserved for young lecturers who, irrespective of their interest in the methods, are not in a position to refuse. I am approaching the other end of my career in teaching and research, and I can report that teaching numerical methods, particularly to students working on their own data for their dissertations, has been one of the most rewarding aspects of a job that I love. If you can break that fear of numbers, and encourage the students to view statistical and other numerical techniques just as tools, that will help them to make more sense of the geographical problems that interest them, you may be surprised at the quality of work that they can produce. You will also have given them a skill set and a confidence in their abilities that they can be proud of and take with them into the job market.

References

Darvill, C.M., Bentley, M.J. and Stokes, C.R. 2015. Geomorphology and weathering characteristics of erratic boulder trains on Tierra del Fuego, southernmost South America: Implications for dating of glacial deposits. *Geomorphology* 228: 382–397.

Goudie, A.S. 2006. The Schmidt hammer in geomorphological research. *Progress in Physical Geography* 30: 703–718.

Shakesby, R.A., Matthews, J.A., Karlén, W. et al. 2011. The Schmidt hammer as a Holocene calibrated-age dating technique: Testing the form of the *R*-value-age relationship and defining the predicted-age errors. *Holocene* 21: 615–628.

Shakesby, R.A., Matthews, J.A. and Owen, G. 2006. The Schmidt hammer as a relative-age dating tool and its potential for calibrated-age dating in Holocene glaciated environments. *Quaternary Science Reviews* 25: 2846–2867.

2

How to Use Statistics

2.1 Hypotheses

The inferential statistical tests that we use in geography are based entirely on the concept of hypothesis testing. When a student comes to ask me what methods are appropriate for their particular problem, my first reply is always 'what are your hypotheses'? Students often argue that they do not have any firm hypotheses, they just want to use some statistical methods to 'explore their data' and try to 'see what it means'. They use terms like wanting to be 'objective'. Personally, I think this is nonsense and that it is actually impossible to do any piece of research that involves gathering data unless you do have hypotheses; it is just that some people hide them away and refuse to acknowledge them. This is a very bad practice, not just because it prevents the use of inferential statistical methods, but because it also hinders progress.

There is nothing magical about a hypothesis, it is simply a clear statement of what you expect to find when you collect your data. When you consider doing a piece of research, the hypotheses are your way of summarising the 'established wisdom' and making some predictions about what you would expect to find if this was correct. The next step is then to collect data in order to test those hypotheses. Without hypotheses, it is not possible to conduct the research because there is no way to decide what to actually do.

If you are conducting a questionnaire survey, for example, you need to decide whom to ask and what questions to ask, and those critical decisions must be guided by something. That something is a set of hypotheses, whether you acknowledge it or not. For example, if you conduct a survey into shopping habits, and decide to make a note of whether the respondents are male or female, that decision is based on the 'hypothesis' that gender influences shopping behaviour. Otherwise, why bother to record gender? Similarly, if you record socioeconomic group or age, it is because you have some reason for believing that those factors may also influence the behaviour you are interested in. For exactly the same reason, you will probably decide not to record hair length, number of remaining teeth, or shoe size. Deciding what not to record is guided by theory, and therefore hypotheses, just as much as deciding what should be recorded. Since it is not possible to collect your data without some underlying hypotheses, why not 'come clean' and reveal them at the start, rather than hiding them away? That way you can use inferential statistics to test them. The alternative is to keep your hypotheses well hidden, collect some data, describe it, and eventually reveal the hypothesis that you had hidden all along. The result is that you end up with the same hypotheses that you started with, and make no progress.

If you have the confidence to reveal your hypotheses right at the start, then you can use your data to test them. However, testing hypotheses using statistical methods is a rather formal procedure, and it is very important that you follow the rules and use the right language. The first thing to stress is that you NEVER say 'therefore I have proved that my hypothesis is correct'. Statistical testing is never about proving hypotheses to be correct, it is always about trying to prove them wrong. I realise that this is an odd concept, so please bear with me!

At the heart of the problem is the concept of inference. We do not want to just describe what we have measured or observed, we want to go further and make more general statements about the things that we did not measure or observe. When it comes to inference, it is usually much easier to show that a hypothesis is wrong than to show that it is right. In fact, proving that a hypothesis is correct is often well-nigh impossible.

A common example used to illustrate this concept, attributed I believe to the philosopher of science Karl Popper, is swans. The example is used in the brilliant book 'What is this thing called science' (Chalmers 1999, 2013) and was introduced to geography by Haines-Young and Petch (1986), great advocates of Popper's 'deductive' approach to hypothesis testing. Consider the hypothesis that 'all swans are white'. That is a reasonable hypothesis that you might well postulate on the basis of prior experience of having seen lots of white swans. You see white swans and describe what you have seen, but your hypothesis goes beyond description of your observations (the sample) and makes an inference about all swans (the population). Now consider how you could prove that hypothesis correct; you would actually have to observe every swan in the world and confirm that it was white. Even after observing many millions of white swans, you could not definitively conclude that all swans are white because there could still be a black one out there yet to be observed. Consider, in contrast, how easy it would be to prove the hypothesis false: all you would need is a single observation of a black (or nonwhite) swan. This huge difference in the ease with which hypotheses that make inferences can be shown to be right (verification) or wrong (falsification) is called the 'asymmetry of verification and falsification'. Basically, it is a waste of time trying to prove hypotheses are correct, you can only test them by trying to prove them wrong.

You would be forgiven for thinking that this is all a bit negative! If all we can do is prove our hypotheses wrong how do we make any progress? The answer is that we never do research that involves a single hypothesis; hypotheses always come in pairs, or multiples of pairs. The one that we suspect may be true is not the one that we use statistical methods to test directly. We actually set up an exact opposite, which we call the null hypothesis, and test that instead.

2.2 The Null Hypothesis

The null hypothesis is a bit of an odd concept. You want to test your 'real' hypothesis, but the 'asymmetry of verification and falsification' tells you that you will probably never be able to prove it right, only prove it wrong. Of course, if your hypothesis really does turn out to be wrong then that is good, but if it is actually close to the truth then it is problematic. The solution is the 'null hypothesis' which is effectively the mirror image, or complete opposite of your real hypothesis. An example might help here.

Imagine that your hypothesis is that there is a gender difference in height among geography students. Let's assume that you are a bit dull and have no idea whether males are likely to be taller or shorter than females, you just think there is likely to be a difference. In this case, the null hypothesis would be that there is no difference. However, if you were a bit less dull, you might propose the hypothesis that males tend to be taller than females. In this case, the null hypothesis would be that males are not taller than females. It is very important to make sure that your null hypothesis really is the exact opposite of your real (statisticians call it the 'alternative') hypothesis. In this case, if your hypothesis is that males are taller than females then null hypotheses that stated 'there is no difference' or 'females tend to be taller' would be wrong.

When it comes to the first example: 'there is a gender difference in height', then the null hypothesis makes the prediction that there will be no difference. When you take an unbiased sample of geography students and measure their heights, there are clearly two ways that this null hypothesis could be wrong: a lot of the males could be taller or a lot of the females could be taller. If you take the second example, where you hypothesize that males will be taller, then there is only one way that the null hypothesis 'males are not taller' can be wrong, and that is if a lot of the males are taller. It does not matter whether the females are the same size as the males, or whether the males are shorter. In the first example, there are two ways to be wrong and in the second, there is only one way to be wrong, and when it comes to statistical testing, this difference is really important. It is the basis of 'two-tail' and 'one-tail' probabilities, which are explained later.

2.3 Bad Hypotheses

A hypothesis has to be a clear statement, but to be useful it also has to be 'testable', and because of the 'asymmetry of verification and falsification' problem 'testable' means that if it is wrong, it should be possible (at least theoretically) to prove it wrong. The concept of being 'testable' is very important, so not everything that looks like a hypothesis is useful for statistical testing. Table 2.1 shows some examples of untestable hypotheses and

TABLE 2.1

Some Examples of Hypotheses That are 'Bad' Because They are Not Testable, Together with the Reason They Cannot Be Tested

Bad 'Hypothesis'	Why It Is Useless
Male students may be a bit taller than female students	Too vague, by adding 'may' you remove the ability to make any predictions
The beetle remains in this peat deposit belong to the order coleoptera	All beetles belong to the order coleoptera, so this statement cannot possibly be wrong, and is therefore untestable
In Swansea, different people have different views about immigration	Too vague, 'different people' is the problem, you need to define the groups more clearly
Professor McCarroll is a miserable tosser	Student feedback is always welcome but 'miserable tosser' is not sufficiently well defined to allow a formal test
God created the Earth	Gods are omnipotent, so there is no arguing with fundamentalists, don't waste your time
The upper diamicton unit at Aberdaron is not the same as the lower till unit	This is too vague. 'Not the same' does not give enough information to make any predictions
The moon is made of cheese	Actually a good testable hypothesis because it can (and has) been shown to be wrong by going there and sampling the rocks
The Vice Chancellor of Swansea University has a tattoo of a Welsh dragon on his right buttock	This is an excellent testable hypothesis but the VC has refused to facilitate the required test

Note: The most common problem for geography students is being too vague.

explains where they go wrong. For geography students, the most common problem is just being a bit vague. For example, the 'hypothesis' that 'males *may* be taller than females' is useless because it cannot be proven wrong. It is like the horoscope that says you may come into money today; it means that you might but you might not. When you formulate a hypothesis think about what predictions it makes. If your hypothesis is that males tend to be taller than females, you can predict that if you take a decent unbiased sample, then within that sample the males will be, on average, taller than the females.

2.4 Multiple Working Hypotheses

One of the problems of critical hypothesis testing is that you run the risk of just falsifying your hypothesis and being left with nothing much to say. The solution is simple; rather than starting with one hypothesis start with several. This is not some kind of trick. Geographers tend to work with problems that involve either the natural environment or humanity, and they are rarely so simple that there is only one potential hypothesis. On the contrary, if the topic that you are interested in is really worth researching, then there are likely to be multiple ways of explaining the existing evidence. Your aim should not be to simply produce a list of potential hypotheses, but to design your research so that it forces those hypotheses to compete (Figure 2.1). You should then focus your efforts on gathering the critical information that is needed to separate competing pairs of hypotheses. At the end of the research, you are unlikely to have completely solved the problem, but you may have been able to exclude several reasonable hypotheses, which adds weight to those that remain. You are likely to find that doing the research adds further insights into the problem, so that the end product of your research is not a definitive statement of the absolute truth, but rather a more refined set of competing hypotheses that remain to be tested.

2.5 Unbiased Sampling

When you have formulated clear and testable hypotheses, and the correct null hypotheses to accompany them, you are ready to design your sampling strategy. There are many ways to do this and whole books are dedicated to research design. However, there is just one guiding principle: *your sampling strategy must not impart a bias in the results*. For example, if you were interested in testing the hypothesis that male students tend to be taller than female students, then a very bad sampling strategy might involve measuring the height of your friends in the men's basketball squad. Clearly in this case, the sample would be biased in terms of the parameter that you are interested in (height) because basketball attracts tall men.

One of the best sampling strategies is 'random sampling', but it is often very difficult to achieve true randomness when you design an experiment, particularly in human geography. Standing in the centre of town on a Wednesday afternoon, for example, and asking 'random' people to answer a questionnaire for you will certainly not provide a true random sample of the residents of that town. The sample will be biased, for example,

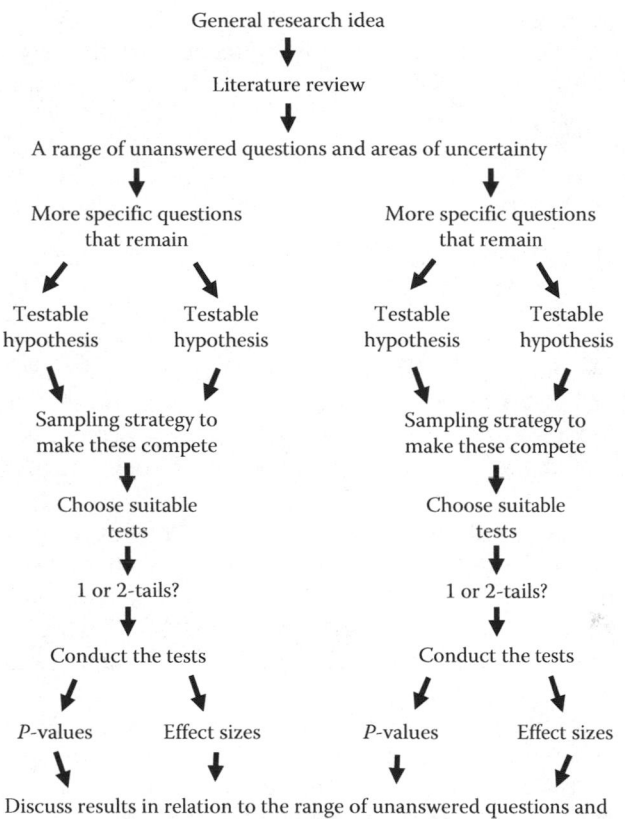

FIGURE 2.1
Reasonable strategy for designing a research project and using inferential statistical tests.

towards people who do not work 'normal' hours, because most workers are not in town on Wednesday afternoon. If you really want to make an inference about the whole population of the town, then a stratified sample might be more appropriate, and the categories that you use to 'stratify' it should be determined by the hypotheses that you have formulated. If, for example, you think that age, gender and ethnic group are likely to influence opinions, then you should ensure that each of these categories is well represented and that there is a good mixture of all three in your sample. For example, if all of the young people you speak to are girls, then you will not be able to separate the effects of age and gender.

No doubt many readers will already have collected all of their data and it is too late to go back and redesign the sampling strategy. There is no need to panic, just take an honest look at the way that you sampled and think about any potential sources of bias and about what kind of population your sample really represents. When you apply statistical tests to your data, and make an inference, just make sure that the population that you make the inference about is really the population that you have sampled from. If there are potential biases because of the way you sampled, then make those clear in the discussion. Your lecturers are aware that you have limited time and limited resources; they do not expect your sampling strategies to be perfect. You will probably gain marks by being honestly critical of your results, and making recommendations for how the approach could be improved.

You will likely be penalised if you do not mention weaknesses in your design or make statements that cannot be supported by your results because there is some source of bias that you could not control. In research, as in life, honesty is the best policy.

2.6 Probability: Is It Just Luck?

When you conduct a statistical test, the end product is usually a 'probability' value. Some statistical tests are used to look at differences between samples, and in those cases, the probability value is essentially the chances that the difference is purely due to luck. Other methods are used to look at the relationship between two variables and again the probability value is the chances that a relationship as strong as the one that is seen in the sample could occur just by luck. Probability (p-values) is normally expressed on a scale of zero to one, and high values (close to one) mean there is a very high probability that luck is the explanation. If the probability value is very low (close to zero), then luck is unlikely to be the explanation. We can never be absolutely sure that chance, or luck, is not the true explanation, we can only say how unlikely it is. If luck is not the likely explanation, then we can be more confident that the difference or relationship that we see in the sample is a true reflection of a similar difference or relationship in the wider population from which the sample was drawn.

Given access to computers and the internet, most of the statistical tests that are covered in this book will return a probability value that is a decimal fraction (on a scale of zero to one). Some software and websites will return values with many decimal places, or in scientific notation. When you present your results, you should not report all of the decimal places, it is better to round to just three. If the p-values are very small, then just report them as less than one in a thousand ($p < 0.001$). In some cases, it is more convenient to look up the probability using tables, and in those cases rather than returning a decimal fraction, the probability values tend to be placed in categories, such as less than 0.05 ($p < 0.05$) or less than 0.01 ($p < 0.01$). It has become a standard practice in geography, and most other disciplines, to regard $p < 0.05$ as the critical threshold for deciding that something is 'statistically significant'. There is no real logic behind this threshold, and you should be wary of making a huge distinction between results that lie just on either side of it. Also, be aware that given a probability of 0.05, there is still a 5%, or one-in-20 chance that the difference or relationship that you see in the sample really is just due to luck, so be careful to moderate your language when you describe results that lie close to the boundary. Table 2.2 gives some suggestions on how to report different thresholds of probability.

An example is perhaps the best way to explain how probability works. Imagine that you want to test the hypothesis that taking caffeine tablets improves exam performance. A simple experiment might involve giving a sample of students an exam when they have not taken caffeine and giving them a similar exam when they have taken the tablets. A good strategy would be to divide the samples in half so that some take the first exam with and others without the tablets, just in case there is some 'learning' effect of taking the exam or it is difficult to make the two exams exactly equal in terms of difficulty. To exclude any placebo effect, we could give the students a tablet of some sort for both tests, but one would not contain any active ingredients. Incidentally, I am not suggesting this as a suitable student topic; you are not actually allowed to experiment on your colleagues with mind-altering drugs.

TABLE 2.2

Probability Values Expressed as Decimal Fractions and as Percentages Together with Some Simple Language for Reporting the Results and Appropriate Descriptors of the Level of Confidence that You Can Have in the Results

Probability	As a Percentage	Reporting This Value in Words	Appropriate Description	Level of Confidence
$p > 0.05$	>5%	More than 5%	Not statistically significant	Unlikely
$p = 0.05$	5%	5% or one in twenty	At the boundary of statistical significance	Possible
$p < 0.05$	<5%	Less than a 5% probability	Statistically significant	Probable or likely
$p < 0.01$	<1%	Less than 1%	Strongly significant	Very likely
$p < 0.001$	<0.01%	Less than one in a thousand	Very strongly significant	Almost certain
$p < 0.0001$	<0.001%	Less than one in ten thousand	Extremely strongly significant	Almost certain

Note: If the p value is even smaller, you still report it as $p < 0.001$ or $p < 0.0001$.

The working hypothesis for this experiment is that taking caffeine tablets improves exam performance, so the appropriate null hypothesis is that the tablets do not improve exam performance. When we use statistical methods to test this null hypothesis, what we are trying to do is to calculate the chances, or probability, that the difference in results that we see in our sample could have occurred purely by luck. In this example, for each participant in the experiment, there are only three possible outcomes: they can perform better when using caffeine, perform the same, or they can perform worse. For simplicity, let us just ignore those cases where the result with and without caffeine was exactly the same so that we effectively have a series of positive or negative results, that is, better or worse.

Let us assume that we have a reasonably unbiased sample of six students, which is too small for a real experiment of course, but it keeps it simple for the purpose of explaining probability. Now assume that all six performed better after taking the caffeine tablet.

In this example, if there is really no advantage to taking the caffeine, then we would have to assume that the only reason that all six students performed better is pure luck. In this case, our sample is very small, only six students, so it is clearly not impossible that, even if the caffeine did not help, all of the students would, just by luck, perform better in the test with the caffeine. The question is not whether it is impossible but just how lucky would you have to be to get a six out of six positive result purely by chance? Thankfully, this is quite easy to calculate because there are just two possible outcomes (better or worse) and so it is similar to tossing a coin, where the two possible outcomes are heads or tails. In this example, just assume that 'heads' equates to performing 'better'.

If you toss one coin, then there are two possible outcomes: a head or a tail. If you were betting on the result, the 'chances' of tossing a head would be one from two, or a half, or 50%. Probabilities are usually expressed not as percentages but as decimal fractions, which effectively means 'out of one' (whereas a percentage value is out of 100). In this case, the probability of tossing a single coin and getting a head is $p = 0.5$, which is the same as a half. Consider now the probability of tossing two coins and getting two heads. Now there are four possible outcomes: two heads (HH), two tails (TT), a tail then a head (TH) or a head then a tail (HT). Since only one of these possible outcomes is the one we predicted, the probability is one from four, or 1/4 which is the same as $p = 0.25$. If we toss three coins, then the number of possible outcomes increases to eight, of which only one fits our prediction of all heads:

HHH, HHT, HTH, THH, HTT, THT, TTH, TTT

In the case of tossing three coins, the probability of obtaining three heads is one in eight, or $p = 0.125$ (one divided by 8). With four coins, the possibilities double again to 16:

HHHH, HHHT, HHTH, HTHH, THHH, HHTT, THTH, HTTH,
THTH, THHT, TTHH, HTTT, THTTT, TTHT, TTTH, TTTT

Again there is only one answer that fits the prediction of all heads so the probability of obtaining four heads from four tosses of a fair coin is one in 16 or $p = 0.0625$. With five coins in play, the number of possibilities doubles again to 32:

HHHHH, HHHHT, HHHTH, HHTHH, HTHHH, THHHH, HHHTT, HHTTH,
HTTHH, TTHHH, THHHT, THHTH, THTHH, HTHHT, HTHTH, HHTHT,
TTTHH, TTHHT, THHTT, HHTTT, HTTTH, HTTHT, HTHTT, THTTH,
THTHT, TTHTH, HTTTT, THTTT, TTHTT, TTTHT, TTTTH, TTTTT

Again of the 32 possible outcomes only one fits our prediction of 'all heads' and so the probability of tossing five heads in a row with a fair coin is 1/32 or $p = 0.03$.

Hopefully a pattern is beginning to emerge. Every time we add another coin the number of possibilities is doubled (Table 2.3), so for six coins, the probability of tossing six heads is one in 64, or $p = 0.016$.

From this example, we can see that although six is a very small sample, it is actually quite unlikely that all six students would perform better with the caffeine tablet 'just by luck'. When we use statistical tests in geography, we usually conclude that something is too unlikely to be due to pure luck if the probability drops below 5% which is one in 20 or $p = 0.05$. Since in this case, the probability that all six performed better purely by luck is much smaller than this, at one in 64 or $p = 0.016$, we would conclude that chance, or luck, is not the likely explanation. In formal language, we would say that we can 'reject the null hypothesis that caffeine tablets do not improve exam performance'. Note that we do not conclude that caffeine tablets do improve performance, because remember that we never try to prove a hypothesis correct (verify), we only try to show how unlikely it is to be true. In this case, all we can say is that the results of the test suggest that pure chance is not the likely explanation for the better performance of the drugged students. However, having excluded luck as the likely explanation, it does give us a bit more confidence in our original hypothesis. A small pilot study like this might be enough to convince a researcher that it is worth pursuing the original hypothesis, perhaps by organising a larger sample of students.

Now consider an alternative outcome to this simple experiment, where five of the six students performed better on the caffeine. In this case, the probability is not one in 64, and it is not two in 64 either. Think again about the coin-tossing example. With six coins, there are 64 combinations of results, and if you predict in advance that all six coins will be heads, then only one result from all 64 fits your prediction. Now what is the probability of tossing five heads and one tail? In fact, if you look at all of the possible outcomes, there are six that fit that prediction. This is because the tail could be obtained on the first toss, the second,

TABLE 2.3

Number of Possible Combinations of Heads and Tails for the Tossing of a Fair Coin Together with the Probability of All Heads or of All the Same (All Heads or All Tails)

Fair Coins Tossed	Possible Combinations	Probability of All Heads	Probability of All the Same
1	2	0.5	1
2	4	0.25	0.5
3	8	0.125	0.25
4	16	0.063	0.125
5	32	0.031	0.063
6	64	0.016	0.031
7	128	0.0078	0.016
8	256	0.0039	0.0078
9	512	0.0020	0.0039
10	1,024	0.00098	0.0020
11	2,048	0.00049	0.00098
12	4,096	0.00024	0.00049
13	8,192	0.00012	0.00024
14	16,384	0.000061	0.00012
15	32,768	0.000031	0.000061
16	65,536	0.000015	0.000031
17	131,072	0.0000076	0.000015
18	262,144	0.0000038	0.0000076
19	524,288	0.0000019	0.0000038
20	1,048,576	0.0000010	0.0000019

Note: The number of possibilities doubles each time another coin is tossed.

the third, etc. This means that the probability of tossing exactly five heads from six tosses of a coin is six in 64. The probability of tossing at least five heads is seven in 64 because then we include the single case where all six tosses are heads.

Looking back at our small pilot study, if only five of the six students performed better on the caffeine tablets, we would conclude that the likelihood of at least five of the six students performing better purely by luck is seven in 64, which is about 11% or $p = 0.11$. That is much larger than the critical threshold of $p = 0.05$ (larger meaning more likely to be just luck) so in this case, we would accept the null hypothesis that 'caffeine does not enhance exam performance'.

When you conduct an experiment like this, where there are only two possible responses, it is always possible to calculate the probability of any given result simply by listing all of the possible outcomes and counting the number that fit the result. Of course, in reality that would be a very tedious procedure and it is not really necessary because all of the possible results have already been calculated and you can just look them up in a table. Alternatively, you can enter the results of your experiment into an online calculator, or spreadsheet, and obtain the probability value. The name used for this procedure is the binomial test, and it is explained later (Chapter 5).

In the coin-tossing example used above, we saw that each time a coin was added to the sequence the number of possible outcomes doubled and therefore, the probability of a single predicted outcome, such as 'all heads', halved. The reason for this is that there are only two possible outcomes each time. It is like playing 'double or quits' by tossing a coin, where each time you place a bet you have a 50/50 chance of winning or losing and so your

TABLE 2.4

Number of Possible Combinations of Numbers for the Rolling of Fair Dice Together with the Probability of All Sixes or of All the Same Number

Number of Dice	Possible Combinations	Probability of All Sixes	Probability of All the Same
1	6	0.17	1
2	36	0.028	0.17
3	216	0.0046	0.028
4	1,296	0.0008	0.005
5	7,776	0.0001	0.001
6	46,656	0.00002	0.0001
7	279,936	0.000004	0.00002
8	1,679,616	0.000001	0.000004
9	10,077,696	0.0000001	0.0000006
10	60,466,176	0.00000002	0.0000001

Note: The number of possibilities increases by a factor of six each time another die is rolled.

winnings are doubled with each successful toss. However, probabilities do not always change by a factor of two, it depends on the number of possible outcomes. If you roll a die, for example, there are six possible outcomes, so with one die there is a 1/6 chance of rolling a 6. If you roll two dice, then there are 36 possible outcomes (six times six) so the chances of rolling two sixes is 1/36, or $p = 0.028$. Adding a third die increases the number of possibilities by a factor of 6 again to 216, so the chances of rolling three sixes with three dice is one in 216 (Table 2.4). When there are several possible outcomes at each step, the probability of predicting a single outcome overall, just by luck, becomes very remote indeed. With ten dice, for example, there are more than 60 million possible results, so the probability of rolling 10 sixes in a row is less than one in 60 million. Think about that next time you take a multiple choice exam, because the chances of getting 10 out of 10 answers right, purely by luck, if every question has five possible answers, is almost 10 million to one!

2.7 One or Two-Tail Testing

The concept of 'one-tail' and 'two-tail' has already been touched upon in the section on hypotheses. A one-tail hypothesis applies when the direction of a prediction is specified in advance and a two-tail hypothesis applies when the direction is not specified. The distinction is critically important because it drastically changes (by a factor of 2 i.e. halves or doubles) the probability when a statistical test is applied. An example is perhaps the best way to explain.

Take the example used above where six students sit an examination both with and without caffeine tablets. In the earlier example, our hypothesis was that caffeine tablets improve performance and so we tested the null hypothesis that caffeine does not improve performance. We used a one-tail hypothesis because we predicted the direction of the effect, which is improvement. In that case, we were interested in the number of students who performed better with caffeine and if only three out of six performed better, we would clearly just accept the null hypothesis because there would be no evidence of improved performance. However, if all six performed worse under the influence of caffeine, we would also accept the null hypothesis because again there would be no evidence for improvement. Now consider a different hypothesis that just states that taking caffeine tablets 'influences' performance, but does not predict the direction of the effect. You might formulate this hypothesis if you were unsure whether the caffeine really helped performance or made it worse. In this case, if three students performed better and three performed worse, you would clearly just accept the null hypothesis that caffeine has no effect on performance. However, there are now two ways that the null hypothesis could be rejected: if a high proportion of students performed better on caffeine or if a high proportion performed worse.

Using the coin-tossing analogy, we can again calculate the probability. In the early example, which was one-tail, we were only interested in the probability associated with tossing several heads (representing students who performed better on caffeine). In the second example, however, which is two-tail, we are interested in the probability of tossing either a lot of heads or a lot of tails. In effect, there are now two ways to 'win'; by tossing a lot of heads or by tossing a lot of tails. In our first example, with a one-tail test, all six students performed better on caffeine and the probability of tossing six heads from six tosses is one in 64. If we use a two-tail hypothesis, that does not predict the direction of the effect, then it is the number of students who get the same result that is important, and it does not matter whether that result is better or worse with caffeine. In this case, we are not interested only in the probability of tossing six heads, we want the probability of tossing 'six the same', and they could be heads or tails. Of the 64 possible combinations with six coin tosses, there are now two that fit our prediction: six heads or six tails. This means that the probability associated with the null hypothesis that there is 'no effect' is not one in 64 but two in 64, which is the same as one in 32 (Table 2.3).

When you use most statistical tests, you have to stipulate in advance whether you want to calculate the 'one-tail' or the 'two-tail' probability. Your decision should be based on looking at your hypothesis, not at the data. This may seem like a trivial distinction but it is actually critically important and quoting one-tail probabilities for a two-tail hypothesis, or vice versa, is a common reason for students to lose marks in project work. Recall that most of the statistical tests that we use in geography are specifically designed for testing hypotheses, they are not 'descriptive statistics' that simply describe data. This means that they have to be used in a formal logical order: first you formulate your hypothesis, then you formulate the appropriate null hypothesis, and only then do you perform the statistical test. The hypothesis must come first and it is the hypothesis, which is either one-tail or two-tail, which determines whether you calculate the one-tail or two-tail probability (Table 2.5). If you are in any doubt at all, then use two-tail hypotheses and calculate two-tail probabilities.

2.8 Effect Size

When we use inferential statistical tests in geography, the end product is a probability value. This usually represents the chances that something that we observe, such as a difference between two samples, or a correlation between two variables, could have arisen purely by chance. When probability values are reported in the published literature, in many fields, not just geography, they are often misinterpreted as measures of the 'strength' or 'importance' of some effect. However, the probability does not actually tell us anything about the magnitude of a difference or the strength of a relationship. If you have a very

TABLE 2.5

Example Hypotheses and the Appropriate Null Hypotheses, Which May Be Directional (One-Tail) or Non-Directional (Two-Tail)

Hypothesis	Null Hypothesis	Tails
There is a gender difference in the heights of geography students	There is no gender difference in the heights of geography students	2
Male geography students tend to be taller than female geography students	Male geography students do not tend to be taller than female geography students	1
Fear of crime among Swansea citizens is contingent on age	Fear of crime among Swansea citizens is not contingent on age	2
Swansea residents over the age of 60 tend to be more afraid of crime than those in the 20–50 age bracket	Swansea residents over the age of 60 do not tend to be more afraid of crime than those in the 20–50 age bracket	1
The upper and lower diamicton units at Aberdaron contain different proportions of erratics	The upper and lower diamicton units at Aberdaron contain the same proportions of erratics	2
The lower diamicton unit at Aberdaron contains more erratics than the upper diamicton unit	The lower diamicton unit at Aberdaron does not contain more erratics than the upper diamicton unit	1
Postgraduate students drink more beer than undergraduate students	Postgraduate students do not drink more beer than undergraduate students	1
Female students underperform in multiple choice examinations	Female students do not underperform in multiple choice examinations	1
Human geographers hate statistics more than physical geographers	Human geographers do not hate statistics more than physical geographers	1
There is a difference in the vehemence with which human and physical geography students hate statistics lectures	Physical and human geography students hate statistics lectures with equal vehemence	2

large sample, then even a small difference or a very weak correlation between two samples is unlikely to occur just by luck, but if you have small samples, then pure chance can result in quite big differences. To understand how important a difference or relationship is, we really need two measures: one that describes the size and one that describes the chances that it could have occurred just by luck.

The difference between effect size and probability values is perhaps best exemplified using scatter plots that show the relationship between two variables (Figure 2.2). In this case, I have used a tree growth index based on measurements from a lot of pine trees from several sites close to the tree line in northern Fennoscandia and some meteorological data from the same region (McCarroll et al. 2013). On scatter plots, it is possible to assess whether there is a correlation between two measurements and a useful 'effect size' in such cases is the Pearson's correlation coefficient, which is often called just the '*r*-value'. Correlation is described in detail in Chapter 10 but for now all we need to know is that the *r*-value is a measure of the strength of the correlation and it ranges between zero and either plus or minus one, depending on the kind of relationship. In Figure 2.2, the growth index is plotted against the mean temperature of September and against the mean temperature of the summer (June to August) and the number of years included is varied from just 15 to the full meteorological record for that region, which is 116 years.

It is well known that trees growing at high latitudes (these are from well north of the Arctic Circle) are sensitive to summer temperature and, as expected, we see a large effect size (*r*-value) when tree growth is plotted against summer temperature, irrespective of

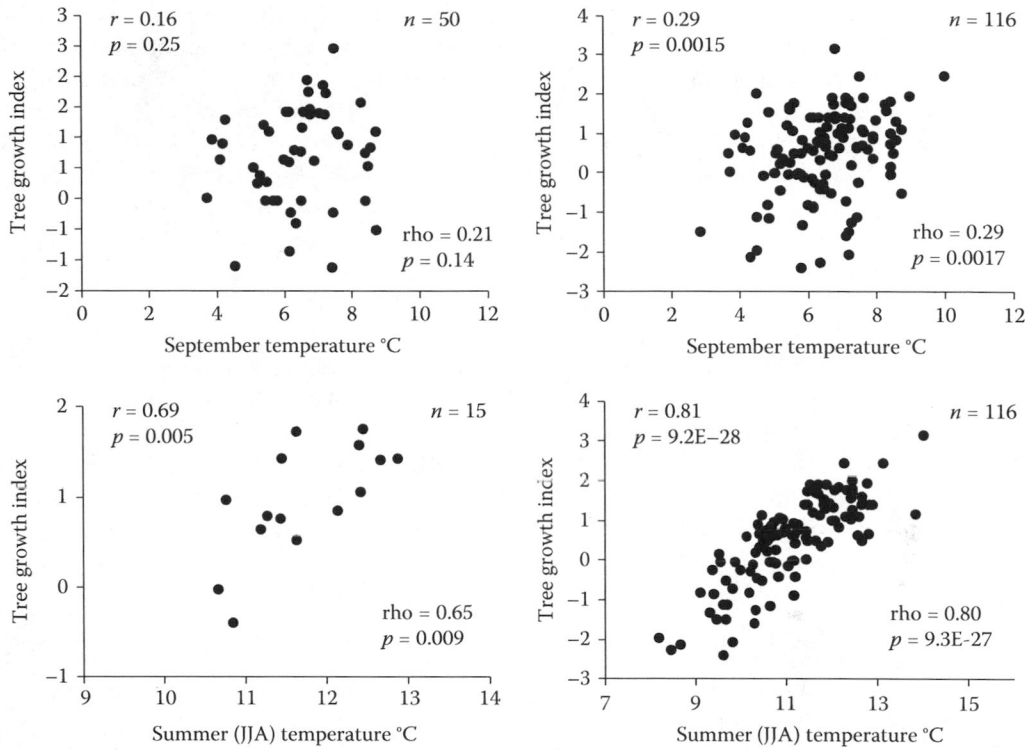

FIGURE 2.2
Scatter graphs to show the relationship between a tree growth index and instrumental temperature readings. The effect size 'r' is the Pearson's correlation coefficient and it is accompanied by a two-tail probability (p). The effect size 'rho' is based on Spearman's rank correlation and is also accompanied by a two-tail probability. The sample size is indicated by n.

the number of years (n) that are included. We would logically expect a weak relationship between tree growth and September temperature because at high latitudes most of the growth is finished by early September and the trees are shutting down for the winter. As expected, the effect size is much lower when growth is plotted against September temperatures. Irrespective of the sample sizes used we can conclude, based only on the effect sizes, that the relationship between tree growth and summer temperature is strong and the relationship between tree growth and September temperature is weak.

When we use the Pearson's correlation coefficient, we can also calculate a probability to go with it. The probability can be either one-tail or two-tail, and in this case, I have used the two-tail probabilities, so the null hypothesis that is being tested is that there is no relationship between tree growth and temperature. The probability, as always, is based on the assumption that the relationship that we see is just due to luck. With a sample of 50 years, the correlation between tree growth and September temperature is just 0.16 and the probability value 'p' is 0.25, which is the same as 25%. This tells us that if we were to take two enormous samples of tree growth and September temperature values, and randomly pick out sets of 50 pairs of values (so that they do not necessarily come from the same year), then about 25% of the time we would expect to produce a correlation as high as the r-value of 0.16 (or as low as −0.16) that we obtained for this sample. Raising the sample size to 116 captures a wider range of values on both axes and the effect size is a little higher

($r = 0.29$), though it would still be regarded as weak. In this case, however, the probability has dropped to 0.0015 or 1.5 in a thousand. This result tells us that although there is only a weak relationship between tree growth and September temperature, that relationship is very unlikely to be due simply to luck. If you were to randomly sample sets of 116 pairs millions of times, you would only expect to obtain an r-value as high as 0.29 (or as low as −0.29) about three times in every two thousand tries. Taking a very large sample has effectively allowed us to identify a weak but significant effect in the population.

If we compare tree growth with summer temperature, using just the last 15 years of the record, then we see a strong effect size but the probability is $p = 0.005$. This is five times higher than the probability that we obtained for September using the full sample of 116 years. That means it is five times more likely to be just luck. This does not mean that the temperature of September is a more important control on tree growth than that of the summer. It is just that with a small sample of just 15, it is a lot easier to obtain a strong cor-relation just by luck. Using the full data set for summer temperature, we obtain a probabil-ity that is such a small number that I have resorted to scientific notation (9.2E–28). The 'E to the minus 28' means that you have to move the decimal place 28 places to the left, which would give us a number that has 27 zeros after the decimal place.

$$p = 9.2\text{E–}28 = 0.000000000000000000000000000092$$

That is a very small number so we can be virtually certain that the strong relationship that we see between tree growth and summer temperature has not occurred just by luck and you would have to take many millions of random samples of 116 pairs before you could reasonably expect to produce an r-value as high as 0.81 (or as low as −0.81, since we are using a two-tail test). The relationship between tree growth and summer temperature at high latitudes and high altitudes is so strong that we can use tree growth to reconstruct temperature far beyond the period covered by instrumental records (McCarroll et al. 2013; Luterbacher et al. 2016).

Effect size is a useful concept because there are times when a relationship is 'statistically significant', in that the probability that it is just due to luck is very low, but so weak that it is not really important or useful. The relationship between tree growth and September temperature in Figure 2.2, for example, is much too weak to be used for reconstructing September temperatures back in time. Confusing statistical significance with effect size is so common that in some disciplines, including psychology, the specialist journals actually require both the statistical significance and the effect size to be quoted. In geography, the concept of effect size seems to have been rather overlooked and that should be redressed.

Where possible I try to give an effect size for each of the tests that are presented. In most cases, the effect size falls on the scale of zero to one, and the best ones are comparable in scale to the Pearson's correlation coefficient, in which case they use the same terminology and are called 'r-values'. With r-values, it is a common practice (Cohen 1988; Field 2013) to consider values of 0.5 or more to represent a large effect and values of less than 0.3 to rep-resent a small effect, with intermediate values considered as a medium effect (Table 2.6). There are other measures of effect size that use different scales, but to avoid confusion I try to avoid them.

If you are using very large secondary data sets, you need to be particularly careful to check and quote the effect size of your results. Be careful not to over-state the importance of some result where the statistical significance is very high (low p-value) but the effect size is very low. When you collect primary data, you are more likely to be dealing with small samples, in which case an effect will have to be reasonably large before it is significant.

TABLE 2.6

Guide to the Strength of the Effect Size Based on r-Values
(Equivalent to Pearson's Correlation Coefficient)

Effect Size	Description
$r < 0.3$	Small or weak effect
$r = 0.3{-}0.5$	Medium effect
$r > 0.5$	Large or strong effect

In this book, I generally recommend the use of non-parametric statistical tests as the first option, and you might well question that advice on the basis that those tests are 'less powerful' than the equivalent parametric tests. However, parametric tests are only more powerful if all of the assumptions are met, and powerful in this sense means that if there is a small effect in the population, the stronger test is more likely to identify it. In most geography student projects, if the effect is so small that a non-parametric test does not recognize it, then it is probably not important enough to worry about anyway.

In Figure 2.2, a second set of effect sizes (rho) is included based on a non-parametric method called Spearman's rank correlation or Spearman's rho (Section 10.5), together with the associated probabilities. Note that for very large sample sizes, the two methods give almost identical results. For smaller sample sizes, the differences are larger but not large enough to change any of the conclusions. For the smallest sample, of just 15 years, the non-parametric method is certainly more appropriate.

References

Chalmers, A. 1999. *What is This Thing Called Science: An Assessment of the Nature and Status of Science and Its Methods* (3rd ed.). Open University Press, Berkshire, 266pp.

Chalmers, A. 2013. *What is This Thing Called Science: An Assessment of the Nature and Status of Science and Its Methods* (4th ed.). (e-book). University of Queensland Press and Open University, Berkshire.

Cohen, J. 1988. *Statistical Power Analysis for the Behavioural Sciences* (2nd ed.). New York: Academic Press.

Field, A. 2013. *Discovering Statistics Using IBM SPSS Statistics* (4th ed.). Sage, London, 916pp.

Haines-Young, R.H. and Petch, J. 1986. *Physical Geography: Its Nature and Methods*. Sage, London, 248pp.

Luterbacher, J., Werner, J.P., Smerdon, J.E. et al. 2016. European summer temperatures since Roman times. *Environmental Research Letters* 11: 024001.

McCarroll, D., Loader, N.J., Jalkanen, R. et al. 2013. A 1200-year multiproxy record of tree growth and summer temperature at the northern pine forest limit of Europe. *Holocene* 23: 471–484.

3

Different Kinds of Data

3.1 Kinds of Data

One of the factors that determine what statistical tests you can use is the kind of data that you have. In some areas of geography you may be measuring something with quite a high degree of precision, so that your data are numbers with perhaps a couple of decimal places, whereas in others you may just be counting the numbers that fall into certain categories, and sometimes those categories have a logical order and sometimes they do not (Table 3.1). Being aware of the kinds of data that you can collect, or just recognising the kind of data that you have, is a critical skill if you want to use statistical tests.

Different types of data can be placed into a logical order defined by the amount of information that they contain, starting with counts in named categories and ending with 'individual measurements'. Generally, the higher the category of data the more tests are available to use. When you have very rich data, like individual measurements, you can still apply statistical methods designed for lower orders of data but then you lose some of the information by degrading your data. For example you might have measured some parameter very carefully, to several decimal places, but then decide to place all the values into five size categories, from very large to very small. Students commonly degrade their data and use weaker statistical tests for no good reason, which is not strictly wrong, just a bit of a waste of effort. Another common mistake is to collect data that is at a low level when, with the same amount of effort, much richer data could be obtained. Some examples are provided in Tables 3.1 and 3.2 and in the following sections.

3.1.1 Nominal

Nominal just means named, so this is the kind of data that represents counts in different categories, where there is no logical basis for placing the categories in rank order. If, for example, you were recording the number of people that fall into three different religious groups it would not make sense, or at least be politically correct, to place the groups in any rank order. Similarly you might be looking at glacial deposits and counting the numbers of different erratics, and again there would probably be no logic in placing the rock types in any specific order. Some data clearly are truly nominal, but much of the data that students treat as nominal is actually ordinal in character.

3.1.2 Ordinal

Ordinal means that it is possible to put the data into a logical rank order. Where the data consist of counts in categories, this means that there is some logical order for the categories.

TABLE 3.1

Different Types of Data with Descriptions and Examples from Human and Physical Geography

Type of Data	Description	Human Geography	Physical Geography
Counts in two nominal categories	Two categories that cannot be placed in any logical order	Yes or no, right or left, religious or atheist, black or white	Local or erratic lithologies, native or exotic plants
Counts in several nominal categories	Several categories that do not fall into a logical order	Religious groups, ethnic groups, political allegiance	Rock types, plant families, colours
Counts in two ordinal categories	Two categories that can be placed in a logical order	Worst and best, higher and lower, bigger or smaller	Above or below a threshold size or weight
Counts in several ordinal categories	Several categories fall into a logical order	Age groups, income bands, socio-economic groups, levels of agreement	Clast roundness classes, size classes
Measurement by rank order: ordinal data	Placing in rank order rather than ascribing absolute measurements	Concepts that are hard to quantify, or where respondents might use a numerical scale in different ways	Climate forcing by volcanic eruptions and solar irradiance: relative order is known but forcing is hard to measure
Individual measurements that form clear groups: counts in ordinal categories	If measurements are not continuous they are treated as counts in categories that fall in a logical order	Number of years in full time education, number of children, number of bedrooms, number of cars	Chemical data from cheap kits with few divisions, measurements using imprecise instruments, counts taken at set time intervals
Individual measurements	Not falling into clear groups	Level of agreement, etc. marked on a line rather than using categories, age, income	Most instrumental measurements

In questionnaire data, for example, respondents may fall into groups according to age and socio-economic status. Although this kind of data is very often treated as nominal, which restricts the number of tests that are available to use, they are actually ordinal and more options are available. In physical geography, size categories are similarly ordinal. Another common kind of data is clast roundness, which is often based on Power's six-point scale, again providing counts in categories that have a natural order, in this case from very angular to well rounded.

When you only have two categories they are usually considered to be nominal, but there are times when it is logical to place them in order (Table 3.2). This can be useful, for example when you can predict the direction of difference that you might expect based on the logical order of the two categories. For example you may be interested in comparing the salaries of a sample of people who either have or have not received a university education. By treating the two categories as ordinal you can make the reasonable prediction that the university-educated group will earn higher salaries, in which case you would calculate one-tail rather than two-tail probabilities.

The data, or information, that you collect within groups or categories can also be ordinal in scale, meaning that they can be placed in rank order. If you collect individual

TABLE 3.2

Some Questions That Might be Asked and the Type of Data That Is Produced

Typical Question	Options for the Answer	Type of Data
To which religion do you ascribe?	Christian, Muslim, Hindu, Jedi, none	Counts in several nominal categories
Do you agree?	Yes or no	Counts in two nominal categories
To what extent do you agree?	Strongly agree, agree, ambivalent, disagree, strongly disagree	Counts in ordinal categories
On a scale of one (weakest) to five, how strongly do you agree?	One to five	Counts in ordinal categories
On a scale of one to one hundred, how strongly do you agree?	Percentages	Individual measurements
Please mark your level of agreement on the scale provided	Continuous, depends on how accurately you measure the results	Individual measurements
Are you old enough to vote?	Yes or no	Counts in two nominal categories
Which of these age categories do you fall into?	Several categories	Counts in ordinal categories
What is your age?	Continuous	Individual measurements
Have you graduated from a university?	Yes or no	Counts in two ordinal categories
What is your highest academic qualification?	School level, undergraduate degree, masters, doctoral	Counts in ordinal categories
Which of these two things do you prefer?	Two choices but one costs more	Counts in two ordinal categories
Place these ten images in order of attractiveness	Rank order, with or without allowing shared ranks	Ordinal level data
Using a scale of 1–20, give each of these images a score for attractiveness	Twenty possibilities, but data may be grouped in fewer options	Could be treated as individual measurements or as counts in categories.

measurements it is of course possible to place them in rank order, but one of the advantages of ordinal level data is that you can often put a series of things in rank order even when it is difficult to make a real measurement. A simple example might be something like beauty, which is rather subjective and difficult to measure. However, if you were interested in how different groups of people perceived the beauty of natural (or urban) landscapes you could give them a series of images, such as a set of postcards, and ask them to put them in rank order of beauty. You would then have ordinal scale data and could test, for example, whether young people from different ethnic backgrounds had similar perceptions of landscape beauty. If you can obtain ordinal data it substantially increases the number of tests that you can use, so it is worth thinking carefully about how to collect your data (Table 3.2).

Sometimes when we make measurements we assume that we are generating individual numbers when in reality the data are just counts in categories. An example of this is the measurement of soil or water chemistry using simple test kits. These kits often involve mixing a few chemicals in a tube, which results in a colour change, and then reading the values using a colour chart. Although the result for each individual test is a number, there are usually so few colour options that the data actually fall into categories. The categories fall in a natural order, so the data are ordinal, but within each category all of the numbers are the same. Such data are not 'individual measurements' in the sense used here, they are ordinal data recording counts in categories.

In Swansea we have a research group that specialises in soil water repellence, or hydrophobicity, and some students perform projects that measure this property by recording the time that it takes for water drops to penetrate the soil surface. Although this sounds like an example of individual measurements, they do not actually sit and observe the exact time of penetration of every drop, instead they count the number of drops that are left after a given number of seconds. Again these data are ordinal in nature, recording counts in categories that can be placed in a logical rank order.

Many of the tests that are described in this book rely on the ability to place data into rank order. Some tests use categories that fall in rank order, and others use individual numbers that can be placed in rank order. If you can collect individual numbers that can be placed in rank order, rather than just counting the number of cases that fall into categories, you will have richer data and more statistical tests will be available for you to use. Where possible, try to obtain individual measurements.

3.1.3 Individual Measurements

By individual measurements I mean numbers that are continuous, in the sense that they can be placed into rank order, without too many shared ranks, rather than data that fall naturally into clear groups. Clast roundness, for example, is often recorded using a six-point scale and every clast falls into one of those six categories so the data are ordinal rather than individual measurements. Clast size, in contrast, is often calculated using measurements of the longest and shortest axes of a clast, and the data do not necessarily fall into groups, so they are individual measurements.

Some data that are typically collected as 'counts in categories' are actually just as easy to record as individual measurements. For example, in a questionnaire survey you might ask respondents to tick one of five boxes labelled from strongly agree to strongly disagree. There is nothing wrong with that and you can use a series of statistical tests on those ordinal 'counts in categories' data. However, an alternative strategy is to provide a line marked strongly agree at one end and strongly disagree at the other and ask the respondents to mark their level of agreement on the line using a cross. By measuring the distance along the line with a ruler you obtain individual measurements and so there are more statistical tests available for you to use. If you make the bar 10 cm long and read off the points in mm you will effectively have 'percentage agreement' values. A potential disadvantage of this approach is that it might make it more challenging for certain participants to make a choice, so it is not always the best option.

Statisticians do not use the term 'individual measurements', I am just using that for simplicity, they use the rather unfamiliar terms 'interval' and 'ratio' to categorise such data. For the tests presented in this book the distinction does not matter, and seems to be a perennial source of confusion. The difference is just that for interval data there is no absolute zero whereas for ratio data there is a logical zero point. Measurements of temperature using degrees Celsius and Fahrenheit are examples of interval data. Distance measurements are ratios because no distance really does mean zero. If you are now totally confused do not worry about it, in practice it really does not make much difference, the important distinction is between individual measurements that are continuous, meaning for our purposes that they can be placed in rank order, and data that fall into groups and therefore represent counts in categories.

3.2 Independent or Linked Data?

For tests that involve comparing more than one sample of individual measurements (or ranks) it is necessary to decide whether your data sets are independent or linked. When you have two sets of data then 'linked' is equivalent to 'paired'. It is an important decision and you have to get it right or you will use the wrong test. Data sets are independent when you want to compare all of the values from one sample with all of the values in another sample or samples and you do not care about the order in which the data are listed. They are linked (or paired) when it is important to compare a particular value in one data set with a particular value in the other data set or sets. Some examples may help.

Imagine you have measured soil compaction at 30 points in a field grazed by cattle and 30 points in a field grazed by sheep. In this case there is no relationship between data point 14 in the first field and data point 14 in the second field, they are independent of each other. If, however, you were interested in soil compaction along a footpath up a mountain you might sample at 30 points up the path and at each point also take a 'control' sample on the verge away from the path. In this case you have 30 samples from the path and 30 samples from the verge, but the samples are paired because they were taken at the same height (altitude) up the mountain. In this case it is important to compare location 14 on the path with its paired location 14 on the verge.

In human geography you might be interested in a possible gender difference in response to some question and take a sample of 30 males and a similar sample of 30 females. In this case there is no reason for linking an individual male respondent with an individual female respondent, so the samples are independent. Alternatively, you might want to assess how some activity changes people's perceptions, opinions, or skill level. For example, you could give some students a small statistics test before they attend a lecture course and then give them a similar test after they have attended the course. In this case you would be interested in how the individual performances have changed, and so the two data sets are linked because each respondent has a value from before the course and a value after the course.

Linked samples are not limited to pairs; there can be several linked samples. For example, you might be interested in whether there is a tendency for student performance to improve during their time at university, in which case you might obtain all of the exam results for a set of 30 students who have already graduated. Assuming there are two exam periods in each of 3 years that would give you 6 sets of 30 exam results and they are linked because each value belongs to one of the 30 individual students. Alternatively you might just want to know whether in a particular calendar year there is a tendency for students in the first year of studies to obtain the lowest marks and for final year students to obtain the highest marks. In this case the samples are independent because no two results belong to the same student.

Deciding whether your data are independent or linked is important not just for choosing the right statistical tests; it also dictates how it is logical to draw graphs to illustrate the data. If data are linked then it makes sense to plot them so that the link between the individual pairs is clear. If the data are independent then you should not plot them in such a way that they appear to be linked. This is a common error in student submissions and likely to lose your marks. If you are comparing two or more sets of linked data then using a line graph works well. If the data are independent then avoid line graphs and plot the data as histograms (Figure 3.1).

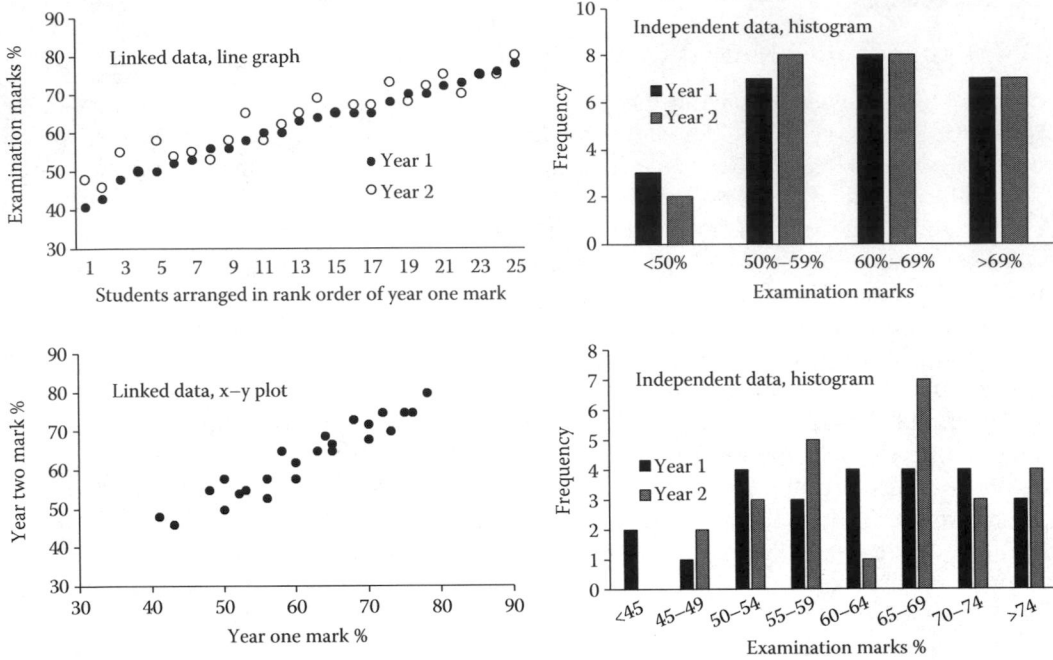

FIGURE 3.1

Different ways of displaying data that are linked (paired) or independent. It makes no sense to plot two independent series on a line graph because it gives the false impression that the points are paired. When using a scattergraph make sure that there is some logic to the choice of axes; the convention is that the independent variable goes on the horizontal (x) axis and the dependent on the vertical (y) axis. In this case it is reasonable to assume that year 2 marks might be influenced by year 1 marks, because knowing that a student performed very well in year 1 suggests they have a good chance of also doing well in year 2, so the year 2 mark is placed on the vertical axis. When plotting a histogram you have to choose the categories or 'bins'. Make sure that they do not overlap.

3.3 Assumptions

By far the most important assumption in using any statistical test is that your sampling strategy has not caused a bias in the results. If, for example, you wanted to test the hypothesis that the clasts (stones) in one glacial till were more rounded than those in another till unit you would not search the first till looking for a location with a lot of rounded stones and then search the other one looking for a location with a lot of angular stones to measure. That would clearly give you a biased result that is completely useless. Your sampling strategy must provide an unbiased sample.

Most statistical textbooks go into considerable detail about the assumptions involved in statistical testing, but that is largely because they tend to focus on tests that are defined as 'parametric'. Those tests calculate probabilities based on the assumption that the data that you have collected come from a population that has specific properties (parameters). The most common assumption involves a 'normal' or Gaussian distribution (the terms 'normal distribution' and 'Gaussian distribution' mean the same thing). Most of the statistical tests that are presented in this book do not make assumptions about the properties of the population and are defined as 'non-parametric' or 'distribution-free'. This does not mean that they have no assumptions at all, and the important assumptions are described for each test, but for most tests it is very likely that the kind of data that students can collect will be perfectly adequate.

The only parametric methods that are covered in detail here are Student's *t*-tests, Pearson's correlation, and regression analysis. Before you apply these methods it is important to check that your data could reasonably be considered to be normally distributed. The basics are described in Section 3.3.1 and more detail is provided in the relevant sections, but it basically means that if you plot the relevant data as a histogram it should look a bit like a bell shape, with most of the data in the middle and decaying in frequency about equally in both directions, so that the mean and median values (Section 4.7) are about the same. It is often assumed that to apply regression your data also have to be normally distributed, but that is not quite true and the assumptions of regression are discussed in Chapter 11.

Personally I think that students, and researchers in general, often worry too much about whether their data are normally distributed or not, and as long they are not too far off, and the data set is large (>30) then it will probably be fine. If you are using very large data sets, with hundreds of values, then normality actually does not matter much and you can go ahead and use parametric tests even on data that is very skewed or flat. Watch out for outliers (extreme values) though, they are the most serious source of error in large data sets.

If you are worried about the assumptions of normality, and particularly if your data sets are not large (>30), I would suggest that you just stick to non-parametric methods where possible. If your data sets are very small you should certainly not use parametric methods. Where there is room for doubt a good strategy is to use both parametric and non-parametric tests and check that they give the same result. For example, if you wanted to know whether two independent samples of 30 individual measurements were different enough to conclude that they had not been taken from the same population, and they look reasonably but not perfectly normally distributed, you could report the results of both Student's *t*-test (parametric test) and the Mann–Whitney *U*-test (non-parametric). You might choose to report that 'the results of Student's *t*-test suggest that the two samples are significantly different ($p < 0.05$) and although they may not be perfectly normally distributed, the significance of the difference is supported by the results of the Mann–Whitney *U*-test ($p < 0.05$)'.

3.3.1 Checking for a 'Normal' or Gaussian Distribution

If you are using non-parametric tests, which includes most of those covered in this book, then your data do not have to be normally distributed. Just check whether they are continuous or not; that is a very important assumption for both parametric and non-parametric tests. If you have individual measurements put them in a spreadsheet and use the 'sort' function to put them in rank order. If a lot of the values are identical, and they form clear groups, then it is probably best to treat them as counts in categories. The categories fall in a logical order, of course, because they are measurements, so you can treat the data as counts in ordinal categories. There are plenty of statistical tests available for this kind of data (Chapters 6 and 7). If there are just a few shared ranks, but not clear groups, then you can assume your data are continuous and treat them as individual measurements.

To check whether a set of continuous measurements is near normally distributed the best option is to draw a histogram that shows the number of values in different size categories. The size categories (often called bins) need to be appropriate for your data; choose them so that there are not too many empty categories. You can use the 'histogram' function in

TABLE 3.3

Selection of Websites That Will Produce Histograms

Provider	Website	Comments
Social Science Statistics	http://www.socscistatistics.com/descriptive/histograms/	Easy to use. Number of bins can be changed. Figure can be cut and pasted.
Free Statistics & Forecasting Software	http://www.wessa.net/rwasp_histogram.wasp	Uses R code you can edit. Replace the data with your own. Options to change bins, colour, axis labels, etc. Good site.
Easycalculation.com	https://www.easycalculation.com/graphs/create-histogram.php	Data must be separated by a comma. Choose number of bins. Graph is a bit ugly.
Shodor	http://www.shodor.org/interactivate/activities/Histogram/	Replace data and click update. Sliding scale for bins is nice. Colourful!
Mathcracker	http://www.mathcracker.com/histogram-maker.php	Data must be separated by commas. A bit clumsy and inflexible.

Excel if you like but I prefer just to put the data in rank order, using sort, and then look over them to choose the appropriate bin sizes. Then I just count the number in each bin, using the count function if it is a big data set. For data sets up to about 100 this is probably the easiest option and you can produce the histogram by inserting a bar graph. You can also calculate some simple descriptive statistics, also in a spreadsheet, like the mean, median, standard deviation (SD), skewness and kurtosis: these are described in Chapter 4. Alternatively import your data into SPSS or *R* Commander (these are described in Chapter 4) and you can produce a histogram and obtain descriptive statistics that way. There are even some free websites that produce a histogram for you (Table 3.3).

When you have plotted the histogram, and obtained the descriptive statistics, compare your results with those in Figure 3.2. If your graph is nice and symmetrical and does not have marked outliers to one side, then the mean and median should fall in the highest part of the graph, which is in the middle, and they should be very similar. How similar depends on the units you are using and on how variable the data are. If you also calculate the SD then you can assume that the difference between the mean and median should be very small relative to the SD. If it is not, your graph will not look symmetrical anyway. The skewness and kurtosis values should be close to zero. If your graph does not look symmetrical then just assume that the data are not normally distributed. There are various methods available for 'transforming' your data so that they are closer to a normal distribution, but I find them terribly confusing so they are beyond the scope of this book. It is easier just to use non-parametric tests, which do not assume a normal distribution. If your sample size is less than about 30 it is safer to stick to non-parametric tests even if your data are near normally distributed.

Having data that are skewed is not a problem for non-parametric tests. Similarly, if the histogram is a bit flat, or very sharply peaked, it does not matter. However, be wary of data sets that have clear outliers. Outliers are not a problem for non-parametric tests but are sometimes an indication of errors, so check them carefully. It is easy to mistype numbers into a spreadsheet. If your data are clearly bimodal it is likely to cause problems for all kinds of tests, so think carefully about why your data look like that. You may be sampling from two different populations by mistake.

If you have entered your data into SPSS or *R* Commander then you may choose to use one of the statistical 'tests for normality' such as the Shapiro–Wilks test (Figure 3.2) or a version of the Kolmogorov–Smirnov test. Personally I would avoid doing this because they tend to produce weird results. They test the null hypothesis that there is no difference between your sample and a perfect normal distribution, so the probability is based on the chances that the difference between your sample and a perfect normal distribution could have occurred just by chance, or luck. The problem is that very small samples are more likely to be different just by luck, and so are more likely to 'pass the test' and be declared not significantly different from normally distributed, whereas huge samples can fail the test even when they are clearly near enough to normally distributed to make no difference.

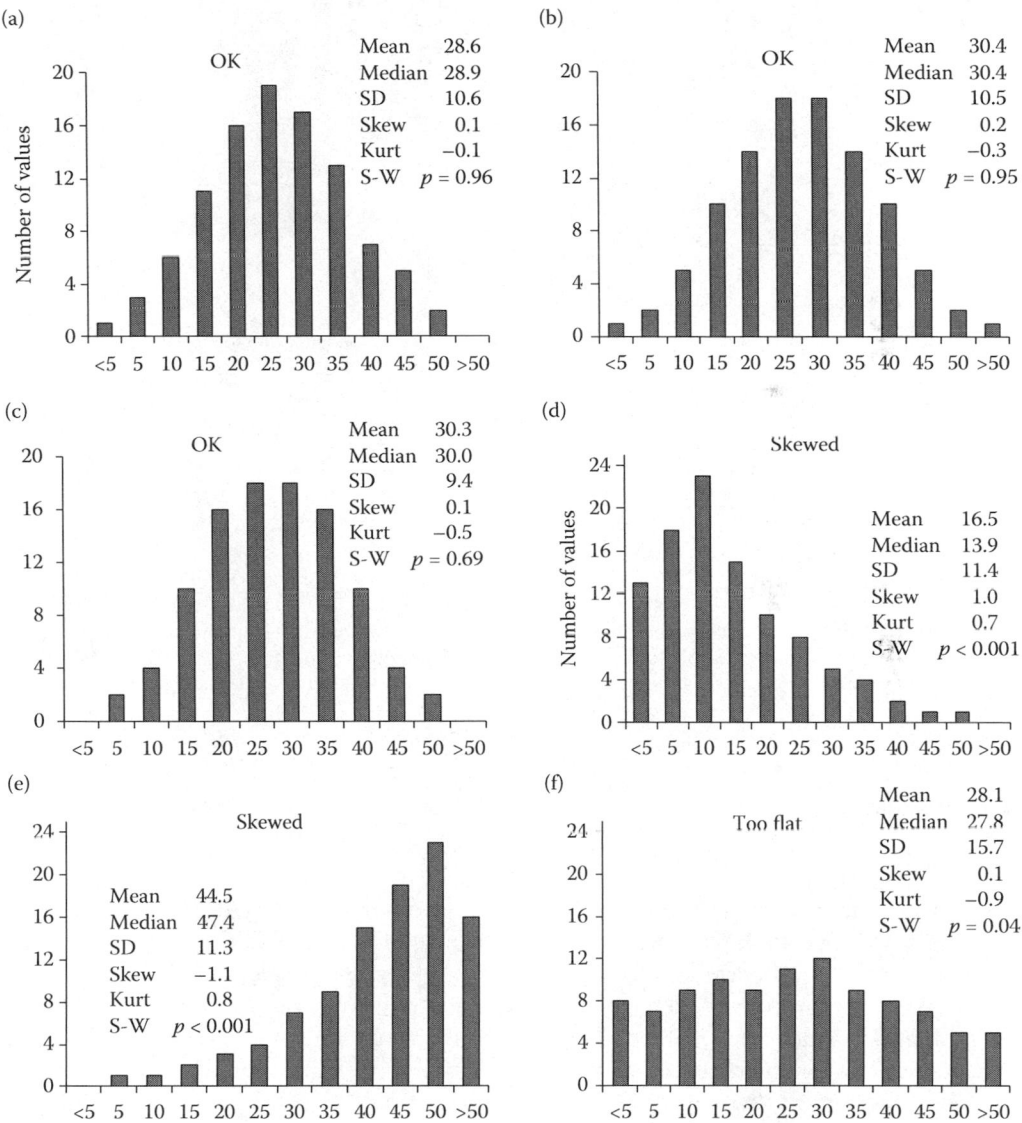

FIGURE 3.2

Histograms showing different shaped distributions. The mean, median, SD, skewness (skew) and kurtosis (kurt) values are shown together with the probability value given by the Shapiro–Wilks test for normality. Probability (*p*-values) below 0.05 is considered to indicate significant difference from a normal (Gaussian) distribution. The first three (a–c) are near enough to normally distributed to be suitable for parametric tests. Note that the mean and median values are very similar. (d) Is too strongly skewed to the left (positive skew) and (e) is too strongly skewed to the right (negative skew). (f) is too flat (negative kurtosis).   (*Continued*)

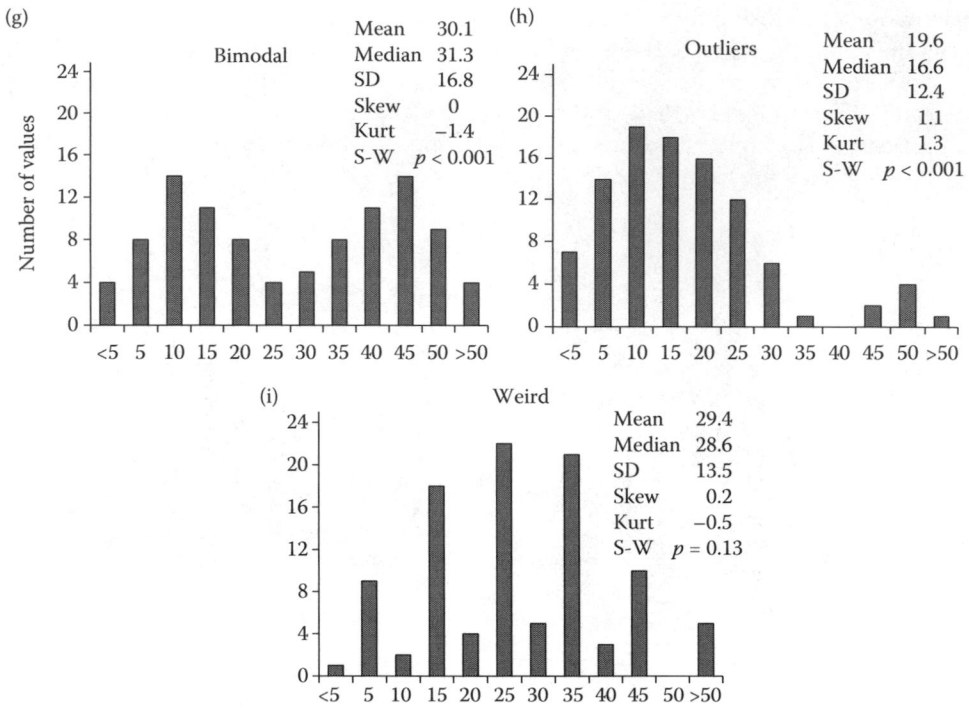

FIGURE 3.2 (Continued)
Histograms showing different shaped distributions. The mean, median, SD, skewness (skew) and kurtosis (kurt) values are shown together with the probability value given by the Shapiro–Wilks test for normality. Probability (*p*-values) below 0.05 is considered to indicate significant difference from a normal (Gaussian) distribution. (g) is bimodal, so the mean and median fall between the two peaks. It probably represents a mixture of two populations. (h) has some high outliers, which is particularly problematic for parametric tests. (i) is very weird with every other category under-represented, which is likely to be due to instrument or operator error.

3.4 Choosing the Right Test

Now that we have access to the internet and to powerful spreadsheet and other software packages, many of which are free, actually performing most statistical tests is not difficult. In many cases it is not necessary to perform any mathematical calculations at all. The difficult step now is finding the most appropriate test for your particular problem and type of data (Tables 3.4 and 3.5) and interpreting the results sensibly.

This book is a practical guide to using statistical tests, so rather than classifying the tests I have arranged them according to the kinds of data that students often collect and the kind of problems that students usually want to use them to address. The simplest and perhaps the most common case is where we want to know whether one sample is different from another. The tests that are suitable for this depend on the type of data that has been collected. For counts in categories data there are two tests included here; Fisher's exact test (Fisher 1922, 1935), which can be used even when sample sizes are very small and the chi-square test for two samples (Pearson 1900) which can deal with large samples. Where you have individual measurements the tests are divided according to whether the data

TABLE 3.4

Tests Covered in This Book Arranged According to Types of Test

Type of Test	Type of Data	Question	Data	Tests
One-sample tests	Counts in two categories	Are proportions as expected?	N	Binomial test or sign test
	Counts in more than two categories	Are proportions as expected?	N	Chi-square one-sample test
			O	Kolmogorov–Smirnov one-sample test
	Binomial sequence	Do two groups fall in random order?	N	One-sample runs test
Two-sample tests for paired samples	Counts in two categories	More change in one direction?	N	McNemar's test for the significance of changes
		Equal proportions?	O	Sign test
	Individual measurements	Is there a shift in where the middle lies?	O	Wilcoxon's matched-pairs signed-rank test
			I	*Paired sample t-test*
		Are the pairs correlated?	O	Spearman's rho
			I	*Pearson's correlation*
		What is the shape of the relationship?	I	*Regression analysis*
Two-sample tests for independent samples	Counts in two categories	Are counts contingent on groups?	N	Risk ratio and odds ratio
				Chi square two-sample test
				Fisher's exact test
	Counts in more than two categories	Are counts contingent on groups?	N	Chi square two-sample test
				Fisher's exact test
	Individual measurements	Do the samples differ?	O	Kolmogorov–Smirnov two-sample test
		Difference in where the middle lies?	O	Mann–Whitney *U*-test
			I	*Student's t-test*
		Equal variability?	O	Siegel–Tukey test
			I	*F-test for equality of variance*
		Any difference	I	Kolmogorov–Smirnov two-sample test for continuous data
Tests for several independent samples	Counts in categories	Are counts contingent on groups?	N	Complex chi square
	Individual measurements	Do groups differ?	N	Kruskal–Wallis *H*-test
		Trend across the groups?	O	Jonckheere–Terpstra trend-test
Tests for several linked samples	Individual measurements	Do groups differ?	N	Friedman's test
		Trend across the groups?	O	Page's trend-test

Note: Data or groups may be nominal (N), ordinal (C) or interval (I) in nature and parametric tests are italicised.

TABLE 3.5

Tests Covered in This Book Arranged According to Types of Data

Type of Data	Samples	Question	Data	Tests
Counts in two categories	Single sample	Are proportions as expected?	N	Binomial test or sign test
		Is order random?		Runs test for randomness
	Two samples paired data	More change in one direction?	N	McNemar's test for the significance of changes
		Equal proportions?	O	Sign test
	Two independent samples	Are counts contingent on groups?	N	Risk ratio and odds ratio
				Chi square two-sample test
				Fisher's exact test
Counts in more than two categories	Single sample	Are proportions as expected?	N	Chi square one-sample test
			O	Kolmogorov–Smirnov one-sample test
	Two independent samples	Are counts contingent on groups?	N	Chi square two-sample test
				Fisher's exact test
		Do the samples differ?	O	Kolmogorov–Smirnov two-sample test
	Several independent samples	Are counts contingent on groups?	N	Complex chi square
Individual measurements	Two samples paired data	Is there a shift in where the middle lies?	O	Wilcoxon's matched-pairs signed-ranks test
			I	*Paired sample t-test*
		Are the pairs correlated?	O	Spearman's rho
			I	*Pearson's correlation*
		What is the shape of the relationship?	I	*Regression*
	Two independent samples	Is there a difference in where the middle lies?	O	Mann–Whitney *U*-test
			I	*Student's t-test*
		Equal variability?	O	Siegel–Tukey test
			I	*F-test for equality of variance*
		Any difference?	I	Kolmogorov–Smirnov continuous test
	Several samples: linked data	Do groups differ?	N	Friedman's test
		Trend across the groups?	O	Page's trend-test
	Several independent samples	Do groups differ?	N	Kruskal–Wallis *H*-test
		Trend across the groups?	O	Jonckheere–Terpstra trend-test

Note: Data or groups may be nominal (N), ordinal (O) or interval (I) in nature and parametric tests are italicised.

are paired (linked) or independent. For paired samples there are three tests; the sign test that can be used even if all you have is nominal data (such as yes/no, higher/lower), the Wilcoxon matched-pairs signed-ranks test (Wilcoxon 1945) that requires some measurement of the difference so that the differences can be placed in rank order, and the paired-samples t-test. Where the measurements are independent the Mann–Whitney U-test (Mann and Whitney 1947) is the non-parametric equivalent of Student's t-test (Student 1908) and they both test for a difference in location (where the middle of the two samples lie). There are also some tests for differences in the variability of the two samples or for any difference.

Where you wish to compare more than two samples a common approach is to use two-sample tests multiple times. So if you had five samples, for example, you might use Student's t-test to compare all possible pairs. This is actually bad practice because it changes the probability of a difference between two samples occurring just by luck. When we use the usual significance level of 5%, ($p = 0.05$) we are accepting that there is still a 5 in 100, or one in 20, chance that the difference that we observe has occurred just by luck and that the two samples are actually drawn from the same population. With five samples (A–E) there are ten possible combinations (AB AC AD AE BC BD BE CD CE DE) which means that you have ten chances of finding a significant difference rather than just one chance. There are different ways of dealing with this problem. One is to adjust the probabilities to take account of the fact that you are using multiple tests and one such approach (the Bonferroni correction: Dunn 1959, 1961) is covered in Section 9.2. The other is to use tests that have been formulated specifically for the case where you wish to compare multiple samples, and those are the ones that are described in Chapter 9. The parametric approach that deals with comparing multiple samples is called analysis of variance (commonly called ANOVA) but there are also non-parametric versions available that can deal with either linked (Friedman's test: Friedman 1937, 1940) or independent data (Kruskal–Wallis H-test: Kruskal and Wallis 1952). Where the data are counts in categories the most common approach is to use the chi-square test, but this test has strict sample size assumptions. Where these are not met it is often possible to use Fisher's exact test.

Looking for trends in data is a common problem for geography students and can apply either to individual measurements or to data that fall into categories. The most common approach to dealing with any data that fall into categories is to use the chi-square test, but that is actually one of the weakest tests and it treats the categories as nominal rather than ordinal, so the order of the categories makes no difference. Where independent sample measurements are sufficiently precise to allow the data within the categories to be placed in rank order a more powerful alternative, which does make use of the order of the categories, is the Jonckheere–Terpstra trend-test (Terpstra 1952; Jonckheere 1954). This test is very rarely used in geography but is ideally suited to a lot of geography student projects where the desire is to check whether there is a general trend across several categories rather than just a difference between the categories. Age groups, income groups and size categories are common examples. Where the data are linked rather than independent the equivalent is Page's trend-test (Page 1963). Where individual measurements are available along a continuum, rather than in categories, then relationships can be investigated using either non-parametric correlation methods based on ranking, two of which are commonly used (Spearman's rho is covered here) or using parametric correlation and regression. You can use these methods to look at trends over space or over time. You have to be very careful, however, because when you take measurements along a spatial transect, or through time, there is often some 'memory effect' between one measurement point and the next, so that each data point may not be entirely independent from the neighbours on either side. This potential complication is discussed in the sections on parametric correlation and regression, which is where it is most important.

When we look at trends we are in effect looking at the relationship between two variables, one is the thing we are measuring and the other might be distance or time. The same methods that are used for looking at trends can also be used for looking at other relationships. The most common examples involve individual measurements in pairs, when the appropriate methods are correlation, which can be non-parametric, based on ranks, or parametric based on the actual measurements. Regression techniques allow you to demonstrate the shape of the relationship, for example whether a unit increase in one parameter generally leads to a particular increase in the other, which is a linear relationship, or whether the magnitude of response tends to increase or decrease as you move up the measurement scale, indicating a non-linear relationship.

Regression methods are particularly useful because they allow you to make predictions based on your data. For example, if you can define the relationship between age and salary for a particular job then you can predict the most likely salary for a person of a given age. You can even use these methods, with great care, to predict into the future. For example if you can define the relationship between average summer temperature and the frequency of droughts then you might be able to predict drought frequency based on predicted changes in temperature. Regression is a complex topic but it is now very easy to perform and, although it falls in a different category to most of the techniques in this book, it is dealt with in some detail because it is so useful and so easy to get wrong.

The remaining category of tests, and that which is dealt with first, is a little odd because it involves just a single sample and it is not so easy to define a single question to which these methods provide an answer. Commonly, if you have collected just a single sample of something then there is already something else with which you wish to compare it. Students often define the other thing in a vague way such as 'what you would normally expect'. A clearer definition is required before statistical tests can be used on single samples and a few examples are provided to help. Common questions include 'is my sample representative of the group I am interested in?' and 'is this sample biased in some way, for example with regard to gender?'

References

Dunn, O. J. 1959. Estimation of the medians for dependent variables. *Annals of Mathematical Statistics* 30: 192–197.

Dunn, O. J. 1961. Multiple comparisons among means. *Journal of the American Statistical Association* 56: 52–64.

Fisher, R. A. 1922. On the interpretation of χ^2 from contingency tables, and the calculation of P. *Journal of the Royal Statistical Society* 85: 87–94.

Fisher, R. A. 1935. *Statistical Methods for Research Workers.* Oliver and Boyd, Edinburgh.

Friedman, M. 1937. The use of ranks to avoid the assumption of normality implicit in the analysis of variance. *Journal of the American Statistical Association* 32: 675–701.

Friedman, M. 1940. A comparison of alternative tests of significance for the problem of m rankings. *The Annals of Mathematical Statistics* 11: 86–92.

Jonckheere, A. R. 1954. A distribution-free k-sample test against ordered alternatives. *Biometrika* 41: 133–145.

Kruskal, W. H. and Wallis, W. A. 1952. Use of ranks in one-criterion variance analysis. *Journal of the American Statistical Association* 47: 583–621.

Mann, H. B. and Whitney, D. R. 1947. On a test of whether one of two random variables is stochastically larger than the other. *Annals of Mathematical Statistics* 18: 50–60.

Page, E. B. 1963. Ordered hypotheses for multiple treatments: a significance test for linear ranks. *Journal of the American Statistical Association* 58: 216–230.

Pearson, K. 1900. On the criterion that a given system of deviations from the probable in the case of a correlated system of variables is such that it can be reasonably supposed to have arisen from random sampling. *Philosophical Magazine Series* 5 50: 157–175.

Student (Gosset, W.S.) 1908. The probable error of a mean. *Biometrica* 6: 1–25.

Terpstra, T. J. 1952. The asymptotic normality and consistency of Kendall's test against trend, when ties are present in one ranking. *Indagationes Mathematicae* 14: 327–333.

Wilcoxon, F. 1945. Individual comparisons by ranking methods. *Biometrics Bulletin* 1: 80–83.

4

Tools of the Trade

4.1 Introduction

A multitude of tools are available to perform statistical tests and I am not going to try to cover many of them here. They fall into four categories and I will just choose one or two examples from each. The categories are using a calculator, a spreadsheet, dedicated statistical software or an online calculator. For spreadsheets I will give instructions for using Excel, because it is so widely available, but if you have no access to Excel you can use 'LibreOffice Calc' which is free. For the dedicated software I will use SPSS (IBM SPSS Statistics), because it is widely used by geographers, and R Commander because it is very powerful and free to use. Online calculators are very convenient but need to be treated with caution, and there is no guarantee that they will remain active after publication of this book. For most tests I provide a table with some sites that I have checked together with some notes on how they perform. I have also produced a set of simple spreadsheet-based calculators that will perform most of the tests in this book. They are no substitute for a real statistics package but they will allow you to apply your chosen test, or a combination of tests, quickly and easily without having to learn how to use any fancy software.

4.2 Arithmetic

I hated mathematics at school. I was in a class where the main job of the teacher was to maintain enough control to stop the pupils from killing and eating each other. I still hate mathematics and I am useless at it, but I am not proud and when I get stuck I just ask someone who is better than me, which is sometimes a colleague but frequently a student. Fortunately you do not need to be good at mathematics to cope with most of the statistical tests that you will need as a geographer. However, you will need to be able to perform some of what is patronisingly referred to as 'simple arithmetic'. I have to confess that I have never found arithmetic simple and spent many years being completely frustrated because I would often get the wrong answer even when the question appeared to be very simple. I have since discovered that I am not alone, and when I teach my first year students I give them a simple test and typically find that between a third and a half of the students cannot reliably perform 'simple arithmetic'. Do you have the same problem?

Consider the following questions:

$2 \times 2 + 2$	is the answer:	A: 6	or	B: 8
$2 + 2 \times 2$	is the answer:	A: 6	or	B: 8
$1 + 3 \times 3$	is the answer:	A: 10	or	B: 12
$(1 + 3) \times 3$	is the answer:	A: 12	or	B: 10
$2 + 4 \times 5$	is the answer:	A: 22	or	B: 30
$(2 + 4) \times 5$	is the answer:	A: 30	or	B: 22

The correct answer in each case is A, not B. If you got them all right well done, you have been well taught. If you got any of them wrong, do not despair or berate yourself. It does not mean you are stupid, you are in the same position as me when I was a student. I had spent all of my years at school sitting through compulsory lessons in mathematics learning about all sorts of equations that have never been of any use to me whatsoever. What they failed to teach me was that mathematics is not written in the same way as English; you cannot simply read the problem from left to right and make each calculation in turn. The different parts of the problem have to be completed in a specific order, irrespective of the order that they are written in. The correct order is defined by the BODMAS rule, which is:

B: Brackets

O: Order (to the power of)

D: Divide

M: Multiply

A: Add

S: Subtract

I have absolutely no memory of having been taught this critical sequence at school and as a result I would get calculations correct if they happened to be written in a convenient order, and otherwise I would get them wrong and could never work out why! Not knowing this sequence has nothing to do with how clever you are. If you have not been taught it you cannot just work it out on the basis of logic because it is just a set of rules that the cursed mathematicians have made up. In the case of $2 + 2 \times 2$, for example, you have to multiply before you add, so it becomes $2 + 4$, not 4×2 and so the correct answer is 6 not 8. I must confess that I did not discover this rule until I was a university lecturer, and it is perhaps the most useful single thing that I have ever learned! If you apply the BODMAS rule you should be able to cope with all of the calculations in this book.

I suspect that, given modern access to computers and the Internet, most readers will not be interested in making any calculation by hand, and I do not really blame them. It is the geographical problems that are interesting, not the mathematics. However, for completeness, and in case your evil professor insists that you make the calculations by hand, I have included the mathematical steps for most tests, but have generally kept them out of the way of the main text by putting them in boxes or at the end of chapters. I have tried to include every step in the calculations, however small, so that even if, like me, you lack confidence in mathematics you should be able to follow the procedure and perform the same steps with your own data.

4.3 Using a Calculator

I conducted my doctoral research in the mountains of Norway, living in a tent in the middle of nowhere, in the days before portable computers. The evenings were long, and the drink prohibitively expensive, so I did most of my data analysis by hand with nothing other than graph paper and a calculator. Unless you are similarly constrained, I do not recommend it. It is a very dull way to spend your time and you are likely to make mistakes, so a spreadsheet is a much better option. Tests that involve putting data in rank order, which applies to many tests in this book, are particularly tedious to complete by hand, and a calculator does not help with the ranking, whereas a spreadsheet will rank your data very easily.

Some tests are relatively easy to complete by hand and if you are dealing with very small samples it may be less bothersome than using a computer. Much more important than actually performing the calculations by hand, however, is an ability to scan your data and make an educated guess at what the result is likely to be. This may seem like a waste of time, given that you do not intend to actually perform the calculation, but it is a very useful skill. It ensures that if you make a mistake somewhere, which is easy to do when using a spreadsheet or when cutting and pasting data, and produce a result that is clearly nonsense, you will spot it rather than including it in your report. For that reason alone it is worth thinking about what each method is based on, so that you understand the logic, if not the mathematics. If you are comparing two samples, for example, to see if they are significantly different, you can look at the two mean values and the spread of results and just decide whether they look very similar or very different. If they look very similar you should expect a result that indicates a small effect size and suggests the difference is not statistically significant. If they are clearly very different you expect a large effect size and result that is strongly significant. If they are a bit different, but not different enough to be really sure, you would expect a result that is close to significant, but could be on either side of the threshold.

4.4 Spreadsheets

My colleagues who are very good at mathematics and statistics generally frown on the use of spreadsheets to perform statistical calculations, preferring dedicated software. However, I find spreadsheets to be very useful in my own research, not least because they allow me to draw the diagrams that I need as well as performing the statistical calculations. By moving data around in a spreadsheet, and experimenting with different graphs, you often spot things you might otherwise have missed and that can lead to new and interesting hypotheses. One problem is that it is easy to make a mistake, by referring to the wrong column for example, and by copying and pasting you can spread mistakes around that are difficult to trace, but as long as you are reasonably careful, and especially if you have enough understanding of the method you are using to be able to estimate the likely result, and thus notice if you make a mistake and produce nonsense, I think spreadsheets are often the most convenient way for students to complete their statistical analysis. A good tip is to use one of the examples from this book when you set up your spreadsheet, so that you can see whether you are getting the right result, then replace the data with your own.

Excel, which is part of Microsoft Office, is perhaps the most widely used spreadsheet package, and if you are a student you may have free access to it. If not I suggest using 'LibreOffice Calc' which is part of the 'LibreOffice' suite of software. It is just as good as Excel for most purposes but is completely free. To download the whole suite, which includes an excellent word processor and software for giving presentations (an alternative to PowerPoint), just go to the website (https://www.libreoffice.org/) and follow the simple instructions. Download is fast, safe and completely free, but if you can afford it consider making a small donation to support this tremendous endeavour. It is a great example of global collaboration.

Most of the functions you will need in Excel or LibreOffice are already there and some of the most useful are listed in Table 4.1. Different versions of Excel seem to vary for no obvious reason other than to make life difficult, so if you have some old or very recent version you may need to search around a bit. In some versions you have the option to use an 'Analysis toolpack' that includes a range of statistical tests, but sometimes this is hidden away as an 'add in'. To access it you have to click on the multicoloured 'Office button' in the top left corner and choose 'Excel Options' which appears as a button on the bottom of the pop-up. When you click this the list on the left includes 'Add-Ins' and there is a box at the bottom that allows you to 'Manage Excel add-ins'. Click on this, tick the analysis tool pack and click OK. Now when you go back to Excel and click on the tab marked 'Data' you should see 'data analysis' as one of the options, probably at the far right of the screen. It is quite a handy add-in, particularly if you are dealing with big or repetitive data sets, but not really essential. I am not aware of a similar add-in for LibreOfficeCalc.

If you are using Excel on a personal computer, I strongly recommend that you download the free statistics 'Resource Pack' provided by Charles Zaiontz (2015) at 'Real Statistics using Excel', who also provides explanations and examples via his excellent website at http://www.real-statistics.com/. This combination of Resource Pack and website is a really fantastic free resource and if you would rather not bother with dedicated statistical packages it allows you to perform almost all of the tests covered in this book just using Excel. Apart from being very good at statistics, Charles is clearly an extremely gifted teacher and his explanations are a model of clarity. To download the Resource Pack go to his website and just follow the instructions. If you are using university computers you probably will not have download rights, so you will have to ask your IT department to download it.

TABLE 4.1

Descriptive Statistics and the Appropriate Functions in Excel and Calc.

Statistic	Function
Average or mean	Average
Median	Median
Maximum value	Max
Minimum value	Min
Standard deviation	Stdev
Variance	Var
Skewness	Skew
Kurtosis	Kurt
Confidence intervals	Confidence.t
Count how many	Count
Sum (add up)	Sum

TABLE 4.2

A Very Inefficient Way of Using a
Spreadsheet to Make Calculations

	A	B
1	4	$= 4 \times 4$
2	6	$= 6 \times 6$
3	8	$= 8 \times 8$
4	10	$= 10 \times 10$
5	12	$= 12 \times 12$

Note: In this case every cell has to be
completed individually.

If you do not have much experience of using spreadsheets they can seem a bit daunting. There is no space here to provide a real primer, and there are plenty of books specifically on Excel (Salkind 2015; Schmuller 2013), but some tips might help. The main advantage of a spreadsheet is that you can make a lot of calculations very efficiently. A common mistake that students make when learning to use them is to complete each cell individually with numbers rather than using and copying equations. For example, if you wanted to find the square of a series of numbers you could type the numbers in one column and in the next column you could type all of the numbers again to get the square (Table 4.2). So, if in column A cells A1 to A3 contained the numbers 4, 6 and 8 you could type in column B ' = 4*4', ' = 6*6' and so on (= means 'this is a calculation' and * is used for multiply, don't include the '' or it will treat everything as text instead of numbers). That would work but it would be tedious.

The alternative is to use a formula that includes reference to the cell where the number that you want is located (Table 4.3). If you do this and then just copy and paste the cell (just click in the little box in the bottom right-hand corner and drag it down) the spreadsheet will assume that you want to perform that calculation on each number in column A and you will get the same results but with less effort. With big data sets that makes a big difference. In this case cell B1 would contain the formula ' = A1^2' which means take whatever is in cell A1 and square it (^ means 'to the power of'). When you copy it down to cell B2 it will automatically change the reference cell from A1 to A2 and so will give you the answer to 6 squared rather than 4 squared again. If you have a go you will see what I mean.

TABLE 4.3

A Much More Efficient Way
of Using a Spreadsheet to
Make Calc.

	A	B
1	4	$= A1^2$
2	6	
3	8	
4	10	
5	12	

Note: By using a formula that refers to
the cell to the left it is only necessary to complete one cell manually and it can then be copied to
fill the rest of the column.

Usually the assumption that you want to perform the same operation but on a different cell each time is justified, but sometimes you want to keep one part of the equation fixed. For example if you wanted to take a single number and multiply it by many different numbers then you can 'protect' one part of the equation by adding a dollar ($) sign. Placing the dollar sign before the letter fixes the column and placing it before the number fixes the row. If you fix both the column and the row then the formula always refers to that single cell even when the formula is copied and pasted elsewhere in the spreadsheet.

One of the most useful operations that a spreadsheet can perform is to re-arrange your data for you, for example by placing it in rank order. This is very helpful for many of the methods used in this book. However, you have to be careful because I know from bitter experience that if you make a mistake you can make a mess of your data and become totally confused. My advice is to keep your raw primary data in a separate spread sheet that you do not use for any data manipulation. That way if it all goes horribly wrong you can start again without having to type it all in again. As you work with your data keep a notebook to hand, or an open word processor file, and keep notes of what you are doing. When you achieve something useful make a note of the name of the spreadsheet and the page name. When you are exploring your data and trying a range of methods, and maybe making a few mistakes as you learn, you can generate a lot of spreadsheets and if you are not organised you will probably get confused. Students often bring their data to me for help and when they open the directory of their memory stick they pore through the long lists of spreadsheet files, all with similar names, and desperately try to remember where the critical bits are hidden. A good policy is to keep two folders: one for 'current versions' and one for 'old versions' of your files.

When it comes to sorting your data in Excel the 'Sort' option is on the 'Data' page. You highlight the data that you want to sort and then choose which column to use as the sorting parameter. For example if you have data in three columns labelled name, exam mark and gender then you could sort those data alphabetically by name, in ascending or descending order of exam mark, or you could separate the male and female students to see if there is a gender difference in the results. This is where it can all go horribly wrong however, because you have to be sure to pick up all of the data that you want to be sorted. A common mistake is to pick up some of it but leave some behind, so that everything just becomes hopelessly muddled. You might pick up the columns for name and mark, for example, but leave the gender behind, in which case gender will not remain linked to the name. One method that helps is always to add a column of numbers that simply lists your data from one to whatever. When you are sorting always include this column, so that if it goes wrong you can re-sort using this column as the sorting parameter.

When starting a new spreadsheet I find it is useful to leave about ten rows free at the top and arrange the data in columns. This gives you space to perform calculations that refer to the columns below and you can see the results and the headers at the same time (Table 4.4). It is easier to draw graphs if the data are arranged in columns. This is also a very efficient layout because if you make a calculation for the data in one column, such as calculating the average, you can just drag and copy that cell and it will perform the same calculation for the other columns. If you copy your data onto new pages or into a new spreadsheet it is a good policy to make sure that the data start on exactly the same line. This means that you can easily copy and paste the whole column and move it around without having to check that you always have the exact cell in the right place. When you copy cells with equations between spreadsheets it really helps a lot if the data do not move up or down the spreadsheet; that is a recipe for total confusion. When working with spreadsheets it really does help to keep everything neat and tidy and to keep clear notes.

TABLE 4.4

Example of a Well Laid-Out Spreadsheet (Middle Section Removed)

	A	B	C	D	E	F	G	H	I	J	K
1											
2		Count	100								
3		Average	28.63								
4		Median	28.85								
5		St dev	10.57								
6		Skew	0.06								
7		Kurtosis	−0.22								
8											
9											
10		ref	A	B	C	D	E	F	G	H	I
11		1	2.4	3.7	8.3	0.0	7.2	0.2	0.1	1.2	3.1
12		2	7.2	9.3	8.8	0.5	11.8	1.2	2.7	1.4	5.3
13		3	8.7	9.6	14.1	0.9	17.0	2.4	3.6	2.2	5.8
14		4	9.3	11.8	14.1	1.7	19.3	3.2	3.6	3.0	6.3
15		5	10.3	12.0	14.5	2.0	21.2	3.2	5.0	3.4	7.2
105		95	47.4	48.8	45.4	37.8	57.2	55.0	54.3	47.3	49.9
106		96	48.2	50.0	45.9	38.3	57.5	55.4	54.4	50.7	55.2
107		97	49.1	50.0	47.9	42.0	57.7	56.2	57.8	51.1	57.4
108		98	49.2	52.2	48.1	44.4	59.5	57.6	58.0	53.3	59.5
109		99	51.3	54.7	50.9	47.3	59.6	58.6	59.3	54.8	59.6
110		100	53.3	56.7	54.7	54.4	59.7	59.1	59.5	59.0	60.0

Note: The data sets are aligned in columns, each starting on the same line, and there is a reference column that can be used to re-order the series if necessary. There is plenty of space above the columns to place the summary statistics. Note that since the series are the same length you only have to add the summary functions above the first column (as shown) then they can just be copied to the right.

4.4.1 Assigning Ranks in a Spreadsheet

One of the useful functions of spreadsheets, especially when you are using non-parametric statistics, is the ability to arrange numbers in rank order, which allows you to replace the original numbers with ranks. One way to do this is to use the 'sort' function, which puts the numbers in rank order, and then add the ranks in a new column. If you do that, however, you must be very careful about the way that you deal with tied ranks. For almost all of the tests in this book there is a simple rule: if the ranks are tied (i.e. two or more numbers are exactly the same) then they each receive the average of the ranks that are tied. So if ranks 4 and 5 are tied they each get a rank of 4.5 and the next one gets a rank of 6. If ranks 4, 5 and 6 are tied they each get a rank of 5 and the next one gets a rank of 7.

In recent versions of Excel a very useful function has been added that allows you to assign ranks to your data without sorting them; it is the RANK.AVG function. Note that this is not the same as the RANK function because the latter does not deal with ties in the correct way. You have to be careful with this function, because it is easy to mess it up. You have to identify which number you want to assign a rank to, which is easy, the problem comes when you define the 'reference' data set. It is important that you get the right reference set and that it remains correct for every cell. For example, if you wish to obtain the ranks for a single column of data, which might be the case if you were performing

TABLE 4.5

Using the RANK.AVG Function to Place Data in
Rank Order

	A	B
1	25	= RANK.AVG(A1,$A$1:$A$5,1)
2	17	
3	21	
4	25	
5	27	

Note: It is important to use dollar signs to fix the reference set from which you wish to draw the ranks.
Otherwise when you copy the equation to other
cells the reference set will move.

Spearman's rank correlation, you would use the formula in Table 4.5. It means assign a rank to the number 25 in cell A1 relative to all of the numbers in the column from A1 to A5. The final number '1' means rank from low to high, if you use a zero it will rank from high to low. Note that it is critical that you include the dollar signs when you define the reference group, so that when you copy this equation down the column the reference set does not change. You have to add the dollar signs by hand, it is not automatic, and if you forget to do that you will produce nonsense.

For some tests you need to obtain a rank for each number in a column but the reference set includes more than one column (Tables 4.6 and 4.7). This is the case for the Mann–Whitney *U*-test, Kruskal–Wallis *H*-test and several others in this book. It is not difficult, you just need to be clear about what is the appropriate reference set. In the example below we want to know the ranks of each number relative to the full data set of three columns (for the Kruskal–Wallis *H*-test) so the reference set includes all three and again dollar signs are used to fix it so that it does not change when the equation is copied down and to the right.

Unfortunately there is, as yet, no function exactly equivalent to 'RANK.AVG' in LibreOffice Calc so it is safer not to use the rank function. Use the 'sort' option to get your data in rank order and then allocate the ranks by hand, which is quite quick. Make sure that you share the ranks correctly, as described above and in Table 4.8. If you get it right

TABLE 4.6

Using the RANK.AVG Function in Excel to Translate Sample Values into Ranks Where the
Reference Set Is All of the Groups

	A	B	C	D	E	F
1	Sample A	Sample B	Sample C	Ranks A	Ranks B	Ranks C
2	17	35	36	= RANK.AVG(A2,$A$2:$C$6,1)		
3	25	34	45			
4	32	38	43			
5	18	26	47			
6	5	41	40			
7						
8		Sum or ranks		= SUM(D2:D6)		

Note: This procedure is used for the Mann–Whitney *U*-test, where you wish to compare two groups, and for the Kruskal–Wallis *H*-test when there are more than two groups. Note that dollar signs are used to fix the reference set so that the contents of cell D2 can be copied to complete the three columns of ranks.

TABLE 4.7

Same as Table 4.5 But Displaying the Results

	A	B	C	D	E	F
1	Sample A	Sample B	Sample C	Ranks A	Ranks B	Ranks C
2	17	35	36	2	8	9
3	25	34	45	4	7	14
4	32	38	43	6	10	13
5	18	26	47	3	5	15
6	5	41	40	1	12	11
7						
8		Sum of ranks		16	42	62

Note: If these three samples had been drawn from the same population, we would expect the ranks to be evenly distributed between the three columns, so that when you add up the sum for each column they should be about the same. The large difference in the sum of ranks suggests that the three samples are not drawn from the same population and that is the basis for the Kruskal–Wallis *H*-test described in Chapter 9.

TABLE 4.8

An Example of How to Deal with Tied Ranks in a Spreadsheet

	A	B	C	D	E
1					
2		Count	Values	Ranks	If
3		1	3.1	1	
4		2	5.3	2	ok
5		3	5.8	3	ok
6		4	6.3	4	ok
7		5	7.2	5.5	ok
8		6	7.2	5.5	TIE
9		7	7.8	7	ok
10		8	8.4	8	ok
11		9	8.5	9	ok
12		10	9.2	10	ok
13		11	12.0	11	ok
14		12	14.4	12	ok
15		13	15.1	14	ok
16		14	15.1	14	TIE
17		15	15.1	14	TIE
18		16	15.4	16	ok
19		17	16.8	17	ok
20		18	17.0	18	ok
21		19	17.3	19	ok
22		20	17.4	20	ok
23		21	17.5	21.5	ok
24		22	17.5	21.5	TIE
25		23	17.7	23	ok
26		24	18.0	24	ok
27		25	18.1	25	ok

Note: The 'count' column helps you to keep on track and the 'IF' function can be used to identify ties automatically.

your last rank should equal the sample size. It helps if you add a 'count' column next to the data. Also with a big data set it helps if you use the 'If' function to help you to identify ties. In Table 4.8, for example, cell E4 has the formula [=IF(C4=C3), 'TIE', 'ok'] which means if the value in column C is the same as the one above write the word TIE, otherwise write ok.

4.5 SPSS

To be honest I am not a great fan of SPSS and I try to avoid using it whenever possible. I like to analyse my data step by step, so that I formulate clear hypotheses and then test them. When I have finished one stage of the analysis I move on to the next, formulating new hypotheses and choosing the appropriate tests. I like to be in control of what is being done to my data and I am not interested in the results of any test that I do not understand. SPSS is set up in such a way that it seems to encourage a sort of shotgun approach, with default settings where everything goes into the mix and gets compared with everything else. Getting all of the data in there in the right format is often more trouble than it is worth and if you go for any of the defaults the amount of output is just bewildering. I find that with students it tends to produce more confusion than enlightenment. SPSS is, however, a very powerful program and can cope with most kinds of analysis and with almost everything that is covered in this book. There are some tests where it is really the easiest and best option. If you are interested in using it to full capacity I recommend the excellent book by Andy Field (2013). If, like me, you are not really interested in using it at all, but just want to perform the appropriate test and get the correct result, then I find that the best approach is to keep it simple and dictate all of the operations rather than letting SPSS make any decisions for you. It is tempting to press a lot of buttons in SPSS and get a lot of results, but unless you really understand what is happening to your data you are very likely to produce nonsense. It is much better to do something relatively simple, but do it well, than to perform a lot of complicated analyses that you can neither explain nor interpret.

I find that the best way to use SPSS is alongside a spreadsheet. This way you can keep the data in the spreadsheet and manipulate it into the format that you need for a particular test, and then cut and paste it across to SPSS. When the data are entered you can name them and then you also have to define what kind of data they are: the options are 'nominal', 'ordinal' or 'scale'. Use nominal for names that cannot be placed in a logical rank order, ordinal for things that can be placed in a sensible rank order and scale for individual measurements. In SPSS it is important that the data are entered in the right format for particular tests and for those tests where SPSS is a good option I explain the format. There are some YouTube videos that I found helpful as well.

4.6 *R Commander*

The 'thing called R', or the 'R-project for statistical computing', is not really a statistical package, it is a 'statistical environment'. Don't worry; I have no idea what that means either. What I do know is that R is free to download and incredibly powerful; much more powerful than anything you can buy. The downside (and it is deep, deep down!) is that R is not operated by clicking with a mouse. When you open it there is virtually nothing on the screen and to get it to do anything you actually have to write lines of code. I realise of course that many people enjoy writing code and are very good at it, but I doubt if any of them are reading this book. I first came across R because I work partly in the field of climate change and many scientists in that field use R to deal with huge data sets. My colleagues who work with global remote sensing data also use R. I wanted to learn to use it but found it very difficult because I have no experience of writing code.

The big breakthrough for me was when someone pointed me towards *R* Commander. This is effectively a plug-in that provides a friendly face for R so that rather than opening a blank screen there are some options to click on that at least give you a chance to get your data in there and make a start. It was designed by John Fox, at McMaster, who should be sainted for his efforts. Apparently he has a book due out in 2016 explaining how to use it. It is actually really easy to use and there are now lots of additional add-ons that extend its capabilities. *R* Commander works a bit like a statistics package, in that it has menus that allow you to choose tests and data sets by clicking with a mouse, but it also has the great advantage that it shows you the lines of code that it is writing and the lines of code that form the output. This makes it very much easier to learn to use R and means that you can alter lines of code to change the output. A simple example is that you can look at the code that tells R to produce a particular graph and change it to alter the colours or symbols. If you hate computers then R, even with *R* Commander, is probably not for you, but if you want to learn a little bit of coding it is a good place to start and adding 'competent at R' to your CV will not do your earning potential any harm either.

To access *R* Commander you first have to download R, which given the amazing things it can do is remarkably easy. First go to the main website (http://www.r-project.org/) and then choose your 'CRAN Mirror' (no, sorry I have no idea either) and choose somewhere not too far from where you live. Then just follow the instructions. It is fast, safe and free. Once it is installed you can open R by clicking on the icon on your desktop or find it in your list of programmes. Opening it gives you a window called 'R console' with some lines of text and at the top of the page a few menu items and icons to click on. The one to choose is 'Packages' then 'Install Packages'. A list called 'CRAN mirror' will appear again so choose somewhere close to where you live. Now you will be faced with a huge list of packages to choose from. These have all been produced by other users, to do all manner of things that are impossible to guess from the name, and so the list is not actually of much use unless you already know what you want. The one you need is called 'Rcmdr'. It should download and then a line of code will appear in the R console to tell you where it is. To use it, however, you now have to 'load' it. You have to do this every time you want to use it. To load it click on 'Packages' again and click 'Load Packages'. Now you will see a much shorter list of packages that I assume are saved on your own computer rather than the remote server. Choose Rcmdr again and click ok and it should open in a new window. At first glance it may not look much friendlier than the original R console, but if you run through the options you will see that there are now a lot of very useful functions available to you at the click of a mouse. Some of the options will appear greyed out, but those will become available when you input the right kind of data.

One of the tricky things about using R is getting your data in and out of there in the right format. For people who use code this may seem very obvious, but for those of us who tend to rely on spreadsheets it is really not obvious at all! *R* Commander will allow you to open one of your spreadsheets in R and use the data, but in my experience that is more trouble than it is worth. It is much easier to prepare your data in a spreadsheet in exactly the right format and then save it as a 'tab delimited text file'. For some tests your data should be arranged in consecutive columns with simple titles (no gaps in the titles) in row one. For other tests you put all of the data in one column and in the next column add something to identify which group each item belongs to. More details are given for the relevant tests. If there are any missing data then go to those cells and add NA which means 'not available'. If some data sets are shorter than others you need to fill in the empty cells with NA so that they are all the same length. When the spreadsheet file is ready to save, use 'save as' and 'other formats'. You are prompted for a name and in the box marked 'Save as type' you can scroll down to 'Text (Tab delimited)'.

Back in *R* Commander go to 'Data' and click on 'import data' choosing the option 'from text file'. This allows you to name the dataset and define some properties, but if you have saved the file as instructed above you do not need to change anything. You can then find the file and just open it. If all is well some script will appear in black in the upper script window, which is the code locating and loading the relevant file, and some script will appear in the lower output window describing the data in some unintelligible language. In the messages box you should see some more useful information in blue, such as the number of rows and columns in the data set. If anything goes wrong an error message will appear in red. A common error tells you that one of the columns does not have enough rows, which is what happens if you leave cells blank rather than filling them with NA. If you click on the tab 'view data set' you can check that you have opened the right file and that everything is correct. You can also edit the data using the 'edit data' tab but I tend to avoid that and do any editing in the spreadsheet version.

4.7 Descriptive Statistics

The statistical methods covered in this book are mostly inferential statistics that are used to test hypotheses. However, there is another class of 'statistics' that are used simply to describe data. The descriptive statistics each have a formal definition so they remove any uncertainty and allow you to summarise a data set very efficiently. The inferential statistical tests also use the descriptive statistics, so it is important that you understand and can calculate them. They are one of the tools that you need. Student's *t*-test, for example, is designed to test whether two samples are significantly different and the specific difference that is being tested is between the means (or averages) whereas the Mann–Whitney *U*-test uses the difference between the medians and the Kolmogorov–Smirnov tests are sensitive to any difference between two samples. All of the descriptive statistics described here can be obtained very easily using spreadsheet functions (Table 4.1). If you have imported your data into SPSS or *R* Commander you can generate a list of descriptive statistics following the instructions in Box 4.1.

4.7.1 Measures of the Middle: Mean, Median and Mode

The mean and median are two different ways of describing where the middle of a data set lies. Tests that look for differences in the 'middle' are sometimes called tests for differences in 'location'. Parametric tests tend to use the mean (or average) whereas most non-parametric tests use the median. The mode is only useful when the data are arranged in groups, and it just means the most popular group. When a data set is large and fits the normal

(Gaussian) distribution very well the mean and median fall very close together and if the data are arranged in groups, to make a histogram for example, the mean and median both fall in the modal group, which is in the middle. Data sets that are not normally distributed, but perhaps a bit skewed to either high or low values, or that have a few extreme values, tend to produce mean and median values that are quite different (Figure 3.2).

The mean of a data set is the same as the average, which means that it is the sum of all of the numbers added together divided by the count of values. Consider this simple example:

$$2\ 5\ 6\ 8\ 9$$

There are five numbers in this data set and when they are added together the sum is 30, so the mean or average is $30 \div 5 = 6$. An alternative to the mean is the median, which simply means the one in the middle. We have five numbers arranged in rank order so the one in the middle is the third one, which is the number six. In this case the mean and the median are the same. Now consider this alternative sample of 5 values:

$$2\ 3\ 4\ 10\ 11$$

These five numbers added together also add up (sum) to 30 so the mean is 6 again. However, the middle value, or median, is now four. When a sample comprises an odd number the median value is clearly just the one in the middle, but if the sample comprises an even number then of course there is no central value. For example:

$$2\ 3\ 5\ 7\ 9\ 10$$

These six numbers add up to 36, so the mean or average is $36 \div 6 = 6$. The middle of the data set lies between numbers 5 and 7, so the median value is taken half way between those two numbers, which gives us 6 again. Alternatively:

$$2\ 3\ 4\ 6\ 10\ 11$$

In this case the sum is again 36, giving a mean of 6 but the middle of the data set lies between 4 and 6, so the median value is 5. The mean and median do not have to be whole numbers (integers). For example in this data set:

$$3\ 4\ 7\ 8\ 12\ 16$$

The six numbers sum to 50 so the mean is $50 \div 6 = 8.3$ and the median lies half way between 7 and 8 so it is 7.5.

The mean of even quite a large data set is easy to calculate by hand if you have a calculator, you just add them all up and divide by the number in the sample. Finding the median value of a large sample, however, requires the data to be placed in rank order, which can be very tedious. Both numbers are very easy to obtain in a spreadsheet and both Excel and LibreOffice use the functions 'average' for the mean and 'median' for the median. In Excel the 'Analysis tools' add-in includes the option 'descriptive statistics', and returns both the mean and the median. If you use this option it is a good idea to keep the results in the same column as the original data because the table that is produced does not retain any reference to the original data, so it is very easy to get confused and forget which table refers to which data set. I find it easier to just use the individual functions and keep all of the descriptive statistics at the top of the spreadsheet and in the same columns as the data to which they refer.

4.7.2 Measures of Spread or Dispersion: Range, Variance and Standard Deviation

A sample can comprise numbers that are all very close together, or they can be very spread out. There are several ways to describe the amount of spread, or dispersion. Perhaps the simplest is to calculate the full range of the data as the difference between the highest and lowest values. In spreadsheets the functions are 'max' and 'min'. There is no individual function for range so just take the minimum value away from the maximum.

The variance of a set of numbers is a measure of how different each value is from the mean. When all of the values are very close together, and therefore all very similar to the mean, the variance is small and if there is a big spread of values the variance is large. Just calculating the difference between each value and the mean, and either summing them or taking an average does not work, however, because numbers higher than the mean will give positive values for the difference and numbers lower than the mean will give negative values and when you add them up you always get zero. One option would be to ignore the sign, but mathematicians do not seem very keen on that simple option preferring to get rid of the sign by squaring all of the differences. This works because when a negative number is squared the result is always positive (−4 times −4 is 16, not −16). The variance of a set of numbers is defined as the average squared difference between each value and the mean. The standard deviation is the square root of the variance. Both of these numbers assume that the numbers represent a whole population, not a sample drawn from a population, so they are called the population variance and population standard deviation and they are not used very often.

When we are using inferential statistics we do not have access to all of the numbers in the population, only to a sample, so we have no way of actually measuring the variance of the population. If we have an unbiased sample, however, we can reasonably assume that the variance of the sample should be similar to the variance of the population. However, estimates based on very large samples are likely to be more reliable that estimates based on very small samples. When we use samples, therefore, we do not use the average squared difference from the mean, which is the sum of squared differences divided by the count, we apply a penalty to small samples by dividing by the count minus one. This 'correction' has a tiny effect on very large samples and a big effect on very small samples. In Excel when you use the 'VAR' and 'STDEV' functions you get the sample variance and sample standard deviation values, which are nearly always the ones that you need. If you are using other software or online calculators and there are options then choose the 'sample' not 'population' versions.

If you have no access to a computer you can calculate the variance and standard deviation by hand, but most modern calculators have a button that will return the variance and/or standard deviation for a set of data. I have included instructions in Box 4.2, just in case your evil professor forces you to calculate the variance or standard deviation by hand and show your workings (in which case they really need to get a life). A spreadsheet is the most convenient method of calculation. Many parametric statistical tests use the sample standard deviation.

4.7.3 Confidence Limits around the Mean

The standard deviation is often used for describing the variability or spread of a sample and it is common to quote both the mean and the standard deviation together. However, there is a much better measure of variability that is more meaningful when combined with the mean (average) because it also takes account of the sample size, and that is the standard error. The standard error of a sample is the standard deviation divided by the square root of the sample size. So if you have a sample size of 25 with a mean of 20 and standard deviation of 2 then the standard error is 0.4 (2 ÷ √25 or 2 ÷ 5). If your data set is not too small you can generally assume that if you were to take lots of samples of the same size, from the same population, then about 95% of the time the average would lie within 2 standard errors of the mean. In this case you could quote the mean as 20 ± 0.8 and be reasonably sure that if you were to take lots of samples, all with a sample size of 25, then 95% of those samples would produce a mean in the range 19.2–20.8. Quoting a mean value plus or minus 2 standard errors is much more useful than just quoting the mean because it provides an instant indication of how uncertain the mean value is. The uncertainty is a function of the variability of the sample but also of the size of the sample.

Strictly speaking, 2 standard errors only give the 95% confidence limits when the sample size is between 28 and 60 and if the sample is very large then a better multiplier is 1.98 rather than 2. When the sample size drops below 28 then the multiplier should be a little larger, so for samples between 14 and 27 use 2.1 standard errors and between 10 and 13 use 2.2 standard errors. It probably does not make much sense to give confidence limits for a mean based on a sample of less than 10. Excel will calculate the 95% confidence limits for you from the standard deviation and sample size, but be sure to choose the option that uses the t-distribution (CONFIDENCE.T). For 95% limits set 'alpha' at 0.05 and for 1% limits set it at 0.01.

It is a common misconception that confidence limits around the mean should only be used when your samples are near normally distributed, but that is not actually true. The real assumption is that if you were to take a lot of estimates of the mean then those estimates would form a normal distribution. Even when your sample and your population are not normally distributed it is usually the case that multiple estimates of the mean will yield a near normal distribution as long as the sample sizes are not too small (>30 is good) and there are no huge outliers.

4.7.4 Measures of the Shape of a Distribution: Skewness and Kurtosis

When you plot your data as a histogram it gives a good impression of how close to the perfect 'normal' or Gaussian distribution it is. The Gaussian distribution is shaped like a bell, with most of the values in the middle and a gradual and symmetrical decay in frequencies to either side. If a distribution is not perfectly symmetrical it is likely to be skewed to one side and the metric that describes that is called 'skewness'. Positive skewness indicates a skew or bias towards low values and a negative skewness value indicates a bias towards higher values. Kurtosis is a metric that quantifies whether a distribution is more peaked (positive values) or is flatter (negative values) than a normal (Gaussian) distribution. A perfect normal or Gaussian distribution has zero skewness and zero kurtosis. Some examples of these values together with histograms of the data they describe are shown in Figure 3.2.

The equations for skewness and kurtosis are quite complicated and difficult to calculate by hand, since they involve cubing (to the power of 3) in the case of skewness and using the power of four in kurtosis. If you are really interested choose the relevant function in Excel ('skew' and 'kurt') and choose 'help on this function' to see the equations. In practice these metrics are rarely used for anything other than adding to a table of descriptive statistics and the most practical way to obtain them is in a spreadsheet.

BOX 4.1   OBTAINING DESCRIPTIVE STATISTICS IN SPSS AND *R* COMMANDER

SPSS

Cut and paste your data, without headers, into the 'data view' (spreadsheet) page then go to 'variable view' where you can give each series a name. If your data are individual measurements then identify the 'measure' as 'scale'. Now click 'Analyze', 'Descriptive statistics' and go to 'Explore' (not 'Descriptives'). Use the arrows to put your data sets into the 'Dependent list' box and click on the button for 'Statistics'. You will get a table with more descriptive statistics than you need or want. You can also choose to see some 'Plots' and the options are 'stem and leaf' and 'histogram'. The histogram is useful for checking if your data are near normally distributed and if there are any big outliers. In the plot options you can also choose 'Normality plots with tests' and receive Q–Q plots (not much use if you are not familiar with them) and results of the Kolmogorov–Smirnov and Shapiro–Wilks tests (see Chapter 3).

_R_ COMMANDER

Open R and load *R* Commander (Section 4.6). Import your data by placing it in the top left corner of a spreadsheet, each data set in a separate column, using simple headers without gaps. Do not leave any blank cells and if there are missing values fill those cells with NA. All data sets have to be the same length so fill any empty values at the bottom of shorter series with NA. Save the file as a tab-delimited text file. In *R* Commander click 'Data', 'Import data' and 'from text file'. On the pop-up page you can name the data set and keep the default options for everything else. A message at the bottom will tell you how many rows and columns there are and you can check the data using the 'View data set' button. Clicking 'Statistics', 'Summaries' and choosing 'Active data set' will give you the mean, median, minimum and maximum for all of your series. To get values for the standard deviation, standard error of the mean, skewness and kurtosis values you have to use 'Numerical summaries' instead, but that will not give you the median. To see a histogram of your data go to 'Graphs' and choose 'Histogram'. Under 'Statistics and Summaries' there is also the option to perform the Shapiro–Wilks test for normality (Chapter 3). If you are familiar with Q–Q plots go to 'Graphs' and choose 'Quantile comparison plot'.

BOX 4.2   CALCULATING THE MEAN, VARIANCE AND STANDARD DEVIATION BY HAND

Table 4.9 shows a sample of ten values and they add up to 30. The mean, or average, is the sum divided by the number of values (count) so in this case the mean is 30 divided by 10 which equals 3.

To calculate the variance and standard deviation you first need to take the difference between each value and the mean and then square it. Then add those values to give the sum of the squared differences (Table 4.9). In this case the sum of the squared differences is 11.88. The population variance is the average squared difference, so it is

11.88 divided by the count, which is 10, giving 1.188. The population standard deviation is the square root of the population variance, so in this case it is 1.09.

However, we rarely want the population variance and standard deviation, because most statistical tests call for the sample variance and sample standard deviation. To calculate the sample variance we divide the sum of the squared differences by the count minus one, so it is 11.88 divided by 9 to give 1.32. The sample standard deviation is the square root of the sample variance, so in this case it is 1.15.

TABLE 4.9

Example of How to Calculate the Variance and Standard Deviation by First Calculating the Sum of the Squared Differences from the Mean

Count	Original Value	Squared Difference from the Mean	=
1	1.4	$(1.4 - 3)^2 = (-1.6)^2 = 2.56$	2.56
2	1.7	$(1.7 - 3)^2 = (-1.3)^2 = 1.69$	1.69
3	2.2	$(2.2 - 3)^2 = (-0.8)^2 = 0.64$	0.64
4	2.5	$(2.5 - 3)^2 = (-0.5)^2 = 0.25$	0.25
5	3.0	$(3 - 3)^2 = (0)^2 = 0$	0
6	3.1	$(3.1 - 3)^2 = (0.1)^2 = 0.01$	0.01
7	3.1	$(3.1 - 3)^2 = (0.1)^2 = 0.01$	0.01
8	3.4	$(3.4 - 3)^2 = (0.4)^2 = 0.16$	0.16
9	4.6	$(4.6 - 3)^2 = (1.6)^2 = 2.56$	2.56
10	5.0	$(5 - 3)^2 = (2)^2 = 4$	4
Sum	30	Sum of squared differences	11.88
Count	10	Count	10
Mean	3	Population variance	1.188

Note: Dividing the sum of squared differences by the count minus one gives the sample variance and the square root of the sample variance is the sample standard deviation.

References

Field, A. 2013. *Discovering statistics using IBM SPSS statistics* (4th ed.). Sage, London, 916pp.
Salkind, N.J. 2015. *Excel statistics: A quick guide* (3rd ed.). Sage, London, 168pp.
Schmuller, J. 2013. *Statistical analysis with Excel for dummies* (3rd ed.). Wiley, Chichester, 528pp.
Zaiontz, C. 2015. *Real statistics using Excel.* www.real-statistics.com.

5

Single Sample Tests

5.1 Introduction

When you have collected a single sample, it is often divided up in some way. For example, you may have conducted a questionnaire via social media and just want to check if there is a gender bias in the respondents, or that ethnic minorities are not underrepresented. You may know, for example, that your target population is split evenly between males and females but of course even if you are able to take a perfectly random sample, it is unlikely that your sample will be split exactly 50:50 just due to the 'luck of the draw'. If you were to take many samples, then if they were truly random about half would have more males and about half would have more females. However, conducting a questionnaire via social media certainly does not produce a random sample, because not everyone is equally likely to respond to a request to complete a student's questionnaire. It would make sense to do a few checks to investigate whether your sample is a bit biased in some way before you go on to analyse the main results. Even if your questionnaire comprises mainly open-ended questions and you intend to use mainly qualitative methods it would make sense to complete this quantitative check.

If you can estimate the proportion of males and females in your target population, then you can look at the proportions in your sample and see if they are close enough for the difference to be 'due to luck'. If the chances that the difference could be due to luck drop below 5% ($p = 0.05$), then you would conclude that it probably is not just luck and that the sample has a gender imbalance. There is a very useful test for this, called the binomial test, which is very easy to apply using online calculators or a spreadsheet. If you have several categories rather than just two, there are two useful tests (Figure 5.1). If your categories cannot be placed in logical order (e.g. religious or ethnic groups), you can use the chi-square one-sample test and if there is some logical order to the categories (e.g. age classes, size categories, socioeconomic groups), you can use the more powerful Kolmogorov–Smirnov one-sample test. The one-sample runs test for randomness allows you to test whether a bivariate sequence (male/female, higher/lower, etc.) falls in random order. It is also used with individual measurements to test for randomness in the order of values that fall above and below the median or for randomness in the order of residuals falling above and below a regression line.

Data can also be re-arranged in order to use simpler tests. For example, a questionnaire may ask people to state their age, so that the data exist as continuous individual measurements, but you might choose to put them into age classes and apply the Kolmogorov–Smirnov one-sample test or just distinguish those above and below retirement age and apply the binomial test. Degrading data in this way involves a loss of information, so the tests become less powerful, but if they are being used to answer a focussed question that may be perfectly acceptable.

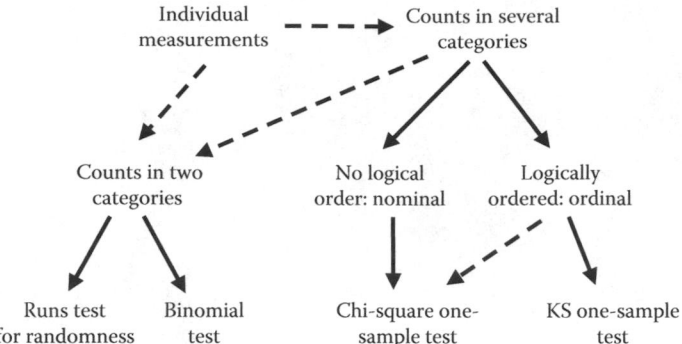

FIGURE 5.1
One-sample tests available for different types of data. Dashed lines indicate a loss of information. KS = Kolmogorov–Smirnov.

5.2 Binomial Test

5.2.1 When It Is Useful

The test is useful when you have counts in two categories and you know the proportions that you would expect to occur. The test returns the probability that the difference between the expected proportions and those observed in your sample could have arisen just by chance. It can be used on all sorts of data (Figure 5.1) from counts in named categories to individual measurements (e.g. to test the proportion in a given size range). The binomial test is used where the population can be divided into two groups. Male/female is an obvious example but there are many potential applications in geography including local people/tourists, working age/retired, local/erratic lithologies, larger/smaller or younger/older than some threshold, etc. Even where the population is logically divided into several groups, you can still use it by choosing the group that interests you and putting all the others in a category called 'other'. For example, if you were to take a sample of clasts (stones) from a glacial deposit that has been well studied, so that the percentage of different rock types is known, you could use the binomial test to see if the proportion of limestone clasts in your sample is as expected. If you have conducted a campus questionnaire, for example, and you are concerned that overseas students may be underrepresented, you can compare the ratio of home to overseas students in your university to the proportions in your sample.

5.2.2 What It Is Based On

The binomial test is based on calculating the exact probability of a truly unbiased sample of a given size yielding a proportion at least as large or at least as small as the one that you have observed (Siegel 1956). It is easiest to understand when the expected proportions are 50:50 because we can then use the coin tossing analogy that was used in Chapter 2 (Section 2.6) to explain probabilities. When the probabilities are 50:50 ($p = 0.5$), the binomial test is exactly the same as the sign test (Section 6.2). When you toss a coin, there is an equal chance of tossing heads or tails, so in an unbiased sample, we would expect about a 50:50 split between heads and tails. Of course with a small sample of tosses, there is still a good chance that the split will not be exactly 50:50, just by chance. If you toss a coin four times, for example, there are 16 possible outcomes:

$$\text{HHHH } \underline{\text{HHHT}} \ \underline{\text{HHTH}} \ \underline{\text{HTHH}} \ \underline{\text{THHH}} \ \text{HHTT HTHT HTTH}$$
$$\text{THTH TTHH TTHT HTTT THTT TTHT TTTH TTTT}$$

If we wish to know the probability of any particular combination of heads and tails from a sample of four tosses, we need only count them from this list. For example, the probability of tossing four heads is one in 16 and the probability of tossing four tails is also one in 16 ($1/16 = 0.0625$ so $p = 0.0625$). That means that the probability of tossing 'four the same', not specifying heads or tails, is two in 16 (or one in eight, which is $p = 0.125$). The probability of tossing at least three heads is five in 16 (they are underlined). The number of possibilities doubles each time you add a coin, so this method quickly becomes rather tedious. With a sample of 20, for example, there are more than a million possible combinations. Statisticians do not actually list the possibilities, they are able to calculate the probability of any given outcome from a sample of any size using an equation. Also, the probability of a given outcome from each step does not have to be one in two ($p = 0.5$), as with coin tossing, the method works for any number of possible outcomes. I am not going to give the equation here, partly because it is rather complicated to use, involving factorials, but mainly because I doubt that anyone actually uses it in practice because you can obtain the correct result by simply entering two numbers in a spreadsheet or online calculator.

5.2.3 How to Do It

5.2.3.1 Online Calculators

The most convenient way to perform the binomial test is using an online calculator (Table 5.1) or the companion site calculator. The calculators are often set up as if you were testing the results of an experiment, but that does not matter. They generally require you to enter the sample size, often called 'number of trials or experiments' and the 'number of successes' which for our purposes is just the number in one of the two groups (e.g. male or female) and fix the 'hypothetical probability of success', which for our purposes is the probability of someone being male (or female) which we are assuming is 50%, which is the same as $p = 0.5$. You can often choose a one- or two-tail test.

5.2.3.2 In a Spreadsheet

Excel includes the function 'BINOMDIST' which performs the binomial test, though it is a bit fiddly to use and quite easy to mess up. It is again set up to test the number of successes in an experiment, so you enter the sample size in the box marked trials and the number you are interested in, for example, males or females, in the box marked 's' (for successes) and when using Excel, it is important to use the *smaller* of the two numbers. Next you define the hypothetical probability, which for a 50% chance is $p = 0.5$ (in which case the binomial test and sign test are identical). Now, you also have to decide whether to write 'true' or 'false' in the box marked 'cumulative'. If you write false, you will get the probability of that exact result, which is rarely what we want. In geographical research, we are usually interested in the probability of an outcome as extreme as, or even more extreme than, the one that we observe. This is what you get when you write 'true'.

TABLE 5.1

Selection of Free Online Calculators That Will Perform the Binomial Test (or the Sign Test)

Provider	Website	Comments
Graphpad	http://graphpad.com/quickcalcs/binomial1.cfm	Easy to use, output is described in the text above
Stattrek	http://stattrek.com/online-calculator/binomial.aspx	To decide which result you need enter one of the examples used above
Vassarstats	http://vassarstats.net/binomialX.html	Output can be a bit confusing so enter one of the examples above to check which one you need. Use the 'exact binomial calculation' where possible
Simple Interactive Statistical Analysis	http://www.quantitativeskills.com/sisa/distributions/binomial.htm	Easy to use but complicated output, use one of the examples to check which one you need
DanielSoper.com	http://www.danielsoper.com/statcalc3/calc.aspx?id=69	Avoid this one, it gives the probability of the exact number rather than the 'at least' or 'at most' probability that we usually need
DanielSoper.com	http://www.danielsoper.com/statcalc3/calc.aspx?id=71	This is the right one to use: it calculates the cumulative binomial probability
Ncalculators	http://ncalculators.com/statistics/binomial-distribution-calculator.htm	Avoid this one, it gives the probability of the exact number rather than the 'at least' or 'at most' probability that we usually need
Easycalculation.com	https://www.easycalculation.com/statistics/binomial-distribution.php	Avoid this one, it gives the probability of the exact number rather than the 'at least' or 'at most' probability that we usually need
Statistics how to	http://www.statisticshowto.com/calculators/binomial-distribution-calculator/	You have to enter the range of results not just one, so for at least four successes enter the range as 4 and zero (the last two entries marked x). The last line of output is the one to use

Unfortunately, the complications are not over. Excel does not give you the option to choose a one- or a two-tail test so you must be very careful when you interpret the results. If you enter a sample size of four and zero successes (always enter the lower value, so in this case zero not four) with a hypothetical probability of 0.5, you will see that the result is $p = 0.0625$, which is the one-tail probability. It is important to remember this because the hypotheses that geography students use this method to test are very often two-tail, in which case this is the wrong answer! Don't panic though; recall that the two-tail probability is double the one-tail probability (Section 2.7), so you just multiply the answer by two when you want the two-tail probability.

5.2.3.3 Companion Site Calculator

I have prepared a simple spreadsheet calculator that will calculate both the one- and two-tail probabilities and give you an effect size. Just enter the number in each of your two groups and then the 'probability', which in the terminology used above is the probability of success. For example, if you want to know if males are underrepresented in your sample, and have reason to expect they form 50% of the population you are sampling from, then enter 0.5. If you were testing the proportion of overseas students, and you know they make up 10% of your population, you would enter 0.1 as the probability.

The output includes both the one- and two-tail probabilities. They are obtained using the BINOMDIST function described above. The one-tail probability is translated into a z-score (from the standard normal distribution) using the NORM.S.INV function and when divided by the square root of the sample size this yields the effect size '*r*' (Section 2.8). When the probability values are small, they may appear in scientific notation, but if you prefer decimal fractions, there are cells to the right that will retain that format.

5.2.3.4 In SPSS

Place your data in a single column, you can use numbers, letters or names (e.g. male/female), but don't include a header, and paste into the 'data view' page. Now go to variable view and give it a name and identify the 'measure' as nominal. In the 'Analyze' menu, choose 'Nonparametric Tests' and 'One Sample'. Check 'Customize analysis' and in 'Fields' move your data set into the 'Test Fields' box. In 'Settings', check the top one; 'Compare observed binary probability to hypothesized (Binomial test)' and then click options. This is where you have to define the expected probability. The default setting is 0.5, which is what you would use if you expect equal proportions in the population (e.g. 50% male and female). Ignore the other options. Click 'Ok' and 'Run'. The output is the two-tail probability. Divide it by two if you are testing a one-tail hypothesis. Double clicking on the output box opens a simple diagram and some more results including a 'standardized test statistic' which is a z-score. You can divide this by the square root of the total sample size to obtain the effect size '*r*'.

5.2.3.5 In R Commander

Open *R* and load *R* Commander. Put your data in a single column with a header and save it as a tab-delimited text file. Use the 'Data' and 'Import data' options to access it. The binomial test is not named as such, it is hidden under 'Statistics' and 'Proportions' and called 'Single sample proportion test'. Under 'Options' set the expected proportion (default is 0.5) and for 'Type of test' choose 'Exact binomial'. The three null hypothesis options effectively allow you to choose a two-tail test (the top one) or a one-tail test. The output includes the number of successes, number of trials and the probability. You can ignore everything else.

5.2.3.6 By Hand

If the expected proportion is $p = 0.5$, which is often the case, it is easy to use tables of critical values of the sign test (Section 12.1). Reject the null hypothesis if the smaller of your two values is less than or equal to the tabulated value for your sample size (*n*). Tables are provided for one- and two-tail tests with probabilities of $p = 0.05$ and $p = 0.01$. If in doubt, use the two-tale tables.

5.2.4 Examples

5.2.4.1 Example: Yes or No Questionnaire Answers

One of the most useful applications of the binomial test is to determine whether there is a significant difference in the number of respondents answering yes or no (or any other two options) to a question. In this case, the null hypothesis is usually that there is no preference in the population and the probability value tells you the chances that the split of answers could have arisen by chance. The simplest option is to use the tables of critical values for the sign test (Section 12.1). For example, in a sample of 50 respondents, if 30 said yes and 20 said no, you would use the two-tail tables where the critical value for $p = 0.05$ is 17. Since the observed value is not 'equal to or smaller than' the critical value, you would accept the null hypothesis and conclude that the observed imbalance in responses could have occurred by chance even if there was no real preference in the population.

5.2.4.2 Example: Is There a Gender Bias in My Sample?

You have conducted a questionnaire using social media but you are concerned that there may be some gender bias in your sample. Let us assume that you did not think about any potential gender bias in advance, but when your sample size reached 30, you noticed that there were 20 females and only 10 males. You had hoped to make inference about users of social media in general, so this potential imbalance in response is a concern. The temptation in a case like this is to start with the hypothesis that women are more likely to complete your survey than men and that men are therefore underrepresented. This would be a one-tail hypothesis. However, this would not really be a fair test, because you only conjured the hypothesis that women are more likely to complete the questionnaire when you looked at the results and noticed that that was the case. If you had noticed that there were more men than women, you might well have posed the hypothesis that men are more likely to complete the questionnaire than women. Since you did not propose the hypothesis in advance of looking at the data (*a priori*), it would be fairer to use a two-tail test. The distinction is important, as we shall see.

The appropriate hypothesis in this case is that there is a gender bias in the sample, and the appropriate null hypothesis is that there is not a gender bias in the sample. The binomial test is appropriate because it will tell you the chances (probability) of a sample of 30 individuals producing a 20:10 split in gender just by luck, even though we can reasonably assume that in the population of social media users the genders are evenly balanced. We want the probability of either 10 males or of 10 females being sampled, not only the probability of 10 males, so we want the two-tail probability.

The companion site calculator returns a two-tail probability of $p = 0.0987$ so there is almost a 10% chance that even when the population has an even mix of genders, a sample of just 30 individuals will give a 20:10 ratio of males to females (or females to males). This is much larger than 5%, so in this case, we would accept the null hypothesis and conclude that the gender imbalance is not statistically significant. The fact that there are more females than males could be just due to chance.

If the sample size was to increase to 60, and you still had the same ratio of females to males (40:20), then you could rerun the test and would find that the two-tail probability is now $p = 0.0135$. This is much less than 5% (1.3%) so in this case, you would reject the null hypothesis and conclude that there is a gender imbalance. The important point here is that it is not just the ratio of males to females in the sample that is important, but also the size of the sample. As the sample size increases, it becomes much more difficult to produce a large gender imbalance just by chance. If you had a sample of 120 with a gender split of 80:40 (same ratio again), you would find that the probability of that occurring just by chance is less than one in a thousand.

5.2.4.3 Example: Have the Limestones Been Removed by Weathering?

Imagine that your life is so empty that you have decided to count the stones in a glacial deposit that has already been very well studied. We already know that this glacial sediment contains 10% limestone clasts, but you are concerned that your sample is taken from too near the surface, so that some of the limestone clasts may have been removed by weathering. Being clever, you proposed this hypothesis in advance because you have read that this can be a problem when you sample close to the surface. You can now formulate the hypothesis that 'limestone clasts are underrepresented in this sample' and test the equivalent null hypothesis which is that limestone clasts are not underrepresented. This calls for a one-tail test. You now look at your results and notice that in your sample of 120 clasts, there are 13 limestones, which is more than 10%. There is no point in now applying the test because you can immediately accept the null hypothesis that limestones are not underrepresented.

5.2.4.4 *Example: Are There Too Few Black Managers in English Football?*

There are 92 football (soccer) teams in the English Premier League and Football League but in 2014 only two of their managers were black. About 25% of the players, however, are nonwhite. Using the BINOMDIST function in Excel, we can enter 2 in the box marked 's' for successes and 96 in the box marked 'trials'. Since 25% of professional players are non-white, we can assume that if you were to randomly sample from that population there is a 1/4 chance of any particular person not being white, so we enter 0.25 as the 'probability s'. We want to know the 'probability that there are at most number s successes' so we enter TRUE in the box marked 'cumulative'. The result appears in scientific notation as 5.47E-10. The E-10 bit means that you have to move the decimal point 10 places to the left. If you find scientific notation a bit confusing just right click on that cell and choose 'Format Cells', then 'Number'. You will need to raise the number of decimal places to 10 to see the full result, which is $p = 0.0000000005$. This tells us that the chances of English football managers representing a truly colour-blind sample of the available talent are less than one in 100 million. That is not quite the same as zero, so I suppose there is still a chance that there is no racism involved. In 2015, one of the black managers was sacked.

5.3 One-Sample Chi-Square (χ^2) Test

5.3.1 Introduction

Also called the 'chi-square goodness of fit test' (Neave and Worthington 1988), it is used to compare counts in nominal categories, with the counts that would be expected on the basis of theory. The one-sample chi-square test is identical to the two-sample test with the exception that the frequencies in one of the samples are based on some theoretical distribution (such as 'all the same') rather than measurements. Since the two-sample case is more commonly used the logic of the test and the critical assumptions are dealt with in Chapter 6. Chi-square is one of the weakest nonparametric tests and it is also one that is easy to get wrong, because it has some rather strict assumptions that have to be met. Don't just jump in here, go to the two-sample case and read about the logic of the test and the assumptions otherwise you are likely to mess it up.

5.3.2 When It Is Useful

When you have a single sample that can be broken down into more than two categories, the chi-square test can be used to test whether the number of occurrences in each category is significantly different from the number that you would expect on the basis of some null hypothesis about the likely frequency. A common null hypothesis is that there will be equal frequencies in each category. The chi-square test, in all of its guises (it can be used with one sample, two samples or many samples), is weak because it does not use much information, just counts in categories. It differs from many other tests in that it treats the categories as nominal, which means that it ignores any logical order that they may fall in. If you have counts in categories that fall into a logical order, the Kolmogorov–Smirnov one-sample test is more powerful (Siegel 1956) and also avoids some of the assumptions.

There are two critical assumptions:

Assumption 1. The frequencies are independent, which means that they sum to the sample size. For example, if you ask 20 people if their favourite animal is a cat, dog or hamster, they can only choose one answer. If they tick all of the animals that they like, the data are not suitable for chi-square because the sum will be more than 20.

Assumption 2. Expected frequencies are large enough. No *expected* frequencies should be less than one and no more than 20% of the categories should have an expected frequency of less than five.

If your data do not fit these assumptions do not just go ahead and use the chi-square test; that is a common error and you will likely be penalised. Either adjust your data, for example, by combining categories (where it is logical to do so) so that the expected frequencies are large enough, or use another test. Where your categories fall in a logical order, the Kolmogorov–Smirnov one-sample test is often appropriate and is not sensitive to the problem of small expected frequencies. Where you only have two categories, or where combining categories effectively leaves you with just two, then it does not make any sense to use chi-square because you can use the binomial test.

5.3.3 What It Is Based On

The one-sample chi-square test compares the observed frequencies (counts) in a set of categories with the number that you would have expected. The chi-square value is effectively a measure of the difference between the observed and expected frequencies and the outcome is a probability value that tells you the likelihood of such a large difference occurring just by chance. The concept of one- or two-tail testing does not apply to this test because it assumes that the categories are nominal, and therefore cannot be placed in rank order. If there is no logical order to the categories, you cannot predict any 'direction' and so a one-tail test would make no sense.

5.3.4 How to Do It

The chi-square test does not involve very complicated mathematics and it is quite feasible to do it by hand, as described in the two-sample case (Section 6.8), and tables of critical values are provided (Section 12.2). However, a spreadsheet makes the calculations easier and avoids the need for tables of significance. For convenience, a simple spreadsheet calculator is provided on the companion site. There are also some online calculators (Table 5.2). Conducting this test in *R* Commander and SPSS is a little trickier.

TABLE 5.2

Selection of Free Online Calculators for the One-Sample Chi-Square Test (Also Called the Chi-Square Goodness of Fit Test)

Provider	Website	Comments
Graphpad Quickcalcs	http://graphpad.com/ quickcalcs/chisquared1/	Good site, easy to use. It says expected values can be added as fractions but I could not get it to work
Vassarstats	http://vassarstats.net/csfit.html	Good site and the option to enter expected values as fractions works well. Gives percentage deviation and standardised residuals (Section 7.2.5)
Social Science Statistics	http://www.socscistatistics. com/tests/goodnessoffit/ default2.aspx	Works eventually but the adverts slow it down terribly. Expected as frequencies or proportions but not fractions
Quantpsy.org	http://www.quantpsy.org/ chisq/chisq.htm	Use the smaller table at the bottom of the page. Expected values as frequencies only. Ignore Yate's correction
Professor Hossein Arsham	https://home.ubalt.edu/ ntsbarsh/business-stat/ otherapplets/goodness.htm	Some advert-free Java script. Works well but you have to enter the degrees of freedom manually (number of categories minus one)
Easycalculation. com	https://www.easycalculation. com/statistics/goodness-of-fit. php	One to avoid. You have to enter degrees of freedom manually and it only gives the chi-square value not the probability
Mathcracker. com	http://www.mathcracker.com/ chi-square-goodness-of-fit.php	Unusual approach. You have to define the level of significance and the result tells you if it is significant or not, but no *p*-value. Accepts fractions and shows the mathematics

5.3.4.1 Companion Site Calculator

The spreadsheet allows you to enter your observed frequencies in up to 10 categories. You must enter the counts in each category so that they add up to the true sample size. You must not use percentages. For the expected values, you can either enter the frequencies directly or you can enter the expected proportions. The latter is useful because it allows you to avoid rounding errors. To enter equal proportions in three categories, for example, enter = 1/3. The equals sign is important. Only fill the boxes for the number of groups that you have, do not fill the rest of the boxes with zero or anything else. You will be warned if the sum of observed and expected values is not the same and if the expected frequencies are not large enough.

The output includes the chi-square statistic, degrees of freedom (number of categories minus one) and the probability. The probability is the chances (out of one) that the difference between the observed frequencies and the expected frequencies could have arisen just by chance. Values of less than 0.05 are normally considered statistically significant. An effect size (Cramer's *V*: see Section 6.8.5) is also provided.

5.3.4.2 In a Spreadsheet

Rather than try to write a complicated equation just complete each step in a new column, as shown in Tables 5.3 and 5.4. First take the difference between the observed and expected frequency, then square it, then divide that number by the expected frequency. The sum of this last column gives you the chi-square statistic. Degrees of freedom are the number of categories minus one. You could now look up the significance level in a table but fortunately

TABLE 5.3

Example of How to Perform the Chi-Square Test Using a Spreadsheet

	A	B	C	D	E
1	Observed frequency	Expected frequency	O-E	$(O-E)^2$	$((O-E)^2)/E$
2	15	10	= A2-B2	= C2^2	= D2/B2
3	10	10	= A3-B3	= C3^2	= D3/B3
4	5	10	= A4-B4	= C4^2	= D4/B4
5					
6				SUM = χ^2	= SUM(E2:E4)
7				Degrees of freedom	2
8				Probability	= CHIDIST(E6, E7)

TABLE 5.4

Solutions to the Equations Shown in Table 5.2

	A	B	C	D	E
1	Observed frequency	Expected frequency	O-E	$(O-E)^2$	$((O-E)^2)/E$
2	15	10	5	25	2.5
3	10	10	0	0	0
4	5	10	−5	25	2.5
5					
6				SUM = χ^2	5
7				Degrees of freedom	2
8				Probability	0.082

Excel (and Calc) contains a function that will give you the probability based on the chi-square value and the degrees of freedom. It is called 'CHIDIST' or 'CHISQ.DIST.RT'.

5.3.4.3 In R Commander

Open *R* and load *R* Commander (Section 4.6). The easiest way to load the data is as a column of 'values' in a text file. For example, if you have a sample of 30 people who chose from three options, then you could use a column of 30 'values' comprising A, B and C or cat, dog, hamster, etc. It is important to give the column a header or the first value will be ignored. Save it as a tab-delimited text file and use the 'Data' and 'Import data' tabs to enter it. Under the 'Statistics' tab ignore the 'Contingency table' options, they are not appropriate for this test. Choose 'Summaries' and 'Frequency distributions' and the pop-up should include the single option of the 'Chi-square goodness-of-fit test (one sample only)'. Check that option and another pop-up shows the expected proportions. The default is equal proportions for all categories, but you can change them. The output includes X-squared, which is the chi-square statistic, degrees of freedom and the probability value. To obtain the effect size 'Cramer's *V*' take the chi-square value, divide it by the sample size and then take the square root.

5.3.4.4 In SPSS

Prepare the data as for *R* Commander and cut and paste, without a header, into the 'Data view page'. Go to 'Variable view' and add a header and identify the data as 'Nominal'.

Now click 'Analyze', 'Nonparametric Tests' and 'One Sample'. Choose 'Customize analysis' as the 'Objective' and in the 'Fields' page move your data into the 'Test Fields' window. On the 'Settings' page, check 'Compare observed probabilities to hypothesized (chi-square test)' and then 'Options'. The default setting assumes an equal probability for all categories and if you want to change that you need to type in the name of each category as it appears in your data and the expected proportion. The 'Hypothesis Test Summary' will appear, but it only gives you the probability rounded to three decimal places. Double click on it to obtain the test statistic (chi-square), degrees of freedom and sample size. SPSS does not calculate Cramer's V so take the chi-square value, divide it by the sample size and then take the square root.

5.3.5 Examples

5.3.5.1 Example: Beautiful Beaches

You are interested in how people perceive and appreciate the natural landscape and since Gower peninsula, just west of Swansea, has many beautiful beaches, including Three-Cliffs Bay, Llangenith and Caswell, you decide to ask a sample of tourists which of these three they prefer. Each person can only choose one. Your null hypothesis is that all three beaches are equally stunning and so you assume an equal likelihood of being chosen. Since the sample comprises 30 respondents, the expected frequencies are 10 in each case. The observed frequencies, however, are 15 for Three-cliffs Bay, 10 for Llangenith and 5 for Caswell. Assuming your sample is unbiased, you can now use the one-sample chi-square test to determine the probability that a difference as large as that observed could have occurred by chance. The numbers from this example are included in the spreadsheet tables above (Tables 5.3 and 5.4). The chi-square statistic is 5 with 2 degrees of freedom and the probability is 0.082. Since this is larger than 0.05, we would conclude that the difference is not statistically significant and would therefore accept the null hypothesis that there is no difference in preference for the three beaches. With a much larger sample, the same proportional split in responses would be statistically significant.

5.3.5.2 Example: Ethnic Groups

You are conducting a survey of local women's fear of violence and have obtained an anonymous sample by leaving a short questionnaire in a range of public places. You obtain the proportion of three ethnic groups in the area from the electoral register. The proportions locally are 60% white European, 20% Afro-Caribbean and 20% Asian. In your sample, you have 160 white European, 30 Afro-Caribbean and 10 Asian whereas the expected numbers are 120, 40 and 40. The resulting chi-square value is 38 and the probability is much less than 0.001 ($p = 4E–09$). You can conclude that your sample is not representative of the local mix of ethnic groups. You need to consider this when discussing the results of your survey.

5.3.5.3 Example: Dolphin Sightings

You have obtained data on dolphin sightings around the Gower coast and intend to investigate whether the location of the dolphins is related to the weather conditions, which might allow you to use the weather forecast to predict the most likely place to see dolphins. However, the sightings are based on reports from the general public via a mobile phone 'app', and you are concerned that there may be a reporting bias because more people are out on the coast at weekends than during the week. The null hypothesis is that dolphins are equally likely to be close to shore on any day of the week, and if significantly more are seen at weekends, then it is likely to be due to a reporting bias. You have reported sightings that suggest the dolphins were near shore on 49 days, of which 14 were Saturdays, 15 were Sundays and the remaining 20 were on weekdays. The expected proportions are 1/7, 1/7 and 5/7. If you enter these expected proportions in the companion site calculator, together with the observed frequencies, the resulting chi-square value is 22.57 with 2 degrees of freedom giving a probability of $p < 0.0001$. There is less than a one in a thousand chance that the difference between the observed and expected frequencies could have arisen by chance. It seems likely that there is indeed a bias towards more sightings at the weekend. This is worrying because there might also be a bias in reporting due to the weather, with more people out on the coast on nice days and fewer out in the rain. Since the intention is to compare sightings with weather conditions that is a serious source of error. The project seems doomed to fail because the reporting data are unlikely to provide an unbiased sample.

5.4 Kolmogorov–Smirnov One-Sample Test

5.4.1 When It Is Useful

This test is useful when you have data in categories and those categories fall in some logical order. Common examples include the categories 'strongly disagree' to 'strongly agree', socioeconomic groups, income groups, size classes and roundness classes. The Kolmogorov–Smirnov one-sample test is similar to the two-sample test but one of the 'samples' is based on a theoretical distribution, such as 'all the same'. This test is more powerful than the equivalent chi-square test because it uses more information (Siegel 1956). It is also less sensitive to sample size, so can be used where the assumptions of chi-square are violated. It is sometimes argued that the Kolmogorov–Smirnov tests (there are three different versions in this book) are inherently two-tail, as for chi-square, but there are cases where one-tail hypotheses are logical, and they are discussed in some detail in the section on the two-sample Kolmogorov–Smirnov test (Section 7.5).

The Kolmogorov–Smirnov one-sample test is often used to test whether a sample of individual measurements is significantly different from a theoretical continuous distribution, such as the normal distribution, but I would caution against using it in this context (Section 3.3). The null hypothesis that is being tested is that the sample is not significantly different from the normal distribution, so it is based on the chances of the observed difference being just due to chance. The problem is that if you have a very small sample any difference is more likely to be due to chance, which can lead to the erroneous conclusion that your tiny sample is normally distributed. If, in contrast, you have a very large sample then the chances of an observed difference arising by luck is very much smaller, so that there is a good chance that you will conclude that a very large sample is not normally distributed when it reality it is perfectly acceptable.

5.4.2 What It Is Based On

The categories are placed in the correct logical order and the counts are first converted into proportions (out of one) and then added together in turn to produce cumulative proportions. A set of cumulative proportions is also produced using the expected frequencies that are based on some theoretical distribution. A common theoretical distribution assumes there is no difference between the categories, in which case they each have an equal proportion (this is called a rectangular distribution). The test statistic 'D' is the largest difference between the two sets of cumulative proportions. The test is sensitive to any difference between the theoretical and observed frequencies. It is easier to understand when you see a worked example.

5.4.3 How to Do It

The Kolmogorov–Smirnov tests are amongst the easiest to complete by hand or in a spreadsheet and there are no complicated calculations. To change counts in each category into proportions divide each count by the overall sample size. If you get it right, they will all add up to one. To produce the cumulative proportions add all the frequencies to the left of a category to the count for that column. If you get it right, the last column will end with a total of one. Now do the same for the expected values. If your null hypothesis is that there will be no difference between the categories, then the expected proportions in each category will be one divided by the number of categories. The test statistic 'D' is the maximum difference between the two cumulative proportions. You now use tables (Section 12.3) to see whether the observed D is larger than a critical value for a particular probability level. Tables are appended that give the critical values for samples sizes up to 50 for $p = 0.05$ and $p = 0.01$, for both one- and two-tail tests. If you are unsure, use the two-tail tables. For samples larger than 50, equations are used to obtain the critical values (Table 5.5).

I could only find one suitable online calculator (http://vassarstats.net/ksm.html), and that can only deal with up to eight categories and is little easier than making the calculations in a spreadsheet or by hand. Similarly, it is not necessary to use SPSS or *R* Commander to perform this simple but useful test.

5.4.4 Examples

5.4.4.1 Example: Are Levels of Agreement Equal?

You have a sample of 30 replies to the question 'to what extent do you agree' and the five possible answers range from strongly agree to strongly disagree. The Kolmogorov–Smirnov one-sample test can check whether the observed frequencies are significantly different from those that you would expect if an equal proportion of people chose each answer. The necessary calculations are shown in Table 5.6. The test statistic 'D' is the maximum difference between the cumulative observed proportions and the cumulative expected proportions, and in this case is 0.333. We are testing for any difference, so the two-tail tables are appropriate and for a sample size of 30, the critical value for $p = 0.05$ is 0.24 and for $p = 0.01$, it is 0.29. Since our calculated value is 'equal to or higher than the

TABLE 5.5

For Large Samples the Critical Values for the Kolmogorov–Smirnov One-Sample Test are Obtained from the Sample Size Using These Equations

Two-Tail Tests		One-Tail Tests	
$p = 0.05$	$p = 0.01$	$p = 0.05$	$p = 0.01$
$1.36/\sqrt{N}$	$1.63/\sqrt{N}$	$1.22/\sqrt{N}$	$1.46/\sqrt{N}$

TABLE 5.6

Example of the Calculations Required for the Kolmogorov–Smirnov One-Sample Test

	Strongly Agree	Agree	Not Sure	Disagree	Strongly Disagree	Sum
Observed counts	0	3	5	10	12	30
Observed proportions	0	0.100	0.167	0.333	0.400	1
Cumulative observed proportions	0	0.100	0.267	0.600	1.000	
Expected proportions	0.2	0.2	0.2	0.2	0.2	
Cumulative expected proportions	0.2	0.4	0.6	0.8	1	
Absolute difference	0.200	0.300	<u>0.333</u>	0.200	0.000	

Note: In this case, the null hypothesis is that there is no preference in terms of answers. The test statistic 'D' is the largest absolute difference between the expected and observed cumulative proportions (underlined).

tabulated value' for $p = 0.01$, we can conclude that there is less than a 1% chance that the difference between the observed and expected proportions could have arisen by chance.

5.4.4.2 Example: Is My Sample Representative?

You have surveyed the opinions of the teaching staff (faculty) at your university using an email request. You have received 200 replies and intend to use the results to make inference about the opinions of all of the teaching staff, so want to make sure that your sample is representative. However, you are concerned that there may be some bias with regard to seniority. The personnel department has given you the percentage of staff in each of six categories and since they represent the stages of a typical academic career they fall in a logical order: postdoctoral researchers (4%), junior lecturers (15%), lecturers (40%), senior lecturers (26%), readers (4%) and professors (11%). You can use the Kolmogorov–Smirnov one-sample test because you have an observed proportion in each category and can calculate the expected proportion based on the figures supplied by the personnel department (Table 5.7).

The test statistic '*D*' is the maximum difference between the cumulative expected proportions and the cumulative observed proportions. The sign of the difference does not

TABLE 5.7

Example of the Calculations Required for the Kolmogorov–Smirnov One-Sample Test

	Post-Doc. Researchers	Junior Lecturers	Lecturers	Senior Lecturers	Readers	Professors	Sum
Observed frequency	14	70	82	28	0	6	200
Observed proportions	0.07	0.35	0.41	0.14	0	0.03	1.00
Cumulative observed proportions	**0.07**	**0.42**	**0.83**	**0.97**	**0.97**	**1**	–
Percent employed	4	15	40	26	4	11	100
Expected proportions	0.04	0.15	0.40	0.26	0.04	0.11	1.0
Cumulative expected proportions	**0.04**	**0.19**	**0.59**	**0.85**	**0.89**	**1.00**	–
Difference	−0.03	−0.23	<u>**−0.24**</u>	−0.12	−0.08	0	

Note: In this case the null hypothesis is that the proportions in the sample are the same as the proportions in the population. The test statistic '*D*' is the largest difference between the expected and observed cumulative proportions (underlined).

matter and can be ignored, so 0.24 is the maximum difference. The table of critical values does not reach beyond 100, so you have to calculate the critical values using the equations in Table 5.5. For a two-tail test with $n = 200$, the critical value for $p = 0.01$ is

$$D_{crit} = \frac{1.63}{\sqrt{N}} = \frac{1.63}{\sqrt{200}} = \frac{1.63}{14.142} = 0.115$$

Since our measured value of D is greater than or equal to the critical value, we can conclude that there is less than a 1% chance that the difference between the proportions has arisen by chance. It seems that the sample is biased.

By comparing the expected and observed proportions, you can see where the differences lie. In this case, it seems that junior staff are overrepresented in the sample and more senior staff are underrepresented. If you went on to use this sample, you would have to discuss this bias and be careful about making bold statements about 'the percentage of academics'. A better approach might be to try to target the more senior staff, effectively shifting your strategy more towards stratified sampling.

5.5 One Sample Runs Test for Randomness

5.5.1 When It Is Useful

It tests whether a bivariate sequence (+/–, male/female, etc.) falls in random order. A sequence can be nonrandom because the elements are too strongly grouped together or because they are too strictly ordered. For example, if you have a long queue of people, the males and females could be randomly arranged, dominated by groups of males and groups of females, or by couples. Where you have individual measurements they can be replaced by plus and minus for values falling above and below the median. In this case, the test can be used to see whether the values are independent of the order in which they were collected. When regression (Chapter 11) is used to fit a line through some points on a scatter graph (*x–y* plot) the runs test can be used to test whether the shape of the line is appropriate (Box 5.1). This test should not be confused with the Wald–Wolfowitz two-sample runs test, which is used for testing for any difference between two independent samples (Neave and Worthington 1988).

5.5.2 What It Is Based On

It works by counting the number of 'runs' of the same type in a sample. A very small number of runs is unlikely to occur by chance, and a very large number of runs is equally unlikely. For example, imagine you have a bag that contains 10 white balls and 10 black balls. If you were to draw the balls randomly from the bag, it is very unlikely that you would draw all 10 black balls followed by all 10 white balls. It is also very unlikely that you would draw them in strict sequence of a black ball followed every time by a white ball. In the first case, the result would be a sequence comprising just two runs and in the second, it would comprise 20 runs.

Clearly, there are only two possible arrangements of these balls that would result in a sequence of just two runs, and only two that would result in a sequence of 20 runs.

In contrast, there are very many ways of arranging the balls so that there are between 8 and 10 runs. It is thus very unlikely that a true random sample drawn in this way would result in as few as two runs or as many as 20 runs, but quite likely that the number of runs will be close to nine. If you know how many black and white balls are in the bag, it is possible to calculate a probability for any given number of runs.

B B B B B B B B B W W W W W W W W W W W = 2 runs

W W W W W W W W W W W B B B B B B B B B B = 2 runs

B W B W B W B W B W B W B W B W B W B W = 20 runs

W B W B W B W B W B W B W B W B W B W B = 20 runs

5.5.3 How to Do It

This test is easy to perform by hand or in a spreadsheet and for unequal sample sizes up to 25 tables can be used to see if the number of runs falls beyond critical values for different sample sizes (Section 12.4). When used on individual measurements, to check the sequence of values above and below the median, the two samples (above and below) are of equal size, and tables can be used for samples of up to 100. With larger samples of unequal size, which might apply if you are using a nominal classification like male/female, there is an equation that allows you to calculate the probability, using a normal approximation (z-score), and the effect size ('r'). For completeness, the procedure is explained in Box 5.2 but on the companion site, there is a simple calculator that will make the calculations for you. If you have downloaded the Real Statistics Resource Pack for Excel, you can use the one-sample runs test option listed under nonparametric tests.

5.5.3.1 Companion Site Calculators

Three calculators are provided on the companion site. The first is designed for use with dichotomous data, where you just have two categories such as male/female, yes/no, etc. (runs (1) binomial). First enter the names of the two groups, as they appear in your data, in the two boxes. Now type or paste your sequence of data (up to 300) in the column marked 'Enter data here'. The spreadsheet will count the number in each category and the number of runs. If you have continuous data, consisting of individual measurements, use the second calculator (runs (2) continuous). The spreadsheet will return the median value and count the number above and below the median. Values that fall exactly on the median are ignored. The number of runs is calculated based on the dichotomised data (above/below the median). In either case, if the size of the larger sample is less than 25, you should now use the tables of critical values (Section 12.4) to determine whether your results are statistically significant. If your samples are larger than 25, you can use the results that appear to the right, which include the one- and two-tail probabilities and the effect size 'r'. The third calculator (runs (3) calculator) allows you to enter values for the size of the two groups and the number of runs and obtain probabilities and effect size. There is no sample size limit for this calculator.

5.5.3.2 In SPSS

To perform the number of runs test for randomness in SPSS, place your data in a single column and enter it, without a header into the 'Data view' page, then go to the 'Variable view and add a header and identify the 'measure' as nominal if it is dichotomous (e.g. male/female) and 'scale' if it is continuous (individual measurements). Click 'Analyze', 'Nonparametric Tests' and 'One Sample'. Choose 'Customize analysis' as the objective, move your data into the 'Test fields' box and on 'Settings' choose the option for 'Test sequence for randomness (Runs test)'. In the 'Options' choose 'there are only two categories in the sample' and the option 'Sample median'. Click 'Run' to obtain the two-tail probability rounded to three decimal places. If it returns $p = 0.000$ report it as $p < 0.001$. Double clicking on the output provides a simple graph and the 'Standardized Test Statistic' which is a z-score. Divide this by the square root of the total sample size to obtain the effect size 'r'. There is no one-tail test option. For high p-values, divide by two to get a one-tail probability. If the two-tail probability is returned as $p = 0.000$, then the one-tail value must be $p < 0.001$.

5.5.3.3 In R Commander

The runs test is not included in the basic version of *R* Commander but it is easy to install and load a 'Plugin' that works well. Open *R* and load *R* Commander in the usual way. Go back to the *R* console and choose 'Packages' and 'Install package'. Choose a 'Cran Mirror' near you and a long list appears. Choose 'RcmdrPlugin UCA' and give it a chance to load. Now when you click on 'Statistics' and 'Nonparametric tests', two new options appear. The 'Randomness test for two level factor' applies to data in two categories (like yes/no) and the 'Randomness test for numeric variable' applies to continuous data as individual measurements. In both cases, the test returns a two-tail probability. Divide by two to get a one-tail probability. If the two-tail probability is returned as $p = 0.000$, then the one-tail value must be $p < 0.001$.

5.5.4 Examples

5.5.4.1 Example: Nominal Data and Small Sample Sizes

You are interested in how people use different spaces at different times and decide to investigate whether students entering the campus bar tend to arrive as single-sex groups or as couples. This is a two-tail test and your null hypothesis is that the gender order will be random. You spend a pleasant evening propping up the bar and recording the gender of people as they enter. The results are 20 males and 18 females and there are 15 runs of the same gender.

MMM,FF,M,FFFF,MMMMMM,FF,MM,FFF, MMM,FFFF,M,F,MM,FF,MM

The tables of critical values for a two-tail test show that 13 runs is the lower critical value and 27 is the upper critical value. Since the number of runs falls between these two critical values, we conclude that there is more than a 5% chance that the gender sequence is random and so accept the null hypothesis.

Conducting the same study at a local restaurant might yield the following sequence:

F,M,FF,M,F,M,F,M,FF,MM,FF,M,F,MM,FF,M,FF,M,FFF,M,F,M,FF

which gives 20 females and 13 males arranged in 23 runs. Given a two-tail test the critical upper boundary for the number of runs at $p = 0.05$ is 23 and at $p = 0.01$ it is 24. Since the observed number of runs is 'equal to or larger than' the critical threshold at $p = 0.05$, we can reject the null hypothesis and conclude that the order of genders is not random, there is a tendency for a female to be followed by a male.

**BOX 5.1   USING THE NUMBER OF RUNS TEST FOR RANDOMNESS
TO ASSESS THE SUITABILITY OF A REGRESSION LINE**

One of the most useful applications of the runs test for randomness is to assess whether a straight line is a good fit for describing the relationship between pairs of data plotted on a scatter graph (Ebdon 1977). A good linear fit is an important assumption of both parametric correlation analysis using Pearson's correlation coefficient (Chapter 10) and of simple linear regression analysis (Chapter 11). You can easily fit a range of different line types through data on an x–y plot (scatter graph), but you should really start by using a straight line. It is only sensible to use curves and other fancy lines when there is some clear theoretical reason for not using a straight line or where a straight line does not fit the data. This test allows you to clearly define whether the straight line (or any other line) fits the data well enough or not. When used in this context, the one-sample number of runs test for randomness is always used as a one-tail test, because we are only interested in whether there are too few runs, not too many.

It is a very simple test to apply, requiring no calculations at all, you just have to read the points on the graph as a sequence, from left to right, and count the number of times that they cross the regression line. A sequence or single point above the line is one run and a following sequence or single point below the line is another run. The two-sample sizes are the number of points above the line and the number below it. If a point sits exactly on the line, you can ignore it. If the regression line fits the data well the points should regularly cross it, giving a lot of runs. If there are very few runs it suggests a very poor fit, which is what typically occurs when a straight line is fitted to a curvilinear relationship.

An example is given in Figure 5.2. When a straight line is used (a), it does not follow the data very well at all, giving three runs with 10 points above the line and 8 below it. We are interested in whether there are too few runs, so using the one-tail table of lower critical values for sample sizes of 10 and 8 gives a critical value at $p = 0.05$ of six runs. Since we only have three runs, we can reject the null hypothesis that the number or runs is large enough to be due to chance. This provides justification for fitting a curved line, in this case (b), it is the 'logarithmic' option in Excel, and there are eight points above, eight points below and two points sitting on the regression line. The one-tail critical value for two samples of 8 at $p = 0.05$ is 5 and we now have eight runs, so we can accept the null hypothesis and conclude that the curved line is acceptable.

When you have a large sample, it can be difficult to apply this test by simply counting the points on a scatter graph, in which case you can define each value as either above or below the line by entering the predicted values for each point based on the equation for your chosen regression line. Then use the 'IF' function to identify values that are higher and lower. A companion site calculator is provided that will do this for you if you enter the data in the correct format. It is described in more detail in Chapter 11. Fitting a straight line (using simple linear regression: Chapter 11) through the Central England spring temperature data used in Figure 5.3, for example, gives samples of 30 points above the line and 20 below it and there are 24 runs, giving a one-tail probability of $p = 0.38$ and a very weak effect size ($r = 0.04$), suggesting that a linear regression line is acceptable.

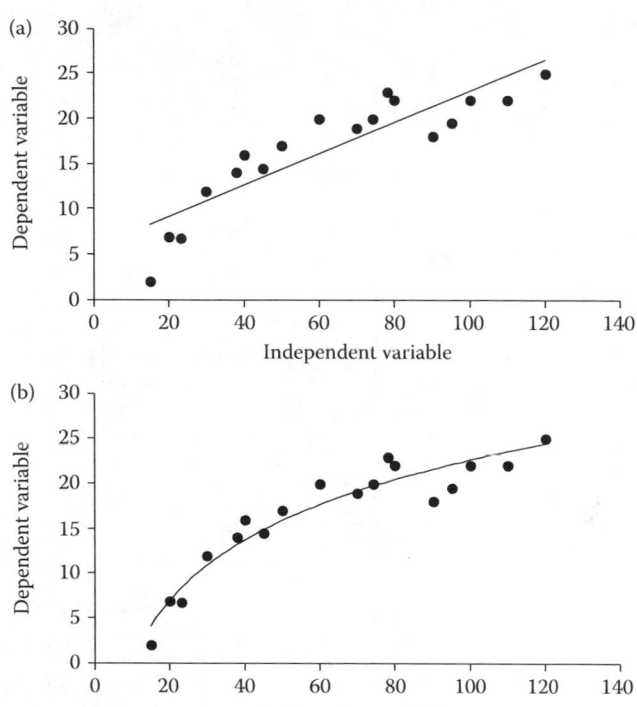

FIGURE 5.2

Two different regression lines fitted to the same set of data. The straight line (a) does not follow the points very well and there are only two places where the points cross the line, giving three runs. The curved line (b) is a much better fit, and the line is crossed seven times giving eight runs.

5.5.4.2 Example: Large Sample of Individual Numbers

When applied to a sample of individual measurements, the method divides the values into a set above the median and a set below the median. Any values that lie exactly on the median are ignored. For continuous data, this should give two samples of equal size, but if there are shared values at the median, there can be a small inequality. Identifying values above and below the median is not difficult in a spreadsheet but if your total sample size is less than 300, the companion site calculator will do that analysis for you. Figure 5.3 shows the effect of this procedure using 50 years of spring temperature data for central England (1966–2015). Note that the magnitude of the difference between the annual value and the median is ignored, so all values effectively have equal weight. In this case, there are 25 values above and below the median and only 14 runs, which is strongly significant (two-tail $p < 0.001$, effect size $r = 0.485$). It is clear from Figure 5.3 that the reason lies in the clear increase in spring temperatures in recent decades.

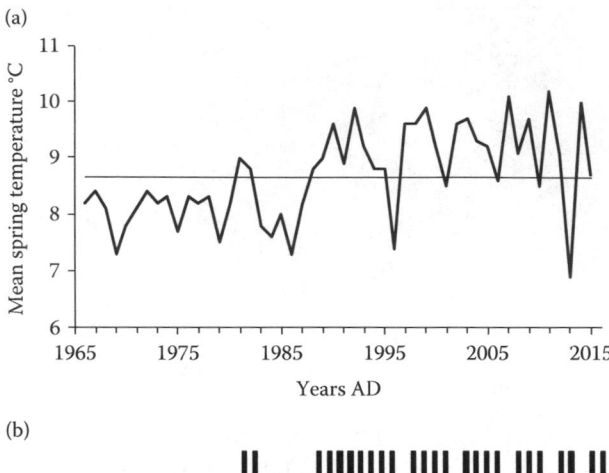

FIGURE 5.3

(a) Evolution of spring (March to May) temperatures for Central England over 50 years, with the median value marked by a horizontal line. (b) the same data translated into +1 for values above the median and –1 for values below the median to graphically display the way that the number of runs test for randomness deals with such data. Note that it is insensitive to the magnitude of the values.

BOX 5.2   RUNS TEST FOR RANDOMNESS USING LARGE SAMPLES: CALCULATING A PROBABILITY USING THE NORMAL APPROXIMATION

On the companion site, there is a simple calculator that will do this for you, but in case you want to see the workings here they are. Where: n_1 is the size of the larger group, n_2 is the size of the smaller group and Runs is the number of runs, the z-score is computed using

$$z = \frac{\text{Runs} - ((2n_1n_2/n_1 + n_2) + 1)}{\sqrt{2n_1n_2(2n_1n_2 - n_1 - n_2)/(n_1 + n_2)^2(n_1 + n_2 - 1)}}$$

Note that the same two calculations appear several times in this equation, and by calculating them first and giving them a code letter (A and B), the equation becomes a little easier to handle. For these calculations, assume that $n_1 = 30$, $n_2 = 20$ and Runs = 15.

$$A = 2n_1n_2 = 2 \times 30 \times 20 = 1200$$

$$B = n_1 + n_2 = 30 + 20 = 50$$

$$z = \frac{\text{Runs} - (A/B + 1)}{\sqrt{A(A - n_1 - n_2)/(B)^2(B - 1)}} = \frac{15 - (1200/50 + 1)}{\sqrt{1200(1200 - 30 - 20)/(50)^2(50 - 1)}}$$

$$= \frac{15 - (24 + 1)}{\sqrt{1200(1150)/2500 \times 49}}$$

$$z = \frac{15 - (24 + 1)}{\sqrt{1200(1150)/2500 \times 49}} = \frac{15 - 25}{\sqrt{1380000/122500}} = \frac{-10}{\sqrt{11.265}} = \frac{-10}{3.356} = -2.979$$

From this z-score, we can calculate a one-tail probability using tables of the standard normal distribution, an online calculator or the NORM.S.DIST function in Excel. In this case, the one-tail probability is $p = 0.0014$ and the two-tail probability is obtained by doubling this value to give $p = 0.0028$.

The effect size 'r' is obtained by dividing the z-score by the square root of the total sample size

$$\text{Effect size } r = \frac{z}{\sqrt{N}} = \frac{-2.979}{\sqrt{50}} = \frac{-2.979}{7.071} = -0.421$$

The sign of the effect size can be ignored, so $r = 0.42$

References

Ebdon, D. 1977. *Statistics in geography: a practical approach.* Oxford: Blackwell, 195pp.
Neave, H.R. and Worthington P.L. 1988. *Distribution-free tests.* London: Unwin-Hyman Ltd., 430pp.
Siegel, S. 1956. *Nonparametric statistics for the behavioural sciences.* New York: McGraw-Hill, 312pp.

6

Two-Sample Tests for Counts in Two Categories

6.1 Introduction

Where data are arranged simply as counts in two categories, there are still some very useful inferential statistical tests available. They are commonly used in medical research, for example, in assessing the outcome of particular treatments. The important distinction in this case is between paired and independent data. Data are paired if an entry in one category is linked to a particular entry in the other category. A common reason for samples being paired is that the same person is being asked a question twice, before and after some event, samples have been collected in pairs from the same place, or at the same time, or the same sample has been measured twice using different methods.

In this chapter, I describe two tests for paired data and two for independent data (Figure 6.1). These tests can also be used when the data are much richer, for example, when there are counts in several classes or where the data consist of individual measurements. In such cases, these methods do not use any of that extra information and may therefore be a bit wasteful relative to some of the tests described later.

6.2 Sign Test

The sign test is just a special case of the binomial test and full details of the way it works and how to conduct it can be found in Chapter 5 (Section 5.2). In the binomial test, the probability of a particular outcome can be varied at will, but in the sign test, the probability of an outcome in either category is always set at 50:50, or $p = 0.5$. The name refers to the fact that is does not use numbers or ranks, just plus or minus signs.

6.2.1 When It Is Useful

The sign test can be used when you have pairs of individual measurements but all you really need is the sign of the difference. Plus does not have to mean 'positive', it is just used to define a direction, so it could be better rather than worse, higher rather than lower or drunker rather than more sober. It is especially useful where the effect that you are interested in is difficult to quantify. You can also use the sign test when you have pairs of individual measurements but are only interested in the sign of the difference. For example, you might want to know whether students tend to improve over the first year at university, so just consider whether they did better or worse in the second exam period. It is quick and

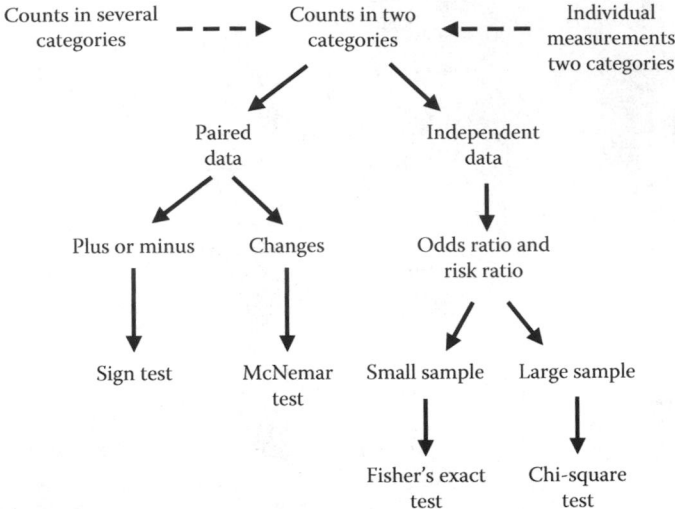

FIGURE 6.1
Tests available for data arranged as counts in two categories. Dashed arrows indicate a loss of information.

relatively easy to use but if you have individual measurements, or where the differences can be placed in rank order, it is rather wasteful of information and you should consider one of the more powerful tests, such as Wilcoxon's matched-pairs signed-ranks test or, if your data are suitable, the paired-sample version of Student's *t*-test.

6.2.2 What It Is Based On

The test calculates the probability of a given ratio of plusses and minuses occurring just by chance for a given sample size and assumes that the probability of a plus or minus occurring just by chance is 50% ($p = 0.5$). When sample sizes are very small, a relatively large difference in the ratio could occur by chance, but as the sample size rises, it becomes increasingly unlikely that a substantial difference is just due to luck. The test calculates the exact probability that an imbalance at least as large as that observed could have occurred just by chance. The probabilities are the same as those that apply when tossing a coin, where there is a 50% ($p = 0.5$) chance of tossing a head or a tail each time (Section 2.6).

6.2.3 How to Do It

The sign test is just a special case of the binomial test, and the options for conducting it are discussed fully in Section 5.2. It can be conducted in a spreadsheet, using online calcula-tors, and in SPSS and *R* Commander. A simple calculator that will return one- and two-tail probabilities and an effect size is provided on the companion site. With small sample sizes, perhaps the simplest way to conduct the sign test is to count the number in the smaller class and use tables of critical values (Section 12.1).

6.2.4 Effect Size

An appropriate effect size is the odds ratio. Since the probability of either outcome is the same ($p = 0.5$ by definition for this test), this is simply the number in the larger class divided by the number in the smaller class. For example, in a sample of 60, you might obtain 40 who answer 'yes' and 20 who answer 'no'. The odds ratio is $40/20 = 2$, which tells us that the participants were twice as likely to say yes as to say no. Alternatively use the more generally applicable effect size 'r', provided by the companion site calculator.

6.2.5 Examples

6.2.5.1 Example: Checking Exam Improvement

The sign test is easier to understand if there are some numbers involved, even though it does not need numbers. Imagine that you want to know whether there is a tendency for geography student grades to increase or decrease between the first and second marking periods at your university. You obtain the marks for an unbiased sample of 25 students and note that 17 improved, six declined and two got identical marks. You have to ignore the two that did not yield a plus or minus, so your sample size becomes 23. You did not predict the direction of difference in advance, so a two-tail test is appropriate. The table in Section 12.1 lists the critical size of the smaller class for a given significance level and sample size (n). For $n = 23$, the critical value for $p = 0.05$ is 6. Since your smaller class is 'equal to or less than' the critical value, you can reject the null hypothesis and conclude that the difference is statistically significant. You might report the results as 'a significant proportion of students showed improvement (sign test, two-tail $p < 0.05$)'. Using the BINOM.DIST function in a spreadsheet gives the probability as $p = 0.017$, but this is a one-tail probability and you need to double it for a two-tail test giving $p = 0.035$.

To use the companion site binomial test calculator, simply enter the numbers in each group (17 and 6) and set the 'probability of success' at 0.5 to obtain the one- and two-tail probabilities. The odds ratio is 2.83 (17/6), so students are 2.83 times more likely to improve than to get worse. You might go on to conclude that the students are getting better, but that is not the only explanation. This experiment assumes that the two exam periods involved exactly the same levels of difficulty and severity of marking, because the underlying assumption is that there is a 50:50 chance of doing better in one rather than the other just by chance. It could be that the evil professors make the first exam difficult just to torture the new students.

6.2.5.2 Example: Crystal Healing

As a keen geologist, you suspect that crystal healing is a load of nonsense. You decide to stand outside of the local alternative medical centre and ask people if they have just had this particular treatment and if so did it make them feel better. Twenty said they felt better, five said worse, two said they felt much the same and several told you to mind your own business. Although you have 27 responses, you cannot use the two who report no difference, so the sample size is 25, with 20 'plusses' and 5 'minuses'. You are testing to see if there is an improvement, so this is a one-tail test. Using the table in Section 12.1, the one-tail critical values for $N = 25$ at $p = 0.05$ and $p = 0.01$ are 7 and 6. Since your smaller value (5) is 'equal to or smaller than the critical value for $p = 0.01$' you can conclude that there is less than a 1% chance that the larger proportion of people reporting that they feel better is just due to chance. The companion site calculator returns the one-tail probability as $p = 0.002$ with a strong effect size ($r = 0.57$). We would have to conclude that having the treatment does make people feel better. Whether it has anything to do with the 'power of the crystals' is another question entirely.

TABLE 6.1

Soil Hardness Values (Artificial) Measured at Eight
Locations on the Path and on the Verge

Pair	Path Value	Verge Value	Sign
A	82	63	+
B	69	42	+
C	73	74	−
D	43	37	+
E	58	51	+
F	56	43	+
G	76	80	−
H	65	62	+

Note: The data are arranged as pairs, they are not independent. The
sign column indicates the direction of the difference.
Reversing the signs makes no difference.

6.2.5.3 Example: Footpath Erosion

You are interested in the effects of trampling on soil compaction because increased compaction tends to cause more overland flow and soil erosion. You measure soil hardness at eight points along a footpath up a mountain and at each measuring station you also take a measurement on the verge, away from the path. You thus have paired data and you are testing the one-tail hypothesis that the path tends to be harder than the verge. The data are presented in Table 6.1. Note that there are eight pairs of measurements and in six cases, the path yields a higher value than the verge. Using a one-tail test, because we predicted in advance that the path values would be higher, the table of critical values for $N = 8$ at $p = 0.05$ is one and in the smaller class in our sample the value is 2. Since two is not 'equal to or smaller than' the critical value, we must accept the null hypothesis that the path is not more compacted than the verge. Entering 2 and 6 into the companion site calculator gives the one-tail probability as $p = 0.145$ with an odds ratio of three.

In this case, where we have pairs of individual measurements, the sign test is rather weak because it is not using all of the information that is available. Note that the two differences that are in the 'wrong' direction (C and G) are quite small. The sign test does not use this information but the Wilcoxon's matched-pairs signed-ranks test does and is thus more powerful. When two tests disagree, you should take the result of the more powerful test, as long as any assumptions are met. Note that if a result is significant using the sign test, it will always be significant using the Wilcoxon's matched-pairs signed-ranks test.

6.3 McNemar's Test for Significance of Changes

Also known as 'McNemar's test for paired proportions', 'McNemar's test for binary matched-pairs data' or simply as 'McNemar's test', it is rarely used in geography although it is well suited to some of the problems that geography students deal with and is very easy to apply.

6.3.1 When It Is Useful

McNemar's test for significance of changes (McNemar 1947) is often used in clinical research to compare the effect of a treatment and in this case, each person effectively acts as their own control, because their condition is assessed before and after the treatment and the medics are typically interested in whether they got better or worse (Fagerland et al. 2014). In human geography, it can be used in a similar way, where you want to examine the effect of some intervention or event. For example, you may be interested to know whether attending a lecture on the evidence for climate change has any effect on people's level of support for 'green' taxes on fuel. In physical geography, it could be used where you are able to use some experimental intervention and can use the same sites or samples before and after. The test only requires counts in two categories and the categories can be nominal (e.g. for/against, better/worse, etc.).

6.3.2 What It Is Based On

The test works by comparing the number of cases that change in one direction with the number that change in the other direction. It ignores those cases where there is no change. Imagine that a sample of 60 students have been given a crystal necklace to wear to see if it improves their health. A nurse examines them before and after a month of crystal adornment (Table 6.2). In this case, we are not interested in the 36 who were healthy both before and after the 'treatment', nor in the 12 people who were sick both before and after, only in the number of healthy people who got sick and the number of sick people who got healthy again. In this case that ratio is 6:6, so it is clear that wearing the crystal had no effect.

Now consider a similar sample of 60 students who were also assessed by the nurse but who were given vitamin supplements rather than a crystal necklace (Table 6.3). In this case, the ratio is more promising. Five healthy people got sick but 20 sick people got healthy. The McNemar test assesses the probability that a difference as large as 5:20 could occur just by chance assuming that the probabilities of getting better or getting worse are the same. This

TABLE 6.2

Contingency Table Where the Entries Are Linked Rather Than Independent

Crystals	Healthy After	Sick After
Healthy before	$A = 36$	$B = 6$
Sick before	$C = 6$	$D = 12$

Note: McNemar's test for the significance of changes uses only the two cells that indicate a change (B and C). In this case, they are the same so the treatment has no effect.

TABLE 6.3

In This Case, the Change in One Direction Is Four Times Larger Than the Change in the Other Direction

Vitamins	Healthy After	Sick After
Healthy before	$A = 30$	$B = 5$
Sick before	$C = 20$	$D = 5$

Note: McNemar's test is used to calculate the probability that a difference of this magnitude could occur by chance.

final assumption is important of course, and you should consider it carefully when you interpret the results, as discussed in the more geographical examples below.

6.3.3 How to Do It

The logic behind the test is similar to that for the binomial test, but it uses the chi-square distribution to calculate the probabilities. In the tables above, we are only interested in the contents of cells B and C, those that show a change, and if the values are not too small the test statistic (confusingly called chi-square) is computed using this formula

$$\chi^2 = \frac{(B-C)^2}{B+C}$$

Note that although this equation uses the chi-square symbol (χ^2), it is not the same as the chi-square test, it just allows you to use the same tables to look up the probability (with degrees of freedom = 1). To avoid any confusion, I have tabulated the relevant values for one- and two-tail tests here (Table 6.4).

For the vitamins, example above the equation becomes

$$\chi^2 = \frac{(B-C)^2}{B+C} = \frac{(5-20)^2}{5+20} = \frac{-15^2}{25} = \frac{225}{25} = 9$$

Consulting Table 6.4 for a one-tail test, which would be appropriate if we predicted better health, the result is significant at $p < 0.01$ and so we would reject the null hypothesis. If you want to know the probability associated with a given χ^2 value, you can use the CHISQ. DIST.RT (or CHIDIST) function in Excel and enter 1 degree of freedom to obtain the two-tail probability ($p = 0.0027$). To obtain the one-tail probability, we half this ($p = 0.00135$).

6.3.4 Effect Size

An appropriate effect size is the odds ratio, which is the number in the larger category divided by the number in the smaller category. Using the example above, it is $20/5 = 4$. In this case, the odds ratio tells us that the participants were four times more likely to get better than to get worse.

6.3.5 Small Samples

With this test, we are only interested in the values in the two cells that represent changes, not in the size of the whole sample. The calculations shown above rely on the assumption that the results follow the chi-square distribution, but when the two numbers are very small that assumption may not hold. There seems to be some uncertainty about how large

TABLE 6.4

Critical Values for McNemar's Test for the Significance of Changes

	One-Tail Test			Two-Tail Test		
p	0.05	0.01	0.001	0.05	0.01	0.001
χ^2	2.71	5.42	9.55	3.84	6.64	10.83

Note: Results are significant if the measured value is equal to or greater than the tabulated value.

the numbers have to be but the most common advice is that they should add up to at least 25. Where the sum is less than this, the advice is often to use the binomial test instead, with the number of changes in either direction as the two values. However, it has been argued that the binomial test tends to be rather conservative and accepts the null hypothesis too easily. There is an alternative 'mid-p' approach (Fagerland et al. 2013), that is supposed to give better results, and it is easy to apply in a spreadsheet, so it is included in the companion site calculator provided for this test along with the standard (sometimes called asymptotic) calculation and the binomial probability.

6.3.6 Correction for Continuity

Many older books recommend applying a 'correction for continuity' to tests that rely upon approximation to a continuous probability distribution (like chi-square) but use discontinuous counts in categories data. The most common example is Yates' correction applied to the chi-square test but a similar correction is often applied to the McNemar test (Edwards 1948). More recent work suggests that these corrections tend to produce results that are too conservative and they are probably best ignored (Howell 2012; Field 2013). However, many online calculators apply such corrections routinely and they are also one of the options in specialist software such as SPSS and *R* Commander, so an explanation is provided (Box 6.1).

6.3.7 How to Do It

6.3.7.1 Companion Site Calculator

To use the companion site calculator for McNemar's test, enter two values: the number of changes in one direction and the number of changes in the other direction. One- and two-tail probabilities are calculated using the chi-square approximation for large samples. For samples where the sum of the two counts is less than 25, it is better to use the 'mid-p' probabilities. Also included are probabilities calculated using the binomial test. The odds ratio is given as the effect size.

6.3.7.2 Online Calculators

Presumably because it is popular in clinical trials, there are plenty of online calculators for McNemar's test (Table 6.5), although the descriptions can be quite challenging. In every case, there are only two numbers that are actually used, the ones that involve a change, so just make sure that you put them in the correct boxes. Running the example given above should help. Unfortunately, all of the sites listed use a correction for continuity, so the results will be a little conservative (*p*-values slightly high).

6.3.7.3 In R Commander

I am not aware of an easy way to run McNemar's test in *R* Commander. There is *R*-code to perform the test but it uses the exact binomial calculation, so you can just use the binomial test as described in Chapter 5.

6.3.7.4 In SPSS

Enter the results in two columns so that the paired results sit next to each other. You can use two words or letters, such as yes and no, or replace with numbers such as zero and one. Click 'Analyze', 'Descriptive Statistics' and 'Crosstabs'. Use the arrows to move one set of data into the 'rows' box and the other into the 'columns' and click 'Statistics'. Choose

TABLE 6.5

Selection of Online Calculators for McNemar's Test for the Significance of Changes

Provider	Website	Comments
Statstodo	http://www.statstodo.com/McNemar_Exp.php	Simple to use, just two numbers to add so not confusing. Uses a Yates' correction of one
Vassar stats	http://vassarstats.net/propcorr.html	Uses the binomial probability. Also gives the odds ratio
Graphpad	http://graphpad.com/quickcalcs/mcnemar1.cfm	Described in terms of risk factor and disease. Uses Yates's correction = 1. Gives odds ratio
CCRB Chinese University of Hong Kong	http://www.cct.cuhk.edu.hk/stat/confidence%20interval/mcnemar%20test.htm	Simple to use, Yates' correction = 1
In silico project support for life sciences	http://in-silico.net/tools/statistics/chi2test	McNemar is an option in the chi-square calculator. Use the top right and bottom left cells. Yates' correction of 0.5

Note: Unfortunately, they all use a correction for continuity so the probability values will be rather conservative.

McNemar's test and click 'Continue' then 'OK'. The output is the two-tail probability which, even for quite large samples is based on the binomial test. You can also access the test via 'Analyze', 'Non-parametric tests' and 'Related samples' but the extra output is not very helpful.

6.3.8 Examples

6.3.8.1 Example: Opinions on Fracking

Fracking, a method of extracting gas from sedimentary rocks, is a very contentious issue in the United Kingdom. Some see the potential gas reserves as an economic windfall whilst others are concerned about the potential environmental impacts. Those with strong opinions are not always well informed. You wish to test the effect of education on the strength of opinion by monitoring opinions before and after a factual and unbiased lecture provided by a local professor of geology. Before the lecture, you ask members of the audience whether they are generally 'for' or 'against' fracking in the United Kingdom. You want to know whether gaining a better understanding of the science will cause people to change their opinions. You are not predicting the direction of greatest change, so this is a two-tail test. The results are shown in Table 6.6.

The two important numbers here are 6 and 18, so you can immediately see that three times as many people were persuaded to change their opinions from 'against' to in favour of fracking (18) as were convinced to change in the opposite direction. You cannot use

TABLE 6.6

Opinions Regarding Fracking for Gas Measured Before and After a Lecture

Fracking	For After	Against After
For before	$A = 43$	$B = 6$
Against before	$C = 18$	$D = 68$

Note: The changes in opposite directions are 6 and 18 and McNemar's test determines the probability that a difference this large could occur by chance.

this information to change your hypothesis to one-tail; that would be unfair. Entering the numbers 18 and 6 into the equation above gives

$$\chi^2 = \frac{(B-C)^2}{B+C} = \frac{(6-18)^2}{6+18} = \frac{-12^2}{24} = \frac{144}{24} = 6$$

To obtain the two-tail probability using Excel, the CHIDIST function is used with one degree of freedom (=CHISQ.DIST.RT,6,1) which yields $p = 0.014$. Using Table 6.4 gives the same result, since 6 falls between the critical values for $p = 0.05$ (3.84) and $p = 0.01$ (6.64). In this case, we can reject the null hypothesis that attending the lecture has no effect on the direction of change in opinions about fracking. Before getting too excited about this significant result, it is worth noting that this test only looks at those who changed their opinions, and in this case that only applies to 24 out of 135 participants, which is less than 18%. You should at least mention that when discussing the results.

6.3.8.2 Example: Land Management

At a local nature reserve, the manager has decided to change the grazing regime, replacing sheep with ponies, with the aim of increasing the abundance of a rare orchid. In a survey in the year before the change, the presence or absence of orchids was reported in 50 marked quadrats. Two years after the change in grazing the presence or absence of orchids is recorded again in the same plots. We predict more sites will gain orchids than lose them, calling for a one-tail test. The results (Table 6.7) are in the direction predicted, with 12 sites gaining orchids and only 4 losing them. Entering the values 4 and 12 into the companion site calculator gives a one-tail probability of $p = 0.024$, so we can reject the null hypothesis. You might conclude that the change in management has been successful, but that is not the only possible explanation. For example, the weather conditions prior to the two measurement periods may have been very different. A better research design would include some 'control' quadrats in locations where the land management has not changed.

6.3.8.3 Example: Golf Green Hydrophobicity (Bad Test)

When golf greens get too dry, they sometimes become hydrophobic, so that water sits on the surface rather than percolating to the grass roots. It can be a serious problem during hot dry summers. You wish to test the effectiveness of a new treatment for hydrophobicity and so tested 30 golf greens during mid-August and then went back in late September. The green keepers promised to use the new product at regular intervals during that time. You are expecting the treatment to make some of the hydrophobic greens hydrophilic, so this is a one-tail hypothesis. At first glance, the results (Table 6.8) look spectacular. All 20

TABLE 6.7

Presence or Absence of Orchids Measured in the Same Places Before and After a Change in Land Management

	Orchids After	No Orchids After
Orchids before	20	4
No orchids before	12	14

Note: Three times more quadrats gained orchids than lost them. McNemar's test determines the probability that a difference this large could occur by chance.

TABLE 6.8

State of Soil Before and After a Treatment for Hydrophobicity

Golf Greens	Hydrophobic After	Hydrophilic After
Hydrophobic before	0	20
Hydrophilic before	0	10

Note: The ratio of changes (20:0) looks very promising but the critical assumption that there is an equal chance of a change in either direction is strongly violated because of the seasonal effect of changes in the weather.

hydrophobic greens have been cured and none of the hydrophilic greens have become hydrophobic. The chances of a ratio of 20:0 occurring just by chance, when there is an equal probability of a positive or negative change occurring, is very remote indeed. The one-tail probability is much less than a thousand to one.

However, in this case, can you really argue that there is an equal chance of those greens becoming more or less hydrophobic between mid-August and late September? On the contrary, even if the greens were not treated at all, it is quite likely that the cooler and wetter weather of September would have resulted in the usual seasonal change from hydrophobic to hydrophilic conditions. The research is poorly planned because it is not possible to hold constant the confounding effect of the seasonal change in the weather. It would have been better to either treat the greens and measure them straight away or take soil samples back to the laboratory and perform the experiment under controlled conditions.

BOX 6.1   CONTINUITY CORRECTION FOR THE McNEMAR TEST FOR SIGNIFICANCE OF CHANGES

Many texts advocate the use of a 'correction for continuity' when you use this test, but there seems to be a lack of agreement with regard to the magnitude of the correction and indeed whether it should be used at all. The equation including a 'Yates' correction (y) is shown below. Note that there are now vertical bars around $B - C$ ($|B - C|$) which mean take the absolute value, or ignore the sign of the difference. Some books and websites advocate using a correction of $y = 0.5$ and others advocate using $y = 1.0$

$$\chi^2 = \frac{(|B-C|-y)^2}{B+C}$$

Using $y = 1$, the result for the land management (orchids) example (6.3.8.2) becomes

$$\chi^2 = \frac{(|B-C|-y)^2}{B+C} = \frac{(|4-12|-1)^2}{4+12} = \frac{(8-1)^2}{16} = \frac{7^2}{16} = \frac{49}{16} = 3.06$$

Or using $y = 0.5$, it becomes

$$\chi^2 = \frac{(|B-C|-y)^2}{B+C} = \frac{(|4-12|-0.5)^2}{4+12} = \frac{(8-0.5)^2}{16} = \frac{7.5^2}{16} = \frac{56.25}{16} = 3.52$$

If you do not use any correction and use the original equation, it becomes

$$\chi^2 = \frac{(B-C)^2}{B+C} = \frac{(4-12)^2}{4+12} = \frac{(-8)^2}{16} = \frac{64}{16} = 4$$

The critical value of χ^2 for a two-tail test at $p = 0.05$ is 3.84, so if we were applying a two-tail test to these data, the decision about whether to use a correction becomes important. Using a correction, we would accept the null hypothesis and without a correction, we would reject it.

If you have the option, I suggest that you do not apply any correction. If the data set is small, so that the two counts add up to less than 25, then consider using the 'mid-p' probability provided by the calculator on the companion website.

6.4 Tests for Independent Samples Arranged as 2×2 Contingency Tables

In geography, it is quite common to collect two samples of data that are arranged as counts in two categories, where the counts are independent of each other. A simple example is testing for a gender difference in response to a question that only has two possible answers, such as yes or no. The result is a 'contingency table' with two columns and two rows. By convention, we put people in the rows and their responses in the columns (Table 6.9). The term 'contingency' refers to the association between the groups (rows) and their responses (columns). If there was a strong tendency for the males to say no and the females to say yes, for example, we would say that the answers were 'contingent on gender'. The definition of contingent is not quite the same as 'dependent'. If one parameter is dependent on another, then we assume that if one changes, then the other will change as well. Contingency implies a less certain relationship, so that a change in one parameter may cause a change in another but it does not mean that such a response is inevitable. If we conclude that opinion is contingent on gender, we are not saying that all women will give the same reply to a given question and all men will give a different answer, only that the difference in the proportions is larger than we would expect to occur purely by chance.

When faced with a 2×2 contingency table, the tendency is to jump straight to the chi-square test. However, unless you have a clear understanding of what is actually being tested, the danger is that you end up with a probability value but no clear idea of what it

TABLE 6.9

Example of a Simple 2×2 Contingency Table

	Yes	No	Sum
Male	–	–	–
Female	–	–	–
Sum	–	–	–

Note: It is important that the answers are mutually exclusive, so that a single individual can only appear in one row and one column.

means. The chi-square test also has some strict sample size assumptions, and 'breaking the rules' of chi-square is a common reason for losing marks. It is better to approach the problem in a series of steps (Figure 6.2), so that you first quantify the difference that you are interested in, before choosing the appropriate test and determining the significance of the observed difference. The appropriate test depends on the size of your sample and the way the counts are arranged in the table, so some calculations are necessary before you can decide whether to use the chi-square test or Fisher's exact test.

There are a variety of measures available to quantify the difference in response of two groups but the most useful are probably the risk ratio and the odds ratio. For both of these metrics, you can also calculate 95% confidence limits. These allow you to go beyond simply describing your sample and to give an estimate of what those values are likely to be in the wider population. Whereas the risk ratio and odds ratio values are based entirely on the magnitude of difference in your sample, the confidence limits also take into account the size of your sample so that the uncertainty declines as the sample size increases.

The risk ratio and odds ratio also allow you to consider whether you should be using a one- or a two-tail test. A one-tail test implies that your hypothesis predicts the direction of difference, but just by looking at four numbers in a 2×2 contingency table, it can be difficult to appreciate how the concept of 'direction' applies. The answer lies not in the raw numbers but in the difference in proportions. For example, if you predict in advance, on the basis of theory rather than by looking at the numbers, that the proportion of men saying no will be larger than the proportion of women saying no then your hypothesis has direction and is one-tail. The risk ratio and odds ratio are based on the difference in proportions, so they tell you if the results are in the direction that you predicted. As always, if you are not sure, use a two-tail test.

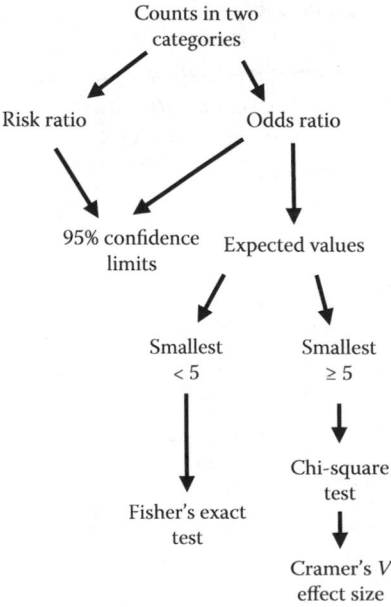

FIGURE 6.2
Logical series of steps for the analysis of a 2×2 contingency table.

6.5 Risk Ratio and Odds Ratio

The risk ratio and odds ratio are measures of 'effect size'. For example, if you are comparing males and females, they tell you the size of the gender difference in response. They also tell you the direction of the difference. Using a medical example helps to explain what is meant by 'risk'.

6.5.1 Risk Ratio

Imagine that we wish to determine whether gender influences your chances of suffering from some medical condition. If we take an unbiased sample of 100 men and 100 women, we can tabulate the results in a 2×2 contingency table (Table 6.10). In this case, it is clear that 40% of the females have the condition but only 20% of the men. It seems sensible to conclude that women are about twice as likely as men to have the condition. This measure of the difference in response is known as the risk ratio and it is simply the proportion (not number) of females who have the condition divided by the proportion of males who have it (Table 6.10). Note, however, that in this case if you were to divide the male risk by the female risk, you would not get 2 you would get 0.5. This means that men are half as likely as women to suffer from the condition. The risk ratio for males is the 'inverse' of the risk ratio for females, which means that one divided by the risk ratio for one group will give you the risk ratio for the other group ($1 \div 2 = 0.5$ and $1 \div 0.5 = 2$).

It is important to remember that the risk ratio relates to the results in one of the columns, and if you use the numbers in the other column, you do not get the same result. In this case, the 'risk' of *not* having the condition is 0.6 for females and 0.8 for males, so the two possible risk ratios in this case are 0.75 and 1.33, so males are 1.33 time more likely than females to not have the condition. If you make the calculation by hand, then it is easy to make sure that you are calculating the number that you really want, the problem comes when you input your data into an online calculator or use some statistical software that yields a single 'risk ratio' that will actually change according to the order in which you arrange your categories.

6.5.2 Odds Ratio

The other metric often used to quantify the difference in response is the 'odds ratio'. The critical thing to understand here is that 'odds' does not mean the same thing as 'proportion'. In this context, the 'odds' of a woman having the condition is not the proportion (or percentage) of women who have it, it is the proportion who have it divided by the proportion who do not have it (Table 6.11). In this case, the result means that if you are female, the

TABLE 6.10

Calculating the Risk Ratio

| | Condition | | | | |
	Absent	Present	Totals	Calculating the Risk Ratio	
Females	60	40	100	Female risk	$40/100 = 0.4$
Males	80	20	100	Male risk	$20/100 = 0.2$
				Risk ratio	$0.4/0.2 = 2.0$
Totals	140	60	200		

Note: In this case, we would conclude that females are twice as likely to have the condition.

TABLE 6.11

Calculating the Odds Ratio

| | Condition | | Totals | Calculating the Odds Ratio | |
	Absent	Present			
Females	60	40	100	Female odds	$40/60 = 0.667$
Males	80	20	100	Male odds	$20/80 = 0.25$
				Odds ratio	$0.667/0.25 = 2.67$
Totals	140	60	200		

Note:　In this case, we would conclude that the odds of a female having the condition are
　　　　 2.67 times higher than the odds of a male having the condition.

odds of you having this condition are 2.67 times higher than they would be if you were male. If you are male, the odds of you having this condition are 0.375 times higher than they would be if you were female. The odds ratio for males is the 'inverse' of the odds ratio for females, which means that one divided by the odds ratio for one group will give you the odds ratio for the other group ($1 \div 2.67 = 0.375$ and $1 \div 0.375 = 2.67$).

The 'risk ratio' and 'odds ratio' are both useful measures of the magnitude of the difference in response of two groups. For geographers, the 'risk ratio' seems a good option because it is easier to understand, perhaps because we are more used to dealing with differences in terms of percentages (and therefore proportions) and the concept of 'odds' is less familiar. In statistics, however, the odds ratio seems to be used most commonly. Just be aware that they are not the same and so it is important to make sure that you use the right language when you report the results (see the different conclusions in the captions for Tables 6.10 and 6.11).

6.6 Confidence Limits of the Odds Ratio and Risk Ratio

An advantage of calculating these 'effect size' metrics is that you can also use them to make inferences about the population rather than just describing the sample. When we calculate an odds ratio, for example, it is correct for that particular sample but we are aware that if we took another sample, the odds ratio is unlikely to be exactly the same. If we take very small samples, then we might expect quite big differences between them, but if our samples are very large, then we would expect the metrics to be much more similar, just because a large sample is more likely to be representative of the 'real' odds ratio in the population. Some software and online calculators will calculate not just the risk ratio and odds ratio of your sample, they will estimate the 95% confidence limits for those metrics. This means that there is a 95% chance that the relevant ratio in the population will lie within these lower and upper values. The magnitude of the odds ratio, for example, is based entirely on the ratios calculated from your contingency table and sample size does not matter, but the 95% confidence limits take into account the size of the sample. Even without performing one of the inferential tests, the confidence limits are very useful because if they cross the number one, it means that the effect is changing direction. A risk ratio or odds ratio of one means that there is no difference between the samples. In the example above, if the confidence limits crossed one, you would conclude that you could not be sure whether females were more likely to have this condition or not.

6.6.1 How to Do It

The calculation of the 95% confidence intervals for the odds ratio and risk ratio is quite complicated, so I will not go through it here. Basically, they use the assumption that both the odds ratio and risk ratio are positively skewed, but normally distributed if you take the natural logarithm. This means that you calculate the confidence limits when the values have been log transformed, but then you return them to the original scale. This explains why the sample estimate of the odds ratio or risk ratio does not lie in the middle of the two confidence limits; it is in the middle when the natural logs are used but then the calculated limits get skewed back when they are returned to the original scale.

6.6.1.1 Companion Site Calculator

Calculators are provided for both the risk ratio and odds ratio and their 95% confidence limits. On both sheets, there is an additional set of calculations to the right that is designed to deal with cases where there is a zero in one or more cells. I have simply added 0.01 to every cell, to avoid division by zero. Of course, if you have very small numbers or zero entries, the confidence limits are likely to be very wide indeed. The same calculations are also included in the companion site calculator for the chi-square test on a 2×2 contingency table.

6.6.1.2 Using an Online Calculator

You can obtain the 95% confidence limits using an online calculator (Table 6.12), but be careful with the risk ratio in particular and make sure that the values you calculate relate to the column that you want to refer to when you report the results. Most of the online calculators cannot cope if there is a zero in any of the cells, because it causes division by zero.

TABLE 6.12

Selection of Free Online Calculators That Returns the Odds Ratio and/or Risk Ratio for a 2×2 Contingency Table

Provider	Website	Comments
Select Statistical Services	http://www.select-statistics.co.uk/confidence-interval-calculator-odds-ratio	Only gives the odds ratio, not the risk ratio. Gives confidence intervals at 90%, 95% and 99% by default but you can also choose. Reverse the columns to change the OR from below one to above one
David JR Hutchon	http://www.hutchon.net/ConfidORnulhypo.htm	Provides an alternative calculation for an estimate of the odds ratio and confidence intervals that still works when there is a zero in one of the cells (Peto odds ratio)
Vassar stats	http://vassarstats.net/odds2x2.html	Returns both the OR and RR with 95% confidence limits. The risk ratio is calculated using the column on the right. Cannot cope if there is a zero in any cell
Medcalc	https://www.medcalc.net/tests/odds_ratio.php	Odd format designed for medics to use but gives an odds ratio and 95% confidence limits even if there is a zero entry. It does this by adding 0.5 to every cell to avoid division by zero. No risk ratio
Interactive Statistical Packages	http://statpages.org/ctab2x2.html	Gives OR and RR (for the left column) with 95% confidence limits, plus lots of other statistics I have never heard of! Does not cope with zero entries

6.6.1.3 In R Commander

Click on 'statistics', choose 'contingency tables' and 'Enter and analyse two-way table'. Enter the values in the 2×2 table that pops up and under 'statistics' click the option for 'Fisher's exact test'. The output will now include both the odds ratio and the 95% confidence limits. The results are not exactly the same as you obtain using other software, apparently because R uses a more sophisticated way of calculating both the odds ratio and the confidence limits. The differences between the methods are small enough to be ignored.

6.6.1.4 In SPSS

If you are using SPSS to analyse a 2×2 contingency table, in order to perform the Fisher's exact test or chi-square test, you can also obtain the odds ratio with 95% confidence limits. The data have to be entered appropriately (see Section 6.9.4) then click 'Analyze', 'Descriptive Statistics' and 'Crosstabs'. Under the 'Statistics' tab, check the 'Risk' box. The results are, confusingly, the odds ratio and not the risk ratio.

6.7 Sample Size Assumptions

Having determined the size and direction of the differences that are of interest, the next step is to check whether the sample size is large enough to allow the chi-square test or whether you need to use Fisher's exact test. The chi-square test is based on measuring the difference between the numbers that are observed in each of the four cells of the contingency table and the numbers that you would expect if the responses were not contingent on the groups. It does not calculate probabilities exactly, it assumes that the calculations fit the chi-square distribution. With large samples that is a reasonable assumption, but with very small samples it is not. For a 2×2 contingency table, the rule is that no expected frequency can be less than five. Note that it is the expected frequencies that matter, not the observed values, so you need to make some calculations in order to decide if your sample is large enough. When the sample size assumptions of chi-square are violated, Fisher's exact test is used.

6.7.1 Calculating Expected Values

The 'expected' values in a contingency table are those that would occur if there was no relationship between the groups in the rows and the responses in the columns. It is easiest to understand if you look at a table where only the marginal totals are filled (Table 6.13). In this case, there are equal numbers of males and females and equal numbers replied yes and no, so if there was no relationship between gender and response, we would expect equal numbers of males and females to answer yes and to answer no, giving expected frequencies of 10 in each cell.

Now consider a more complicated case where the totals of the rows and columns are not all the same (Table 6.14). In this case, there are twice as many males as females, so the ratio of males to females is 2:1. We know that 15 people answered yes, and if we assume that gender makes no difference to response, then we would assume that the ratio of males to females answering yes will be the same as the proportion of males to females overall. We would allocate two-thirds of the yes answers to the males (10) and one-third to the females (5). The column total for the answer no is also 15, so we enter 10 and 5 again. The logic is

TABLE 6.13

Expected Frequencies are Calculated Using the Marginal Totals (Sum of Rows and Columns)

	Marginal Totals				Expected		
	Yes	No	Sum		Yes	No	Sum
Male			20	Male	10	10	20
Female			20	Female	10	10	20
Sum	20	20	40	Sum	20	20	40

Note: In this simple example, the totals are all the same so the expected frequencies are all the same.

TABLE 6.14

In This Example, There Are Twice as Many Males as Females

	Marginal Totals				Expected		
	Yes	No	Sum		Yes	No	Sum
Male			20	Male	10	10	20
Female			10	Female	5	5	10
Sum	15	15	30	Sum	15	15	30

Note: If drinks preference was not contingent on gender, we would expect the same ratio (2:1) to occur in each column.

that the numbers appearing in the column totals are shared out in proportion to the overall number in each group (the row totals).

In reality of course the ratio of one group to the other can be a much less convenient number, so that even if you understand the logic, it can be difficult to see how to divide up the expected numbers in the four cells so that they match the row and column totals. Fortunately, there is a simple rule

$$Expected\ cell\ value = \frac{column\ total \times row\ total}{Grand\ total}$$

The calculations for Table 6.14 are shown in Table 6.15 and the same calculations for a much less convenient arrangement are shown in Table 6.16. Note that there is no requirement for the expected values to be integers; they are usually fractions. Also, it is clear from Tables 6.15 and 6.16 that you cannot decide whether it is appropriate to use chi-square simply by looking at the total sample size; you have to calculate the expected values. In the case of Table 6.15 with a total sample size of 30, the smallest expected frequency is five, so

TABLE 6.15

Expected Value for Each Cell is Obtained by Multiplying the Two Marginal Totals (Row and Column) for That Cell and Dividing the Result by the Grand Total

	Yes	No	Sum
Male	$(15 \times 20)/30 = 10$	$(15 \times 20)/30 = 10$	20
Female	$(15 \times 10)/30 = 5$	$(15 \times 10)/30 = 5$	10
Sum	15	15	30

TABLE 6.16

Calculating the Expected Counts Looks Laborious, But You Only Have to Do It for One Cell

	Yes	No	Sum
Male	$(10 \times 20)/35 = 5.71$	$(25 \times 20)/35 = 14.29$	20
Female	$(10 \times 15)/35 = 4.29$	$(25 \times 15)/35 = 10.71$	15
Sum	10	25	35

Note: The other values can be obtained by subtraction from the marginal totals.

the assumptions of the chi-square test are met. In the case of Table 6.16, even though the total sample size is larger, at 35, the arrangement of the marginal totals is such that one of the cells has an expected frequency of less than five, so the chi-square test cannot be used.

Calculating four expected values like this may seem like a bit of a nuisance, but if you have ever completed a Sudoku puzzle, you may notice that you only really have to make one calculation. The other three can be calculated by subtraction from the marginal totals. When you conduct a chi-square test on a 2×2 contingency table, it is said to have 'one degree of freedom', which is because if you know one value in the table, you can work out all of the others just from the marginal totals.

Checking the sample size assumptions for the chi-square test, by calculating the expected frequencies and checking that none are less than five, is a critical step in interpreting a 2×2 contingency table. Do not be tempted to skip it by blindly entering your data into a chi-square online calculator or other software. You can make the calculations by hand or in a spreadsheet and the chi-square calculator supplied on the companion website will make the calculations for you and warn if any expected values are less than five.

6.8 Chi-Square Test for a 2×2 Contingency Table

The chi-square test can be used on contingency tables with any number of rows and columns, but it is convenient to distinguish the case of a 2×2 table because in this special case, the test is easier to explain, compute and interpret. It is also possible to consider testing one-tail hypotheses, which is not true when the method is applied to more complex tables.

6.8.1 When It Is Useful

In a general sense, the chi-square test is concerned with contingency, in that it tests the hypothesis that the responses in the columns are contingent (loosely similar to dependent) upon the groups in the rows. In the simple case of a 2×2 table, however, it can also be seen as a test of the significance of the difference in response of the two groups.

Although chi-square tests are non-parametric, in this case there are two very important assumptions:

1. The entries in the table must be independent
2. The smallest expected frequency must be at least five

Independence means that each number in the table represents a single choice between the rows and a single choice between the columns. In this case, the column totals and the row totals each add up to the total sample size. If that is not true, because participants were allowed to tick more than one box, for example, then the calculations make no sense. If you have violated this assumption all is not lost, look at the last example (Section 6.9.1: common error and a solution). Note that the sample size assumption refers to the expected frequency, not the observed frequency. Instructions for calculating the expected frequencies are given above (Section 6.7.1). If one or more cells has an expected frequency of less than five, use Fisher's exact test.

6.8.2 What It Is Based On

The chi-square statistic is simply a way of comparing the counts in categories that you observe with the counts that you would expect and it summarizes the differences in a way that takes into account the size of the sample. The 'expected' frequencies are calculated in different ways depending on the type of chi-square test that you are using. In the one-sample chi-square test, for example (Section 5.3), they are based on some theoretical distribution (which might simply be 'all the same'). In most cases, however, the expected values are calculated directly from the data.

The equation is

$$chi\text{-}square = \text{the sum of} \frac{(Observed - Expected)^2}{Expected}$$

or

$$\chi^2 = \sum \frac{(O - E)^2}{E}$$

Taking the difference between the observed and expected counts in a single category tells you the difference between what you expected and what you measured, but if you were to sum those values for your contingency table, the result would be zero, because some differences are positive and others negative. By squaring the differences, you get rid of the sign. The clever bit is that each squared difference is divided by the expected frequency for that cell. This effectively removes the effect of sample size. This means that when these values are calculated for each cell and added together (the Greek capital letter Esh (Σ) means take the sum of), you have an overall measure for your contingency table of the magnitude of difference between the counts that you would theoretically expect and the counts that you actually have, and that measure is not sensitive to the size of the samples.

6.8.3 Calculating Chi-Square

Given a 2×2 contingency table, it is not too difficult to calculate the chi-square statistic by hand and use a table of critical values to check for significance. Using the example in Table 6.17 (the same marginal totals were used to calculate expected values in Table 6.15), we can see that there is a clear difference in the responses of the two genders. The risk ratio tells us that males are more than three times more likely than females to say yes ($13 \div 20 = 0.65$ and $2 \div 10 = 0.2$ so the risk ratio (for saying yes) $= 0.65 \div 0.2 = 3.25$).

TABLE 6.17

Observed and Expected Frequencies for a 2 × 2 Contingency Table
Used to Calculate the Chi-Square Statistic

	Observed				**Expected**		
	Yes	No	Sum		Yes	No	Sum
Male	13	7	20	Male	10	10	20
Female	2	8	10	Female	5	5	10
Sum	15	15	30	Sum	15	15	30

Given the observed and expected frequencies in Table 6.17, the chi-square statistic is

$$\chi^2 = \sum \frac{(O-E)^2}{E}$$

There are four cells so

$$\chi^2 = \frac{(O-E)^2}{E} + \frac{(O-E)^2}{E} + \frac{(O-E)^2}{E} + \frac{(O-E)^2}{E}$$

$$\chi^2 = \frac{(13-10)^2}{10} + \frac{(7-10)^2}{10} + \frac{(2-5)^2}{5} + \frac{(8-5)^2}{5}$$

$$\chi^2 = \frac{(3)^2}{10} + \frac{(-3)^2}{10} + \frac{(-3)^2}{5} + \frac{(3)^2}{5}$$

$$\chi^2 = \frac{9}{10} + \frac{9}{10} + \frac{9}{5} + \frac{9}{5}$$

$$\chi^2 = 0.9 + 0.9 + 1.8 + 1.8 = 5.4$$

For a 2 × 2 contingency table, the critical values are obtained from the chi-square distribution with one degree of freedom and for convenience the critical values of interest are summarized in Table 6.18. For a two-tail test, the critical value at $p = 0.05$ is 3.84 and at $p = 0.01$ it is 6.64. Since our measured chi-square value is equal to or greater than the tabulated value, we can reject the null hypothesis and conclude that the gender difference in response is statistically significant at $p < 0.05$ (but not at $p < 0.01$). The results could be summarised as 'the gender difference in response is statistically significant ($\chi^2 = 5.4$, two-tail $p < 0.05$) with males being 3.25 times more likely to agree'. Alternatively, use the 'CHIDIST' or 'CHISQ.DIST.RT' functions in a spreadsheet to give the two-tail probability as $p = 0.02$.

TABLE 6.18

Critical Values for the Chi-Square Test for a 2 × 2 Contingency Table

	Two-Tail Tests			**One-Tail Tests**		
p	0.05	0.01	0.001	0.05	0.01	0.001
χ^2	3.841	6.635	10.828	2.706	5.412	9.550

Note: These values are taken from the chi-square distribution with one degree of freedom.

6.8.4 Continuity Correction (Yates's Correction)

Many older text books recommend a 'continuity correction' be applied when calculating the chi-square statistic, particularly when it is applied to a 2×2 contingency table. Applying such a correction, often called Yates's correction, raises the probability values so that they are closer to those that are obtained using Fisher's exact test. Recent advice is that both Fisher's exact test and continuity-corrected chi-square tests are too conservative (Lydersen et al. 2009), so that they tend to accept the null hypothesis too easily and you obtain fewer significant results than you should. Where you have the choice, choose the option without a continuity correction.

6.8.5 Effect Size: Cramer's *V*

For a 2×2 contingency table, the most useful effect sizes are the risk ratio and odds ratio (Section 6.5) because they give you direct information about the size of the difference in response and can be interpreted easily in the context of the study. For example, you might use the risk ratio to say that males were three times more likely than females to say no. A disadvantage of those ratios is that they do not fall on a scale of zero to one, so they are not convenient for comparing the effect size of the chi-square test with effect sizes obtained from other statistical tests. Also the two ratios are specific to particular columns, they do not summarise the overall level of difference or contingency in the table.

A simple metric that both provides an overall summary of the level of contingency and falls conveniently on a scale of zero to one is Cramer's *V* (also called phi), which for a 2×2 contingency table is obtained by dividing the chi-square value by the sample size and then taking the square root

$$\text{Cramer's } V = \sqrt{\frac{\chi^2}{n}}$$

The most commonly used measure of the strength (not significance) of correlation is called Pearson's correlation coefficient, commonly referred to simply as the '*r*-value', and *r*-values fall on the scale zero to one (positive or negative), where zero represents no correlation at all and one represents a perfect correlation. Cramer's *V* can be regarded as the contingency table equivalent of the *r*-value, with the exception that it is always positive. If it is zero, then there is no difference between observed and expected values, which means that there is no association at all between the groups and the responses. If it approaches the number one, then there is a very strong association so that it is likely that the responses are very strongly contingent on the groups. Where effect sizes fall in the range zero to one, and can be compared directly to Pearson's *r*-value, a common shorthand is to regard values of less than 0.3 as indicating a small effect, and those of 0.5 or over as a large effect, with values in between regarded as 'medium'.

Cramer's *V* is easiest to understand when it is applied to a simple 2×2 contingency table, as in Table 6.19. In the first example (A), it is clear that drinking preference is not contingent on gender at all. Half of the males prefer beer and half prefer wine, and the same is true for the females. In this case, Cramer's $V = 0$ and you would conclude that there is no difference between the genders. In the second case (B), all of the males prefer wine and all of the females prefer beer, so drinking preference is clearly very strongly contingent on gender, to the extent that if you know the gender, you also know the drink preference. This is the largest difference that is possible. This is equivalent to a perfect correlation so Cramer's $V = 1$.

TABLE 6.19

Two Most Extreme Cases for a 2×2 Contingency Table

A	Wine	Beer	Totals	B	Wine	Beer	Totals
Male	20	20	40	Male	40	0	40
Female	30	30	60	Female	0	60	60
Totals	50	50	100	totals	40	60	100

Note: In the first case (A), there is no difference between the groups, giving a Cramer's $V = 0$ and in the second case, the difference is as large as it can be, so Cramer's $V = 1$.

Cramer's V can be interpreted in much the same way as Pearson's r-value and the R^2 value obtained in regression analysis (R^2 is the square of Pearson's r-value). In regression analysis (Chapter 11), we assume that one variable is dependent on the other and R^2 equals the proportion of variance in the dependent variable that is explained by the variance in the independent variable. For example, if we assume that salary is partly dependent on age, and the Pearson's correlation coefficient is $r = 0.5$, we would conclude that age 'explains' 25% of the variability in salary ($0.5^2 = 0.25$ as a proportion so $0.25 \times 100 = 25\%$). Treating Cramer's V in the same way, if a table arranged like the one above yielded a value of $V = 0.8$, we would conclude that 64% of the variability in drinks preference is 'explained by' gender ($0.8^2 = 0.64$ out of one, so times 100 gives 64%).

6.9 Conducting the Chi-Square Test

6.9.1 Companion Site Calculator

The companion site calculator for a chi-square test on a 2×2 contingency table allows you to simply enter the four observed frequencies (raw numbers, not percentages). The expected frequencies are calculated automatically and you are warned if any fall below the critical threshold of five. As well as the chi-square statistic and one- and two-tail probabilities you will obtain the risk ratio and odds ratio, together with their 95% confidence limits and Cramer's V as an effect size.

6.9.2 Online Calculators

There are lots of online calculators available for analysing 2×2 contingency tables (Table 6.20) using the chi-square test. Be careful to check the sample size requirements.

6.9.3 In *R* Commander

Choose 'Statistics' then 'Contingency tables' then 'Enter and analyse two-way table'. A 2×2 contingency table will pop up and you need only enter the four numbers and under the 'Statistics' tab choose chi-square test and check the button for 'Print expected frequencies'. There are no warnings, so check the expected frequencies are all at least five, and if not go back and choose Fisher's exact test. The output includes the chi-square value and two-tail probability. Cramer's V is not calculated.

TABLE 6.20

Selection of Free Online Calculators for the Chi-Square Test on a 2×2 Contingency Table

Provider	Website	Comments
GraphPad software	Graphpad.com/quickcalcs/ contingency1/	Easy to use, choose one- or two-tail *p*-values. Copes with zero entries and large samples. Chi-square with or without Yates' correction and Fisher's exact test. No sample size warning or expected counts
Vassar stats	Vassarstats.net/odds2x2.html	Gives one- and two-tail probabilities, expected counts and both odds ratio and risk ratio with 95% confidence limits. RR is calculated on left column. Chi-square with and without Yates' correction, not returned when assumptions breached. Phi is Cramer's *V*
SISA quantitative skills.com	http://www.quantitativeskills. com/sisa/statistics/twoby2. htm	Gives chi-square result first (use Pearson's value for large samples) with a link to 'Fisher's exact test'. The relevant *p*-values are starred. Lots of other numbers including expected counts, risk ratio and odds ratio and Cramer's *V*
Social Science Statistics	http://www.socscistatistics. com/tests/chisquare2/ Default2.aspx	Enter names of categories first, or just click next, then numbers. No sample size warning. No Cramer's *V* or other results

6.9.4 In SPSS

The problem with running a chi-square test in SPSS is getting the data in there in the right format. There are two options: long and thin or 'weighted cases'. In the long thin version, you produce a column to represent the rows and another column to represent the columns of your 2×2 table and the number of rows is equal to the total sample size. The result is shown in the first two columns of Table 6.21, but in this case, the frequency column would be omitted and there would be 30 rows, one for each individual. Unless you happen to have your data in this format, this is a very tedious procedure. The other option is to enter the data as three columns, as in Table 6.21. In this case, you have to inform SPSS that the frequency column represents counts, otherwise it will just treat the entries as numbers and assume the sample size is 4 rather than 30. To do this, click 'Data' and choose the bottom option which is 'Weight cases'. Move your frequency data into the box marked 'Frequency variable' and click ok.

Once the data are entered appropriately, running chi-square is easy, if you know where to find it. Click 'Analyze', 'Descriptive Statistics' and 'Crosstabs'. Identify which data are to go in the rows (by convention we put people in there) and columns and on the 'Statistics' tab check 'Chi-Square' and 'Phi and Cramer's *V*' then OK. The output includes the 'Pearson

TABLE 6.21

'Weighted Cases' Option for Entering Data into SPSS

Gender	Answer	Frequency
Male	Yes	13
Male	No	7
Female	Yes	2
Female	No	8

Note: You must inform SPSS that the frequency column represents counts by using the 'Weight cases' option under the 'Data' button.

Chi-Square value', degrees of freedom (1) and the two-tail probability labelled 'Asymp. Sig (two-tail)' which means it is an 'asymptotic' estimate (based on a probability distribution) rather than an exact calculation. Below the output box, there is some text that tells you if there are any cells with expected frequency less than five and the size of the minimum expected frequency. If the minimum is less than five, use the Fisher's exact test results, which also appear. You can ignore the other output.

6.9.5 Examples

6.9.5.1 Example: Organic Produce

You are interested in motivations for buying organically grown vegetables and decide to take an unbiased sample of students and ask them whether they regularly buy organic produce or not. One of the questions of interest is whether there is any gender difference in their shopping preference. Your results can be presented in a 2×2 contingency table (Table 6.22)

Entering the data into SPSS gives a chi-square value of 11.71 with a two-tail significance of 0.001 and Cramer's V is 0.267. We can reject the null hypothesis that propensity to purchase organic produce is not contingent on gender but in this case, a Cramer's V value of less than 0.3 tells us that shopping preference is only weakly contingent on gender. The strong statistical significance of the result is due to the large sample size. We might report the results as: 'shopping preference was weakly but significantly contingent on gender ($\chi^2 = 11.71$, $p < 0.001$, Cramer's $V = 0.27$)'.

It is clear from the tabulated results that a larger proportion of females said yes (more than half) than males (less than a third). In order to describe this imbalance more precisely, we can use the risk ratio calculated on the column marked yes. The 'risk' of females regularly buying organic produce is $41 \div 73 = 0.56$ and the risk for males is $27 \div 91 = 0.30$, so the risk ratio that we want is $0.56 \div 0.30 = 1.87$. This tells us that in this sample females are 1.87 times more likely than males to buy organic produce regularly. To infer what this means for the population, it is helpful to calculate the 95% confidence intervals for the risk ratio using the companion site calculator, which in this case yields values of 1.3–2.8. You can conclude that there is a 95% chance that the risk ratio in the population to which your sample relates lies between those two values.

6.9.5.2 Example: Erratic Content

You are interested in glacial deposits and wish to determine whether, in a particular coastal exposure, a lower diamicton (poorly sorted sediment) contains more erratics than an upper diamicton. You take an unbiased sample of clasts (stones) more than 2 cm in

TABLE 6.22

Example of a 2×2 Contingency Table Suitable for the Chi-Square Test

	Yes	No	Totals
Male	27	64	91
Female	41	32	73
Totals	68	96	164

Note: In this case investigating gender differences in propensity to buy organic vegetables.

TABLE 6.23

Contingency Table Giving Counts of Local and Erratic Lithologies (Rock Types) in Two Sediment Layers

	Erratic	Local	Totals
Upper	12	38	50
Lower	8	42	50
Totals	20	80	100

diameter and classify them as either local or erratic and work until you have 50 identified clasts from each of the units. Your results can now be tabulated in a 2×2 contingency table (Table 6.23).

In this case, the chi-square value is one and the p-value is 0.317 with a Cramer's V of 0.1. We can conclude that the difference between the two deposits, in terms of erratic content, is not statistically significant ($p > 0.05$).

6.9.5.3 Example: Large Sample Parametric Approach

In a large survey, one question asks people if they agree or disagree with a simple statement like, 'man-made global warming is real' (Table 6.24). The chi-square test can be used to test the significance of the difference in opinions between the two groups and in this case, chi-square = 4.04 and the two-tail $p = 0.044$. The data in this case are being treated as counts in two mutually exclusive categories. However, another way to look at a big sample like this is to consider the two categories as just a coarse-scale measurement of a variable that is continuous. In this case, we might assume that, within the population, there is a very wide range of opinions, from extremely strongly agree to extremely strongly disagree. Where this is the case it is sometimes argued that it is acceptable to treat the data as if they are numbers, rather than as counts in categories, and to use parametric statistics, such as Student's t-test. To do this, you assign numbers to the two possible answers (typically 0 and 1 or 1 and 2) and treat them as individual measurements. As long as the two categories can logically be interpreted as representing a scale of some kind (Carifio and Perla 2007), and the sample sizes are large, this procedure is surprisingly robust (Norman 2010; Fagerland 2012).

In analysing large-scale questionnaires, where data are collected as counts in categories, it is a common practice to treat categorical data in this way and it allows the answers derived from different types of questions, with different numbers of possible answers, for example, to be combined. In this case, if the independent samples Student's t-test

TABLE 6.24

When the Sample Size Is Very Large, and the Two Answers Can be Considered Part of a Scale, It is a Common Practice to Assign Numbers to the Two Answers and Use Parametric Tests Such as Student's t-Test

	Agree	Disagree	Sum	Results
Male	120	80	200	$\chi^2 = 4.04$, two-tail $p = 0.044$
Female	100	100	200	$t = -2.015$, two-tail $p = 0.045$
Sum	220	180	400	

Note: You should not do this with small samples.

(Section 8.6) is used to compare the male and female results the probability that the difference has arisen by chance agrees very closely with that derived using the chi-square test ($t = -2.015$, two-tail $p = 0.045$).

6.9.5.4 Example: Common Error with a Solution

You are interested in whether there is a difference in the pet-keeping habits of human and physical geography students. You ask 20 of each to fill in a short survey asking which side of geography they identify with and to tick a box if they have a dog and tick another box if they have a cat (Table 6.25). This kind of data cannot be analysed using the inferential statistics described here because the categories are not mutually exclusive. The classification of the people is ok, because you have to choose either physical or human geography, but for the pets, you could tick the box for cat and the box for dog, so they are not mutually exclusive and that breaks the rules. Note that the row totals are not 20, so they do not reflect the true sample size.

If you have collected data like this all is not lost, you can recast the data into two different tables: one for each question. For example, since 8 of 20 physical geographers have a dog, you can assume that the other 12 do not. In the recast tables (Table 6.26), the row totals are 20 and it is valid to use the chi-square or Fisher's exact tests. Strictly speaking,

TABLE 6.25

These Data Cannot be Analysed Using the Chi-Square or Fisher's Exact Tests Using This Table Because Individuals Were Allowed to Tick More Than One Box, so the Row Totals are Not Equal to the Sample Size

| | Q. Do You Have a Pet? | | |
	Dog	Cat	Total
Physical	8	14	22
Human	12	12	24
Total	20	26	46

TABLE 6.26

Same Data Plotted in Table 6.25 but Here Re-Cast to Show the Numbers with and without a Dog and the Numbers with and without a Cat

| | Q. Do You Have a Dog? | | | |
	Yes	No	Total	Results
Physical	8	12	20	$\chi^2 = 1.6$, $p = 0.32$
Human	12	8	20	Cramer's $V = 0.20$
Total	20	20	40	

| | Q. Do You Have a Cat? | | | |
	No	Yes	Total	Results
Physical	6	14	20	$\chi^2 = 0.44$, $p = 0.51$
Human	8	12	20	Cramer's $V = 0.1$

Note: The data are now independent, the row totals are equal to the sample sizes and it is valid to use the chi-square test.

you should apply a correction (e.g. the Bonferroni correction: Section 9.2) to take account of the fact that you are now conducting two tests rather than one, so that the chances of one of them being significant is increased. In reality, most geographers would not apply a correction and when the results are not statistically significant ($p > 0.05$), as here, there is no point.

6.10 Fisher's Exact Test

This test is named after the famous statistician Sir Ronald Aylmer Fisher, who is said to have devised it to test whether one of his colleagues really could tell the difference between cups of tea prepared with the milk added before or after the tea (Salsburg 2001). It is called an exact test because it calculates the probability directly rather than relying on an approximation to a specific distribution.

6.10.1 When It Is Useful

Fisher's exact test is normally applied to 2×2 contingency tables and, as for chi-square, it is important that the data are independent, which means that an individual can only appear in one column and one row. Asking if someone prefers dogs or cats is fine, but asking them to tick the animal or animals that they like violates the assumptions because the total of the rows and columns would not be equal to the sample size. There is no minimum sample size, so it is particularly useful where the sample size assumptions of the chi-square test are violated. Although there is no theoretical maximum size limit, the computation becomes difficult and most software will default to the chi-square test for large samples.

The problem with Fisher's exact test is that it tends to be too conservative, so the p-values tend to be on the low side. It was devised for use when the total of both the rows and the columns are set in advance, so the calculations are 'conditional' on the row and column totals. However, when we produce 2×2 contingency tables in geographical research, this is rarely true. You might deliberately set the sample sizes in the rows, for example, by asking equal numbers of males and females, you might only fix the overall total, by stopping when you have 100 questionnaires completed for example, or you might not control anything and just use as many replies as you get. In all of these cases, the marginal totals are not defined in advance and so the probabilities obtained from Fisher's exact test are likely to be a little too high. There are some fancy methods available to deal with this problem, but none of them are easy to access. My advice is to only use Fisher's exact test when the sample sizes are too small to allow you to use the chi-square test. If you do use Fisher's test on small samples and obtain probabilities that are close to significant (e.g. around 0.06 rather than 0.05), then mention that the test tends to be conservative and discuss the results in that light.

Fisher's exact test is normally used to test two-tail hypotheses but, as for the chi-square test for a 2×2 contingency table, it is reasonable to also test one-tail hypotheses if you are predicting the direction of difference in advance. However, Fisher's test is unusual in that the one-tail probability is not necessarily half of the two-tail probability. This means that if you use a method that only gives the two-tail value, such as *R* Commander, you cannot simply halve it to get the one-tail value.

6.10.2 What It Is Based On

Apparently, it is based on the 'hypergeometric distribution', but I will not try to explain the mathematics. Nor will I give the equation that is used in the calculation because it involves multiplying factorials, so even with quite small numbers is laborious to calculate, which is probably why it has so rarely been used in geography. In essence, the test calculates the probability that a difference as large as or larger than that which you observe could have occurred purely by chance.

6.10.3 How to Do It

It is probably possible to compute Fisher's exact test in a spreadsheet by writing your own equations but it is certainly more trouble than it is worth. The most convenient method is to use the 'Real Statistics Resource Pack' for Excel, an online calculator or specialist software, including SPSS and *R* Commander.

6.10.3.1 Real Statistics Resource Pack for Excel

The free 'Resource Pack' (Zaiontz 2015: Section 4.4) includes the function 'FISHERTEST' which is clearly explained on the website (http://www.real-statistics.com/chi-square-and-f-distributions/fishers-exact-test/). All you do is identify the block of four cells that make up your contingency table and choose a one- or two-tail test. If in any doubt, use a two-tail test. The result is a probability.

6.10.3.2 Online Calculators

Online calculators (Table 6.27) differ in the way that you enter the data and some cannot cope with zero values in any of the cells. With Fisher's exact test, it is perfectly valid to have a cell with a zero count, so it is just a weakness with those calculators. Most sites offer both a one- and a two-tail probability. As always, you cannot just choose the one you like, your test must fit your hypothesis. If you are only given a two-tail probability, you cannot simply halve it to obtain the one-tail probability. If in doubt, use a two-tail test. Some sites will also give you the odds ratio and associated 95% error limits (Sections 6.5 and 6.6) but very few will give you the risk ratio and associated confidence limits. These can be calculated using the companion site calculator.

TABLE 6.27

Selection of Free Online Calculators for Fisher's Exact Test

Provider	Website	Comments
Daniel Soper	www.danielsoper.com/statcalc3/calc.aspx?id=29	Odd vertical format, be careful how you enter the data. Crashes if you enter zero
Quantpsy.org	Quantpsy.org/fisher/fisher.htm	Good explanation of how the test works. Calculator only works with small numbers, accepts zero
In Silico project support for life sciences	http://in-silico.net/tools/statistics/fisher_exact_test	Odd arrangement with groups in columns. Can cope with zero entries. Maximum cell frequency is 99

6.10.3.3 In R Commander

Fisher's exact test is one of the easiest to perform in *R* Commander because you do not even have to import any data. Just choose 'Statistics' then 'Contingency tables' then 'Enter and analyse two-way table'. A 2×2 contingency table will pop up and you need only enter the four numbers and choose Fisher's exact test. The output includes the two-tail probability value and the odds ratio with 95% confidence intervals. Unfortunately, no one-tail probability is provided and in this case, it is not safe to assume that it is half of the two-tail probability.

6.10.3.4 In SPSS

Follow the instructions as for the chi-square test for a 2×2 contingency table (Section 6.9). The output includes one- and two-tail probabilities for Fisher's exact test.

6.10.4 Examples

6.10.4.1 Example: Small Sample of Snails

You are interested in determining whether two limestone-rich slope deposits represent different climatic conditions. One is slightly more reddish in colour, which is often an indicator of warmer conditions, whilst the other is very pale, often indicating cold conditions. You decide to sample the snail shells in both units to test your hypothesis that the redder unit was deposited under warmer conditions. You take an equally large sample of sediment from each unit back to the laboratory and sieve it carefully to remove all of the snail shells. Most of the snails are not very sensitive to temperature but there are some species that are known to be cold tolerant and others warm tolerant. There are not many of these, so you have small sample sizes for the comparison (Table 6.28). A sample this small would break the assumptions of the chi-square test but Fisher's exact test is ideal. Your hypothesis was that there would be a greater proportion of 'warm' snails in the reddish unit and that is true, so you go ahead with a one-tail test.

For Fisher's exact test, the one-tail probability is slightly less than 0.01. Even though the sample sizes here are very small, we can conclude that there is less than a 1% chance that such a big difference, with the bias towards warm snails in the reddish layer, as predicted, would have occurred just by chance. We might report the results as: 'the ratio of warm-tolerant to cold-tolerant snails in the reddish layer was significantly higher than that in the pale layer (Fisher's exact test, one-tail $p < 0.01$)'.

TABLE 6.28

Small Samples Such As Those Shown Here Produce Expected Frequencies That Are Too Small for the Chi-Square Test, but There Is No Such Restriction on the Use of Fisher's Exact Test

	Snail Tolerant To		
	Warm	**Cold**	**Totals**
Reddish	7	3	10
Pale	1	9	10
Totals	8	12	20

TABLE 6.29

Contingency Table with One Cell That Gives an Expected Frequency of Less
Than Five

	Births	Deaths	Sum
Funded	19	3	22
Unfunded	5	5	10
Sum	24	8	32

Note: The data should be analysed using Fisher's exact test rather than the chi-square test.

6.10.4.2 Example: Start-Up Companies

A small town was granted public funding to improve the facilities for small companies, with the aim of increasing the number of new businesses and helping them to survive. You are interested in whether this has resulted in a better ratio of start-up companies to business failures (births and deaths) over the 3 years after the scheme ended. You use a similar small town that did not receive such funding for comparison (Table 6.29). Given the low counts for company 'deaths' the lowest expected value falls below five (2.5), so Fisher's exact test is used. Since we predict a healthier ratio of births to deaths for the funded town, we can use a one-tail test. The resulting one-tail probability is $p = 0.04$, so we can conclude that the funded town has fared significantly better. Of course, you would have to consider the evidence that the difference between the two towns is due to the funding, perhaps using qualitative rather than quantitative methods.

References

Carifio, J. and Perla, R.J. 2007. Ten common misunderstandings, misconceptions, persistent myths and urban legends about Likert scales and Likert response formats and their antidotes. *Journal of Social Sciences* 3(3): 106–116.

Edwards, A.L. 1948. Note on the 'correction for continuity' in testing the significance of the difference between correlated proportions. *Psychometrika* 13: 185–187.

Fagerland, M.W. 2012. Exact and mid-p confidence intervals for the odds ratio. *Stata Journal* 12: 505–514.

Fagerland, M.W., Lydersen, S. and Laake, P. 2013. The McNemar test for binary matched-pairs data: Mid-p and asymptotic are better than exact conditional. *Bmc Medical Research Methodology* 13: 8.

Fagerland, M.W., Lydersen, S. and Laake, P. 2014. Recommended tests and confidence intervals for paired binomial proportions. *Statistics in Medicine* 33: 2850–2875.

Field, A. 2013. *Discovering Statistics Using IBM SPSS Statistics* (4th ed.). London: Sage, 916pp.

Howell, D.C. 2012. *Statistical Methods for Psychology* (8th ed.). Belmont, California: Wadsworth.

Lydersen, S., Fagerland, MW. and Laake, P. 2009. Recommended tests for association in 2 x 2 tables. *Statistics in Medicine* 28: 1159–1175.

McNemar, Q. 1947. Note on the sampling error of the difference between correlated proportions or percentages. *Psychometrika* 12: 153–157.

Norman, G. 2010. Likert scales, levels of measurement and the 'laws' of statistics. *Advances in Health Science Educational Theory and Practice* 15: 625–632.

Salsburg, D. 2001. *The Lady Tasting Tea: How Statistics Revolutionized Science in the Twentieth Century.* New York: Henry Holt and Company.

Zaiontz, C. 2015. *Real Statistics Using Excel.* www.real-statistics.com

7

Two-Sample Tests for Counts in Several Categories

7.1 Introduction

A common type of problem for geography students is comparing two samples where each sample comprises counts in several categories (i.e. more than two). The critical distinction in this case is whether the samples have a logical order or not (Figure 7.1). Where the categories do not fall into any logical order, and sample size is large enough to meet the strict assumptions, the appropriate test is chi-square. Students (and academics) often use chi-square in a rather simplistic way, to give a yes/no answer to a complicated question, but it is possible to extract a lot more information. As well as deciding whether there is a significant difference between two samples overall, for example, it is possible to say which of the categories show large differences. However, as sample size drops, or if there are big differences in the counts between categories, it becomes increasingly difficult to meet the strict sample size requirements of chi-square. In such circumstances, there are two options; either combine some of the categories so that the minimum size requirements are met, or use versions of Fisher's exact test that allow several categories.

Where there is a logical order to the categories the chi-square and Fisher's exact tests ignore it, which is a waste of information. A good alternative is the Kolmogorov–Smirnov two-sample test, which takes account of the order of the categories and is also much less sensitive to small sample sizes or to big disparities in counts between categories. Websites sometimes warn that the Kolmogorov–Smirnov two-sample test should only be used on continuous data and not on counts in categories, but that is a misunderstanding. The test assumes that the thing that is being measured has an underlying continuous distribution; the fact that it is measured using counts in categories does not invalidate the test. Clast roundness, for example, is often measured using Power's scale, which comprises six classes, but the roundness of stones is a continuum, from extremely angular to extremely well-rounded. Similarly, questionnaire responses may be recorded in classes labelled from strongly disagree to strongly agree, but we can assume that underlying this classification, there is a continuum of levels of agreement.

Where data are arranged in several categories that fall in a logical order, and the sample size is very large, it is also possible to treat the data as if they were continuous, rather than categorical, and to apply parametric statistical methods such as Student's t-test. It is easy to do this by allocating a score to each category. The merits and problems of this approach have been discussed at length in the literature, particularly in relation to 'Likert-like' scales such as strongly agree to strongly disagree. Some authors argue that because the data are ordinal, discontinuous and often skewed, only non-parametric methods should be used. Others argue that the categories are just a convenient way of obtaining information and that the underlying scale of strength of feeling is continuous. The issue is discussed in

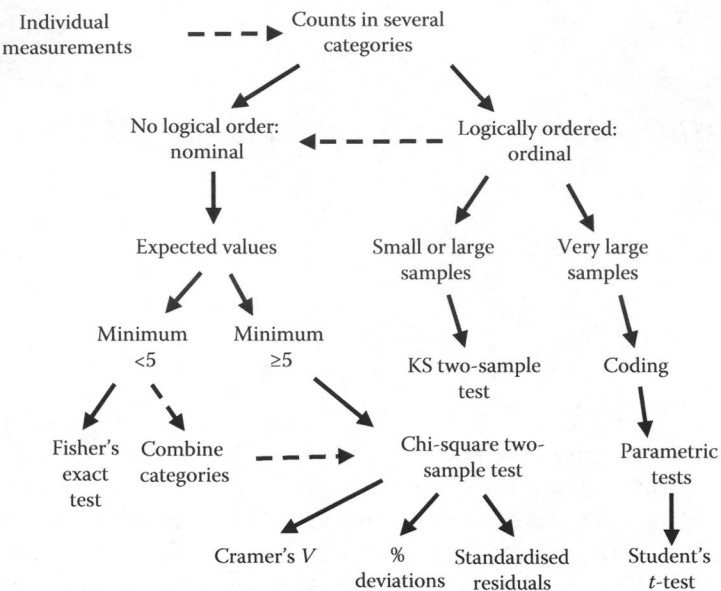

FIGURE 7.1
Logical series of steps for the comparison of two samples comprising counts in several categories. Dashed lines indicate a loss of information.

more detail below, but the bottom line is that people tend to worry too much about the assumptions of parametric tests. If you have very large samples, they work very well even if the data are discontinuous (i.e. fall in categories) and are skewed. If you have small samples, you should use non-parametric tests and the Kolmogorov–Smirnov two-sample test is a good option because it respects the order of the categories. What constitutes a large enough sample for parametric tests is not easy to define, and depends on the number of categories and degree of skewness, but at least 50 in each sample is probably a reasonable guide.

7.2 Two-Sample Chi-Square Test

Although chi-square tests can be used on more complicated tables, where there are several groups, it is useful to treat the two-sample case separately because in this case, the test also serves as a measure of the difference between the two groups.

7.2.1 When It Is Useful

The chi-square test is used to measure the degree to which the 'responses' in the columns are contingent on (loosely meaning dependent upon) the groups (rows). The chi-square test treats the categories as nominal, which means that the order that they are arranged in is irrelevant. If there is a logical order to the categories, it does not use that information.

7.2.2 What It Is Based On

The chi-square statistic is a way of comparing the counts in categories that you observe with the counts that you would expect and it summarises the differences in a way that takes into account the size of the sample. The 'expected' frequencies are calculated in different ways depending on the type of chi-square test that you are using. In the one-sample chi-square test, for example (Section 5.3), they are based on some theoretical distribution (which might simply be 'all the same'). In the two-sample chi-square test, however, the expected values are calculated directly from the data.

The equation for the chi-square statistics is

$$chi\text{-}square = the\,sum\,of\, \frac{(Observed - Expected)^2}{Expected}$$

or

$$\chi^2 = \sum \frac{(O - E)^2}{E}$$

Taking the difference between the observed and expected counts in a single category tells you the difference between what you expected and what you measured, squaring the difference gets rid of the sign and dividing by the expected frequency effectively removes the effect of sample size. When the values for each cell are summed, it provides an overall measure of the magnitude of difference between the counts that you would theoretically expect and the counts that you actually have, and that measure is not sensitive to the size of the samples.

7.2.3 Expected Frequencies

Where we wish to compare two samples, with more than two categories, the null hypothesis is that the counts in the categories are not contingent on the groups. The expected values are those that would occur if the null hypothesis was true. For example, if we were using chi-square to determine whether there was a gender difference in the drinking preferences of students, we might ask males and females to choose their favourite from a list of four choices (Table 7.1). In this case, the null hypothesis would be that drinks choice is not contingent on gender. Note that this is an inherently two-tail hypothesis because it is

TABLE 7.1

Contingency Table Suitable for Analysis Using the Two-Sample Chi-Square Test

	Wine	Beer	Cider	Other	Totals
Males	15	30	20	35	100
Females	45	10	20	25	100
Totals	60	40	40	60	200

Note: Each respondent can fall in only one group in the rows and make a single choice from the columns, so that the totals in the rows and columns faithfully record the true sample size.

TABLE 7.2

Contingency Table with the Observed Values Removed to Show How the Expected Values (on the Right) Are Calculated

	Wine	Beer	Cider	Other	Totals	Wine	Beer	Cider	Other	Totals
Male					100	30	20	20	30	100
Female					100	30	20	20	30	100
Totals	60	40	40	60	200	60	40	40	60	200

Note: In this case, there are equal numbers of males and females so the sum of each column is allocated in equal proportions to the two groups.

TABLE 7.3

Contingency Table with the Observed Values Removed to Show How the Expected Values Are Calculated

	Wine	Beer	Cider	Other	Totals	Wine	Beer	Cider	Other	Totals
Male					200	60	40	40	60	200
Female					100	30	20	20	30	100
Totals	90	60	60	90	300	90	60	60	90	300

Note: In this case, there are twice as many males as females, so the sum of each column is allocated so that 2/3 go to the males and 1/3 to the females.

not possible to predict a direction when there are more than two possible responses that cannot be placed in any logical order.

To better understand the concept of 'expected' values in a contingency table, it helps to look at one that has the marginal totals but not the actual counts in each category (Table 7.2). In this case, if we are testing the null hypothesis that drinking preference is not contingent on gender, then we have to arrange the 'expected' values as if it makes no difference whether you are male or female. In this table, there are an equal number of males and females, so if 60 people prefer wine, we can simply allocate 30 to the males and 30 to the females, and so on (Table 7.2). When the totals for the two samples are the same, it is a simple procedure.

When the totals are not the same (Table 7.3), it requires a bit more thought. The 90 wine lovers cannot be split into half males and half females, because it is clear from the row totals that there are twice as many males as females. Two-thirds of the wine lovers must, therefore, be allocated to the males and one-third to the females. The 90 wine lovers are split into 60 males and 30 females, and so on.

It is already getting quite complicated, but when we get to a table where the numbers are less convenient, it all becomes a bit scary! In Table 7.4, for example, the totals are very

TABLE 7.4

In This Contingency Table, the Marginal Totals Make It Difficult to See How to Allocate the Column Totals So that the Correct Proportions Go to Each Group

	Wine	Beer	Cider	Other	Totals
Male					138
Female					141
Totals	65	74	65	75	279

Note: Fortunately, there is a simple rule.

TABLE 7.5

Very Simple Table Illustrating How to Calculate the Expected Frequencies

	Wine	Beer	Cider	Other	Total
Male	$\dfrac{6\times10}{20}=3$	$\dfrac{4\times10}{20}=2$	$\dfrac{4\times10}{20}=2$	$\dfrac{6\times10}{20}=3$	10
Female	$\dfrac{6\times10}{20}=3$	$\dfrac{4\times10}{20}=2$	$\dfrac{4\times10}{20}=2$	$\dfrac{6\times10}{20}=3$	10
Totals	6	4	4	6	20

Note: For each cell, multiply the row total by the column total then divide by the grand total.

inconvenient. Clearly you would need a brain the size of a planet and a doctoral degree in Sudoku to fill this one in. Actually that is not true, because there is a simple rule that allows you to distribute the column totals so that the correct proportions go to the two groups. For each individual cell, you multiply the column total by the row total and then divide by the grand total. A very simple example is shown in Table 7.5.

Using the same method for the more complicated Table 7.4, in the first cell, the expected value is the column total (65) multiplied by the row total (138) and the answer is then divided by the grand total (279)

$$\frac{Column\ total\times Row\ total}{Grand\ total}=\frac{65\times138}{279}=\frac{8970}{279}=32.15$$

The same procedure is used for each cell, so that the entry for each cell depends on its own column total and its own row total, so in this case, they are all different (Table 7.6). Notice that the numbers of males and females in this case are almost but not quite the same, so the expected values in each cell are almost, but not quite, half of the column totals. Expected values do not need to be integers, they are usually fractions.

Having calculated the expected values (Table 7.7) they can be combined with the observed values (the counts in categories that you actually recorded) to calculate the chi-square statistic. For example, for the observed (and expected) values shown in Table 7.1, the chi-square value would be:

$$\chi^2 = \frac{(15-30)^2}{30} + \frac{(30-20)^2}{20} + \frac{(20-20)^2}{20} + \frac{(35-30)^2}{30}\ continued.$$

$$+\frac{(45-30)^2}{30} + \frac{(10-20)^2}{20} + \frac{(20-20)^2}{20} + \frac{(25-30)^2}{30}$$

$$\chi^2 = \frac{(-15)^2}{30} + \frac{(10)^2}{20} + \frac{(0)^2}{20} + \frac{(5)^2}{30} + \frac{(15)^2}{30} + \frac{(-10)^2}{20} + \frac{(0)^2}{20} + \frac{(-5)^2}{30}$$

$$\chi^2 = \frac{225}{30} + \frac{100}{20} + \frac{0}{20} + \frac{25}{30} + \frac{225}{30} + \frac{100}{20} + \frac{0}{20} + \frac{25}{30}$$

$$\chi^2 = 7.5 + 5 + 0 + 0.83 + 7.5 + 5 + 0 + 0.83 = 26.67$$

The chi-square value provides a measure of the total difference between the observed counts and the expected counts. The chi-square 'test' now checks whether that difference is larger than you would expect to occur just by chance. To achieve this, one more piece of

TABLE 7.6

Expected Values for a Contingency Table with Row Totals That Are Almost But Not Quite Equal

	Wine	Beer	Cider	Other	Totals
Males	32.15	36.60	32.15	37.10	138
Females	32.85	37.40	32.85	37.90	141
Totals	65	74	65	75	279

TABLE 7.7

Contingency Table with Observed and Expected (in Brackets) Frequencies

	Wine	Beer	Cider	Other	Totals
Males	15 (30)	30 (20)	20 (20)	35 (30)	100
Females	45 (30)	10 (20)	20 (20)	25 (30)	100
Totals	60	40	40	60	200

Note: These numbers are used in the worked example showing how to calculate chi-square.

information is required and that is the 'degrees of freedom'. A simple example of this concept is a bag with the numbers 1–10 in it. If you take out one number at random, you cannot predict what it will be because there are 10 possibilities, and if you pull out a second, there are still nine possibilities. However, when you have pulled out nine of the numbers, you can already be absolutely certain which number is left in the bag. That means that with 10 numbers in a bag, there are nine degrees of freedom. However, in the context of a contingency table, the degrees of freedom are not obvious unless you see an example. Taking a 2×2 contingency table, with the marginal totals revealed (Table 7.8), although there are four empty cells, only one needs to be completed in order to deduce what the other three must be. They can be computed simply by subtraction from the marginal totals. For a 2×2 contingency table, therefore, there is only one degree of freedom because when one cell has been completed, the others are also known.

Where there are more than two columns and/or rows the rule is that the number of degrees of freedom is equal to the number of rows minus one times the number of columns minus one. So for a 2×4 contingency table, there are $1 \times 3 = 3$ degrees of freedom, meaning that if numbers have been entered in three columns then all of the other numbers can be deduced with certainty (Table 7.9).

TABLE 7.8

In a 2×2 Contingency Table, Only One Expected Value Needs to Be Known in Order to Deduce the Other Three

	Category 1	Category 2	Total
Group A	8		10
Group B			20
Total	18	12	30

Note: In this case, there is one degree of freedom.

TABLE 7.9

In a 2 × 4 Contingency Table, Values in Three Columns Need to
Be Known in Order to Deduce the Others by Subtraction from
the Marginal Totals

	Wine	Beer	Cider	Other	Totals
Males	30		20		100
Females		30			100
Totals	60	40	40	60	200

Note: In this case, there are three degrees of freedom.

Having calculated chi-square and the number of degrees of freedom, the level of significance can be obtained either from tables of the 'chi-square distribution' or by entering those numbers into a spreadsheet and using the 'CHIDIST' function. Tables of the full chi-square distribution can be quite confusing, so for convenience the critical values for degrees of freedom up to 50 and probabilities of $p = 0.05$, $p = 0.01$ and $p = 0.001$ are compiled in Section 12.2.

In the worked example above (Table 7.7), the chi-square value was 26.67, with three degrees of freedom. Using the tables of critical values (Section 12.2), the critical value at $p = 0.001$ for three degrees of freedom is 16.266, and since the measured value is 'equal to or larger than the tabulated value', we can conclude that the difference between the two groups is strongly significant ($p < 0.001$). Using the CHIDIST function in Excel, we find that the probability is given in scientific notation as 6.9E-06 which means you have to move the decimal places six points to the left. Alternatively right click on the cell and use 'Format Cells' to change it to a decimal. All we really need to know is that the probability is less than one in a thousand, so reporting $p < 0.001$ is perfectly adequate. The probability values for a two-sample chi-square test with more than two categories are always two-tail because direction of difference makes no sense when applied to categories that cannot be placed in a logical order.

7.2.4 Sample Size Assumptions

A serious problem with the chi-square test is that it does not work with small sample sizes. Unlike Fisher's exact test, probabilities are not calculated mathematically, they are obtained from a theoretical distribution the shape of which depends on the degrees of freedom. When sample sizes are large, the fit to the distribution is excellent, but with small samples that may not be true. To meet the assumptions using a 2 × 2 contingency table, all of the expected values should be at least five. In more complicated tables, the rule of thumb is that none of the expected values should be less than one and no more than 20% of the expected values can be less than five.

To determine whether your samples are too small to use the two-sample chi-square test, you first have to calculate the expected values as shown above, which is not difficult in a spreadsheet and some online calculators will do it for you. Where the assumptions are violated, there are two options: use Fisher's exact test, which is only feasible if all of the numbers are small, or combine categories to raise the low expected frequencies until the assumptions are met.

Using a chi-square approach to compare the lithologies in two glacial deposits, for example, might result in observed counts as shown in Table 7.10. Note that despite the large sample size overall, the big disparities between lithologies means that six of the cells contain

TABLE 7.10

This Contingency Table Cannot Be Used to Compute Chi-Square Because Too Many of the Expected Values (in Brackets) Are below 5

	Red Sandstone	Yellow Sandstone	Basalt	Pale Granite	Porphyritic Granite	Flint	Other	Total
Upper till	4	25	15	23	1	2	30	100
	(3.5)	(28.5)	(17.5)	(21.5)	(3)	(1)	(25)	
Lower till	3	32	20	20	5	0	20	100
	(3.5)	(28.5)	(17.5)	(21.5)	(3)	(1)	(25)	
Total	7	57	35	43	6	2	50	200

Note: The large counts in some cells would make Fisher's exact test very difficult to compute, so the best option is to combine some of the categories.

expected values (in brackets) of less than five, so chi-square would be invalid. Fisher's test is not a realistic option because the numbers in the other cells are so large that the calculations would be very complex indeed and most software would crash. The sensible solution is to combine cells, on the assumption that there is really some logic to combining them. In this case, a good solution is to combine the two types of sandstone, and the two types of granite. It does not make sense (geologically) to combine flint with any of the other rock types listed here, so they can just go into the category marked 'other'. The result is a smaller table of just four lithologies (Table 7.11) with expected values well above five.

Applying the two-sample chi-square test to Table 7.11 gives a χ^2 value of 4.07 with three degrees of freedom and a probability of $p = 0.254$. Since the p-value is much larger than 0.05, we can accept the null hypothesis that the frequency of occurrence of different lithologies is not contingent on which till layer they were sampled from and therefore conclude that the observed differences in lithologies between these two till layers are not statistically significant.

Note that in the example above, there was no logical order to the categories, they are nominal in the sense that they can be named but not placed in rank order. Very often, however, the categories that we use in geography do form a logical order. A common example in human geography occurs where a questionnaire includes five possible answers indicating level of agreement with some statement. For example, you might ask representative samples of human and physical geography students the extent to which they agree with the statement 'numerical methods are interesting' (Table 7.12).

In this case, chi-square is invalid because six of the 10 cells have expected values below 5. The problem could be solved by merging adjacent categories, but it is an unsatisfactory

TABLE 7.11

Same Data as Shown in Table 7.10 But after Combining Some of the Categories with Small Counts in Them So That the Expected Counts (in Brackets) Are All Higher Than Five

	Sandstone	Basalt	Granite	Other	Total
Upper till	29	15	24	32	100
	(32)	(17.5)	(24.5)	(26)	
Lower till	35	20	25	20	100
	(32)	(17.5)	(24.5)	(26)	
Total	64	35	49	52	200

Note: The chi-square test can now be used.

TABLE 7.12

Example of a Contingency Table Where the Categories Fall in a Logical Order

	Strongly Disagree	Disagree	Ambivalent	Agree	Strongly Agree	Totals
Human geographers	15 (12)	32 (33.3)	7 (4.9)	4 (4.9)	2 (4.9)	60
Physical geographers	7 (10)	29 (27.7)	2 (4.1)	5 (4.1)	7 (4.1)	50
Totals	22	61	9	9	9	110

Note: The chi-square test does not recognise this order, it treats the categories as nominal and would give the same result if the columns were re-arranged.

solution. Although chi-square is often applied to data like this in geography, it is not good practice. Chi-square is based on the sum of the differences between observed and expected counts in each cell and the order in which the columns are arranged makes no difference. The categories are treated as nominal. Where the categories are ordinal, and fall in a logical order, as here, the Kolmogorov–Smirnov two-sample test is a better alternative because it respects the order of the categories and also has no sample size restrictions, so the low counts in some categories do not matter.

7.2.5 Strength of the Relationship (Cramer's *V*)

The end product of a two-sample chi-square test is a probability value that represents the chances that a difference at least as large as that between the observed and expected counts could occur just by luck if there was no relationship between the categories in the rows and the categories in the columns. That probability value depends on the sample size as well as on the magnitude of the difference, so that with a small sample the difference has to be large to be significant, but with very large samples even quite small differences can be significant. This means that the probability value does not tell us the magnitude of the difference between the observed and expected values, just how unlikely that difference is to be due to luck. When dealing with a 2×2 contingency table, the simple solution is to use either the risk ratio or the odds ratio as a measure of the 'effect size', and those measures are discussed in Chapter 5. Where two samples each have several categories, those ratios can still provide some information about individual pairs of columns, but they cannot provide a measure of the overall magnitude of the difference for the whole table.

Of course the chi-square statistic itself provides a measure of total difference, but it uses a very inconvenient scale and chi-square values obtained from, for example, a 2×3 table cannot be compared directly with those from a 2×4 table. Fortunately, there is a statistic called 'Cramer's *V*' (or sometimes 'phi') that does summarise the effect size for a contingency table and which, very conveniently, has a possible range from zero to one. Since the two-sample chi-square test is concerned with the degree to which the counts in the columns are contingent upon the categories in the rows, Cramer's *V* can be considered a measure of the degree of contingency. It is conceptually similar to Pearson's correlation coefficient that is used as a measure of the degree of correlation when data are available as individual measurements (Section 10.3) and can be interpreted in much the same way. The square of Pearson's correlation coefficient (R^2) is a measure of the proportion of variance in one variable that is explained by variance in the other. The square of Cramer's *V* can be treated in the same way, providing a measure of the proportion of the variability in the responses (columns) that is explained by the groups.

When Cramer's *V* is zero there is no difference between observed and expected values, which means that there is no association at all between the groups and the responses. If it approaches the number one, then there is a very strong association, so that the responses are very strongly contingent on the groups. Where effect sizes fall in the range zero to one, and can be compared directly to Pearson's *r*-value, a common shorthand is to regard values of less than 0.3 as indicating a small effect, and those of 0.5 or over as a large effect, with values in between regarded as 'medium'.

Cramer's *V* and the *p*-value thus provide different but complimentary information; Cramer's *V* records the strength of the association between the variables and the *p*-value indicates how unlikely it is that such a strong association could occur in a sample of this size just by chance, if there was no real association in the population.

7.2.5.1 Calculating Cramer's V

When dealing with a contingency table that only has two rows, as with the data used for the chi-square two-sample test, Cramer's *V* is calculated by dividing chi-square by the sample size and then taking the square root

$$\text{Cramer's } V = \sqrt{\frac{\chi^2}{n}}$$

For more complicated contingency tables, with several rows and columns, the sample size is first multiplied by either the number of rows minus one, or the number of columns minus one, whichever is the smaller.

7.2.5.2 Where Are the Differences?

Using the combination of Cramer's *V* and the *p*-value derived from a two-sample chi-square test, we can come to some conclusions about the strength of the association that we can see in our sample and the chances of it being 'real' in the sense that it is unlikely to have occurred just by chance. However, this tells us about the relationship overall, for example, in the case above, it tells us about the strength and significance of the relationship between drinks preference and gender, and those values are based on the differences between the observed and expected values. However, those metrics do not tell us anything about differences between the categories. In the case above, for example, it might be interesting to know which particular drinks choices are most contingent on gender and which are not.

There are several methods available to break down a contingency table and look at the information in a bit more detail, and two of them are particularly useful: percentage deviations and standardised residuals. The calculation for chi-square uses the difference between the observed and the expected counts in each cell, and if the totals for each group are the same, then they are easy to interpret, however if the groups differ in size, it is more confusing. The solution is to represent the difference between the observed and expected value as a percentage of the expected value. The equation is

$$\text{Percentage deviation} = \frac{Observed - Expected}{Expected} \times 100$$

TABLE 7.13

Observed and Expected (in Brackets) Values Used for a Chi-Square Test Together with the Percentage Deviations

	Beer	Wine	Whisky	Totals
Scouts	20	20	10	50
	(15)	(15)	(20)	
	+33.3%	+33.3%	−50%	
Guides	10	10	30	50
	(15)	(15)	(20)	
	−33.3%	−33.3%	+50%	
Totals	30	30	40	100

Chi-square = 16.67; *p*-value = 0.0002; Cramer's *V* = 0.41

An example of percentage deviation values is given in Table 7.13, where the chi-square result suggests that scouts and guides have different drinks preferences. If there was no difference, we would expect equal proportions of scouts and guides to prefer whisky, but the number of scouts who preferred whiskey was 50% smaller than the expected value and for the guides, it was 50% larger. The percentage deviations for beer and wine are smaller, a third more or less than the expected values in each case. This result illustrates that the difference in drinking preferences is not evenly balanced between the three drinks categories, with whisky contributing more to the overall difference than either wine or beer.

Standardised residuals are even more useful because they tell us something about the statistical significance of the individual differences. The standardised residual (*z*) for a cell is the observed value minus the expected value then divided by the square root of the expected value

$$z = \frac{Observed - Expected}{\sqrt{Expected}}$$

The values are given a positive sign if observed is higher than expected and a negative sign if the observed values are lower than expected. Standardised residuals for the example used in Table 7.13 are shown in Table 7.14. These values are a way of breaking up the chi-square statistic to demonstrate how much is contributed by each cell. It works because the sum of the squares of all of these numbers add up the chi-square value (16.67 in this case). If the samples are not very small, we can assume that the standardised residuals (*z*-scores) follow a normal (Gaussian) distribution with a mean of zero and a standard deviation of one. This provides a simple test of the significance of the difference in any of the cells. If the values lie at or beyond plus or minus 1.96, then they are statistically

TABLE 7.14

Standardised Residuals Calculated for the Data Shown in Table 7.13

	Beer	Wine	Whisky	Totals
Scouts	+1.29	+1.29	+2.24	50
Guides	−1.29	−1.29	−2.24	50
Totals	30	30	40	100

Note: When these values are squared and summed, they equal the chi-square value, so they show the relative contribution of each cell to the overall result.

significant at $p < 0.05$ (two-tail) and if they lie beyond 2.58, they are statistically significant at $p < 0.01$ (two-tail). In this case, we can see that the differences in beer and wine drinking are not statistically significant and it is the guides' (well known) preference for whisky that is the main cause of the difference between the two groups.

7.2.6 How to Conduct a Two-Sample Chi-Square Test

The calculations required to conduct a two-sample chi-square test are not difficult, so it can be done by hand using a calculator, but it is tedious and easy to make a mistake.

7.2.6.1 Companion Site Calculators

Spreadsheet calculators are supplied for two-sample tests with between 2 and 12 columns. Enter the original counts, not percentages. The spreadsheet will calculate the row and column totals and use these to produce a table of the expected values. If any of the expected values are less than five, you will see a warning. If you have broken the rules on sample size, do not just carry on regardless. Try to resolve the issue by combining categories or using another test. Also provided are the components of chi-square, the percentage deviations and the normalised differences. The chi-square value is given together with the degrees of freedom, statistical significance and Cramer's V as a measure of the effect size.

7.2.6.2 Online Calculators

There are lots of online calculators for chi-square tests (Table 7.15), but there are big differences in the amount of output. Many do not show the expected frequencies or warn when the sample size assumptions are violated, so use with care.

7.2.6.3 R Commander

Chi-square is quick and simple to compute in *R* Commander but the output is limited. Choose 'Statistics', 'Contingency tables' and 'Enter and analyse two-way table' options. You can then set the size of your contingency table to two rows (for a two-sample test) and up to 10 columns. There are options to include percentages of either the row totals, column totals or overall total, and a table of the expected values. You should check this to ensure that the sample size assumptions are met because there is no automatic warning. You can also choose to obtain a table of the 'components of chi-square'. These are the values that when added up give you the chi-square value. Taking the square root of these values gives you the normalised residuals. Remember to add the sign, so if the observed value is higher than the expected, they are positive and if the observed value is lower than the expected, they are negative.

The output from *R* Commander includes the χ^2 value (it cannot cope with Greek letters so uses X), degrees of freedom and the probability. These values are computed even when sample sizes are far too small, so check the expected frequencies. If they are too low, choose the 'Fisher's exact' option to obtain an exact probability. If some cells have high counts, the error message 'LDSTP is too small for this problem. Try increasing the size of the workspace' may appear. This tends to happen due to tiny samples in one or two columns and huge samples in others. The only options are to combine categories or remove the columns with the small samples.

TABLE 7.15

Selection of Free Online Calculators for the Two-Sample Chi-Square Test

Provider	Website	Comments
Vassar stats	http://vassarstats.net/newcs.html	Up to five rows and columns. Gives Cramer's V and tables of percentage deviations and standardised residuals. Warns when sample size assumptions are broken. Remember to click 'reset' between tests
Kristopher J. Preacher, Vanderbilt University	http://www.quantpsy.org/chisq/chisq.htm	Allows up to 10 rows and columns. No warnings so be careful. Only gives χ^2, df and p-value. Ignore the Yates' correction
Social Science Statistics	http://www.socscistatistics.com/tests/chisquare2/default2.aspx	Up to five rows and columns. No warnings so be careful. Only gives χ^2, and p-value
http://www.physics.csbsju.edu/	http://www.physics.csbsju.edu/stats/contingency_NROW_NCOLUMN_form.html	Up to nine rows and columns. No warnings so be careful. Only gives χ^2, and p-value
Lawrence Turner	http://turner.faculty.swau.edu/mathematics/math241/materials/contablecalc/	Up to six columns/rows. Gives χ^2, df and p-value and also expected values and components of chi-square, all in one table you can cut and paste into a spreadsheet. Take square root of the components of chi-square to get standardised residuals
In Silico Project support for life sciences	http://in-silico.net/tools/statistics/chi2test	Up to 10 rows/columns but data have to be uploaded as a text file and there are no warnings. Only gives χ^2, df and p-value. Allows one-tail, which is very odd. One to avoid perhaps
Simple Interactive Statistical Analysis	http://www.quantitativeskills.com/sisa/tableprocs/table2xn.htm	Two columns and up to 120 rows. Data can be cut and pasted in from a spreadsheet. Gives χ^2, df and p-value. No warnings so be careful. A lot of choices, tick 'show tables' to get the expected values and chi-square to get the results. Pearson's is the one to use

7.2.6.4 *In SPSS*

To analyse contingency tables in SPSS, the data need to be entered in an appropriate way, which is not very obvious. Follow the instructions in Section 6.9. Once the data are entered appropriately, click 'Analyze', 'Descriptive Statistics' and 'Crosstabs'. Identify which data are to go in the rows (by convention we put people in there) and which in the columns and on the 'Statistics' tab check 'Chi-Square' and 'Phi and Cramer's V' then OK. The output includes the 'Pearson Chi-Square value', degrees of freedom and the two-tail probability labelled 'Asymp. Sig (2-tail)' which means it is an 'asymptotic' estimate (based on a probability distribution) rather than an exact calculation. Below the output box, there is some text that tells you if there are any cells with expected frequency less than five and the size of the minimum expected frequency. You can ignore the other output.

7.2.7 Examples

7.2.7.1 *Example: Garden Visitors*

You are interested in how different elements of the natural and cultural landscape are experienced and wish to compare visitors to a botanical garden with those to a nearby

TABLE 7.16

Contingency Table Classifying Visitors to Two Tourist Attractions

	Family Groups	Young Couples	Mature Couples	Adult Groups	Retired	Totals
Botanical gardens	24	10	43	32	62	171
Castle and grounds	10	4	25	10	68	117
Totals	34	14	68	42	130	288

Note: The totals are very different, so the differences between the groups are not obvious. The counts for 'retired' are similar but they represent very different proportions of the two groups.

TABLE 7.17

Normalised Residuals Give a Clearer Picture of the Differences Than the Raw Counts

	Family Groups	Young Couples	Mature Couples	Adult Groups	Retired
Botanical gardens	+0.85	+0.59	+0.41	+1.41	−1.73
Castle and grounds	−1.03	−0.71	−0.50	−1.71	+2.09

Note: The largest residual indicates disproportionately large numbers of retired people visiting the castle and grounds.

castle and grounds. As part of the study, you observe people arriving and place them into one of five categories (Table 7.16). Applying the two-sample chi-square test, the results are $\chi^2 = 15.31$ with four degrees of freedom, $p = 0.004$, Cramer's $V = 0.23$.

The most important number here is the p-value of 0.004 which means there is a 0.4% chance that the difference between these two samples could arise just by chance. However, Cramer's V is less than 0.3, which means that the level of contingency is low. This suggests that although the difference is statistically significant, it is small. Looking at the table of normalised residuals (Table 7.17) is helpful. Here, we can see that the largest difference occurs in the retired category, and the only value that crosses the threshold of 1.96 is produced by the anomalously high number of retired visitors to the castle and grounds. Seeing the results, you recall that large numbers of retired people arrived by coach.

In this case, despite the strongly significant difference overall, it might be worth re-running the analysis without the retired category. Removing the retired column changes the marginal totals, and expected values, so one of the cells now has an expected frequency of less than five. The rule is that no cell can have an expected frequency of less than one and no more than 20% of the cells can have an expected frequency of less than five. There are six cells, so one cell represents 1/6 or 16.7%, so the assumptions are not violated. The results of a chi-square test are $\chi^2 = 2.15$, $df = 3$, $p = 0.54$, Cramer's $V = 0.12$. Now, it is clear that the significant difference between the two locations was entirely caused by the numbers of retired people. When they are removed, the difference is very small indeed (Cramer's V is very low) and there is more than a 50% chance ($p = 0.54$ is 54%) that it could have arisen purely by chance.

7.2.7.2 Example: Snails (Small Sample)

You wish to determine whether two limestone breccia (rubble) deposits represent different environments. Pollen does not preserve well in such deposits but there are a few snail shells present and they fall into three clear categories. Before trying to identify them to species level, you just want a quick check to see if the proportions of the three types are significantly different. However, you have found very few snails (Table 7.18).

TABLE 7.18

These Samples Are Too Small for a Chi-Square Test But Suitable for Fisher's Exact Test

	Big and Robust	Small and Disc Shaped	Small and Conical	Totals
Unit A	7	2	3	12
	(3.36)	(3.36)	(5.28)	
Unit B	0	5	8	13
	(3.64)	(3.64)	(5.72)	
Totals	7	7	11	25

In this case, chi-square is not a realistic option because the sample sizes are too small, with four of the six cells yielding expected values of less than five. Combining categories is not helpful in this case. You could search for more snails or use Fisher's exact test. This example is used again in Section 7.3.3, and the probability is $p = 0.006$, so despite the small sample sizes, there is only a 0.6% chance that the difference in snail groups has arisen just by luck.

7.2.7.3 Example: Misuse of Chi-Square

You are interested in biodiversity, so you decide to compare the plant-life present in a wildlife park and in the grounds of a nearby derelict factory. The two study sites are similar in size and both surrounded by housing estates. You search both areas carefully and record the total number of species present in each of three groups (Table 7.19). Running a chi-square test would give $\chi^2 = 1.44$, $df = 2$, $p = 0.49$, Cramer's $V = 0.13$. Since the p-value is much higher than 0.05, we would accept the null hypothesis that there is no difference between the wildlife park and the derelict factory grounds.

This is actually a terrible use of the chi-square test. Recall that we are interested in biodiversity, so we want to know if there are *more* plant species present in one site. The problem is that the two-sample chi-square test is not designed to look for a difference in the absolute numbers for two samples, it looks at the difference in the *proportion* of each sample that falls into each group. If the number of species in each group in the wildlife park were to be doubled, for example, suggesting much greater biodiversity, the chi-square test would still suggest that the two samples are not statistically significant. This is because doubling the numbers for the wildlife park does not change the proportion that falls into each category.

TABLE 7.19

This Contingency Table Shows the Counts of Species in Four Groups Identified at Two Sites

	Trees and Shrubs	Flowering Plants	Grasses and Ferns	Totals
Wildlife Park	13	25	8	46
	(12.4)	(23.3)	(10.3)	
Derelict factory	10	18	11	39
	(10.6)	(19.7)	(8.7)	
Totals	23	43	19	85

Note: If the intention is to compare the proportions of plants occurring in each category, then chi-square is appropriate. If the real interest lies in comparing the absolute numbers of species, then chi-square is not appropriate.

7.2.7.4 Example: Common Mistake and a Solution

You are interested in why people choose to buy organically grown produce and suspect that there may be differences in the attitudes of those who make their purchases at farmers' markets and those who buy from shops. You decide to conduct a questionnaire and ask 30 customers at each venue 'What were your reasons for purchasing organically grown produce today?' The results are presented in Table 7.20.

You must not use chi-square on a table like this! The true sample size is 30 for each group, but in the table, the totals come out as 90 and 79. This is because each person was allowed to tick several boxes. The mathematics used to calculate the probabilities in a chi-square test rely on the assumption that the row and column totals represent the true sample size, which requires that all of the values are independent. That means that one person from the farmers market row is allowed only one entry in one column. In this case, the same person effectively appears in several or even all of the columns.

In retrospect, this was probably not the best way to collect data for this comparison. One option would be to ask them to identify only their main reason for purchasing organic produce today, which would allow them to tick only one box, and then you could use the chi-square test on your contingency table. However, the four questions are not really mutually exclusive, and you may be genuinely interested in getting an answer to all of them, so distorting your sampling strategy just to satisfy chi-square would probably not be the best strategy. A good alternative might be to ask them to rank those answers from one to four in terms of how important they are. That would allow you to place the four options in rank order for each customer and use Friedman's test (Section 9.9) to see whether there is a difference in the way that the two sets of customers arranged the four possible options.

If you have already collected data like this, and are just trying to find a suitable way to test for differences, all is not lost. One option is to look at each question in turn and note how many respondents did mark it as important and how many did not. Now, you have a set of 2×2 tables where the row and column totals are an honest reflection of the true sample size. Taking the answer 'better taste', as an example (Table 7.21) gives a 2×2 contingency table that satisfies the assumptions of the chi-square test, giving: $\chi^2 = 5.69$, $df = 1$, $p = 0.017$, Cramer's $V = 0.35$. The risk ratio for yes is 1.50 with 95% confidence limits of 1.09–2.06. We can conclude that there is a 1.7% chance that the difference in response to this question could have arisen by chance ($p = 0.017$). Cramer's V tells us that there is a medium degree of contingency and the risk ratio tells us that customers at the farmers' market were one and a half times more likely to identify better taste as a reason for buying organic produce compared to customers at the shop. Given the size of the samples, there is a 95% chance that the equivalent risk ratio in the population lies between 1.09 and 2.06.

For the question 'better quality' (Table 7.22), there is clearly no difference between the two groups ($\chi^2 = 0.08$, $df = 1$, $p = 0.78$, Cramer's $V = 0.017$). Cramer's V shows a very low level

TABLE 7.20

These Data Are Not Suitable for the Chi-Square Test Because the Sum of the Rows Is Not an Honest Reflection of the True Sample Size

	Better Taste	Better Quality	Fairer to Producer	Better for Environment	Totals
Farmers' market	27	22	14	27	90
Shop	18	20	16	25	79
Totals	45	42	30	52	169

Note: This happens when respondents are allowed to tick more than one box.

TABLE 7.21

Data from Table 7.20 Can be Disaggregated and Divided into a Separate 2 × 2 Contingency Table for Each Question; This One Contains the Numbers for 'Better Taste'

	Yes	No	Totals
Farmers' market	27	3	30
	(22.5)	(7.5)	
Shop	18	12	30
	(22.5)	(7.5)	
Totals	45	15	60

Note: The numbers answering 'no' are derived by subtraction from the known total (30) for each group.

TABLE 7.22

For This Question (Better Quality), the Ratios for the Two Groups are Very Similar and the Difference is Not Statistically Significant

	Yes	No	Totals
Farmers' market	22	8	30
	(21)	(9)	
Shop	20	10	30
	(21)	(9)	
Totals	42	18	60

of contingency, and the *p*-value indicates a 78% chance that the observed difference could have arisen due to chance ($p = 0.78$).

Strictly speaking, if you split up your sample like this and conduct several tests, rather than just one, you should correct the probabilities to take account of the increased chances of a significant result when you conduct several tests. The Bonferroni correction is the easiest option and it is described in Section 9.2. In reality, geographers rarely apply a correction and treat the tests as independent.

7.3 Fisher's Exact Test for More Than Two Categories

Although Fisher originally proposed his exact test to deal with 2 × 2 contingency tables, it was quickly demonstrated (by Freeman and Halton 1951) that the logic and mathematics could be extended to cover more than two rows or columns. The only limitation is the computational complexity, and most software will only allow you to calculate the exact value if the sample size is not very large, otherwise they either crash or default to chi-square.

7.3.1 When It Is Useful

The test is most useful when your sample sizes are too small for chi-square. Some calculators offer both one- and two-tail options, but a one-tail test is difficult to justify when you have three or more nominal categories, so use the two-tail value. As with the 2×2 Fishers exact test, this extended version is likely to be rather conservative and to produce p-values that are on the high side. For samples that meet the sample size requirements the chi-square test is preferable.

7.3.2 How to Do It

It is not really practical to perform this test by hand or in a spreadsheet. The best options are an online calculator, *R* Commander or SPSS. Some online calculators that will allow more than two columns are listed in Table 7.23. It is just a matter of entering the counts in each category and recording the probability value.

7.3.2.1 In R Commander

Choose 'Statistics', 'Contingency tables' and 'Enter and analyse two-way table'. Choose how many rows and columns and enter the counts. Click on Statistics and choose Fisher's exact test. If you do not click anything else, you will just get the two-tail p-value, which is exactly what you need. If your samples are too large, you will get the error message (in red at the bottom) 'LDSTP is too small for this problem', which means that the calculations are so onerous that your computer cannot cope.

TABLE 7.23

Selection of Free Online Calculators That Will Perform Fisher's Exact Test when There Are More Than Two Columns

Provider	Website	Comments
Vassar stats	http://vassarstats.net/fisher2x3.html	2×3 calculator, sample size up to 300. Gives two p-values, the top one is most useful (Pa). Also gives chi-square if sample sizes allow
Daniel Soper	http://www.danielsoper.com/statcalc3/calc.aspx?id=58	2×3 table, but you enter your data in a column so be careful. Copes with large samples (>300). No default to chi-square
BGI Cognitive genomics	https://www.cog-genomics.org/software/stats	2×3 table, click next to p-value to get the probability. Copes with large numbers
In Silico project support for life sciences	http://in-silico.net/tools/statistics/fisher_exact_test	2×3 table, arranged as three rows and two columns, but that does not influence the result. For tail/side click 'both'. Copes with zero entries. Copes with large numbers, no default to chi-square
Vassar stats	http://vassarstats.net/index.html	2×4 calculator, sample size up to 120. Gives two p-values, the top one is most useful (Pa). Also gives chi-square if sample sizes allow
In silico	http://in-silico.net/tools/statistics/fisher_exact_test	2×4 table, arranged as four rows and two columns, but that does not influence the result. For tail/side click 'both'. Copes with zero entries. Crashes with large numbers, no default to chi-square
Simple Interactive Statistical Analysis	http://www.quantitativeskills.com/sisa/statistics/fiveby2.htm	2×5 table. Odd format with seven rows, but just enter data in the first five (or less). The p-value you need is the larger of the two-tail values (not the point value). Gives chi-square even with very small numbers, so ignore that value

7.3.2.2 SPSS

To run Fisher's exact test in SPSS, first enter the data so that it is recognised as a contingency table, using the instructions in Section 6.9. Go to 'Analyze', 'Descriptive Statistics' and 'Crosstabs'. Choose which data you want to appear in the rows and columns, then click the 'Exact' button. Choose the option 'Exact', then click 'Continue' and 'OK'. The output includes the two-tail exact probability for Fisher's exact test. This procedure works even with large numbers in some cells.

7.3.3 Examples

7.3.3.1 Example: Snails (Small Sample)

You wish to determine whether two limestone head (rubble) deposits represent different environments. Pollen does not preserve well in such deposits but there are a few snail shells present and they fall into three clear categories. Before trying to identify them to species level, you want to check if the proportions of the three types are significantly different. However, you have found very few snails (Table 7.24). Sample sizes are far too small for chi-square but Fisher's exact test is appropriate. The result is $p = 0.006$, so despite the very small sample sizes, it is very unlikely that the difference in the mix of snail types between these two units is due to chance.

7.3.3.2 Example: Small Questionnaire

You are interested in differences in fear of crime amongst female students and wish to know whether British Muslim women, whose dress identifies them as such, are more afraid of crime than British non-Muslim women of similar age. You intend to use mainly qualitative methods, but have conducted a quick pilot study to see whether there is a difference that is larger than would be expected due to chance. You ask each person how they would feel if they found themselves alone in the city centre at midnight (Table 7.25).

TABLE 7.24

Contingency Tables with Small Counts, in This Case of Snail Shapes, Cannot Be Analysed Using Chi-Square Because the Expected Values Are Too Low

	Big and Robust	Small and Disc Shaped	Small and Conical
Unit A	7	2	3
Unit B	0	5	8

Note: Fisher's exact test has no minimum sample size requirements and is the appropriate alternative.

TABLE 7.25

Contingency Table Using Four Categories That Fall in a Logical Order

	Relaxed	Worried	Scared	Terrified	Totals
Muslim	0	4	8	8	20
Non-Muslim	4	6	7	3	20
Totals	4	10	15	11	40

Note: Chi-square and Fisher's exact tests treat these categories as nominal, ignoring the order in which they fall.

TABLE 7.26

Subsample of the Information Presented in Table 7.26

	Relaxed	Terrified	Totals
Muslim	0	8	8
Non-Muslim	4	3	7
Totals	4	11	15

Note: Care must be taken in interpreting the significance of differences like this that are identified by looking at the data rather than as part of the experimental design.

Fisher's exact test gives $p = 0.087$, so we can accept the null hypothesis that degree of fear is not contingent on the religion of the young women.

Note, however, that this value represents the overall differences for the table and it is a two-tail probability. Fisher's exact test treats the categories as nominal, so the order in which they are arranged is irrelevant to the result and that makes a one-tail test difficult to justify. However, looking in more detail, it is clear that in the central categories (worried and scared), the differences are quite small, but in the extreme categories, relaxed and terrified, the differences are larger. In this case, it might be worth checking to see whether there is a significant difference in the extreme reactions. The data can be recast into a 2×2 contingency table that only represents those who chose an extreme (Table 7.26).

The sample size is now very small, with only 15 women in total, but this is not an impediment to applying the Fisher's exact test, this is precisely when it is most useful. Given a 2×2 table, it is possible to apply a one-tail test. That is appropriate in this case because your original intention was to test whether the Muslim women were 'more afraid'. The result is statistically significant ($p = 0.026$). For completeness, you could compare the two middle groups as well, which would give you a nonsignificant result ($p > 0.05$). From these results, you might conclude that most of the young women would feel either worried or scared to be alone in town at night, but a disproportionately large number of young Muslim women would feel terrified and a disproportionately small number of Muslim women would feel relaxed. These results might provide some useful insights into how to conduct the qualitative part of the study.

Applying chi-square or Fisher's exact tests to categories that are ordinal rather than nominal, and thus have a logical order, is inherently problematic because the order of the categories is irrelevant in these tests. It is also generally a bad practice to take the responses that are intended to represent some underlying continuum of strength of opinion and to treat the categories individually. However, when you are faced with small samples like this, I think that using these methods to investigate the data in a bit more detail can be justified, at least as an exploratory tool, perhaps as a guide to gathering a larger sample or to inform the qualitative part of a study. You should be careful about placing too much emphasis on the probability values when you dissect your results in this way though, because you become exposed to the 'family-wise errors' that come from applying multiple tests (see Section 9.2).

7.3.3.3 Example: Failed Test and a Solution

You are investigating changes in past climate and environment using the foraminifera (they look like microscopic snails) in marine clay deposits and wish to determine whether the mix of species in two layers is significantly different. You have identified 10 different

TABLE 7.27

Contingency Table That Is Difficult to Analyses Because It Has Too Many Low Counts for Chi-Square But Some Large Counts That Make Fisher's Exact Test Very Difficult to Run

	A	B	C	D	E	F	G	H	I	J	Totals
Clay 1	36	46	28	8	8	2	3	3	0	1	135
Clay 2	53	40	56	7	3	4	3	2	4	2	174
Totals	89	86	84	15	11	6	6	5	4	3	309

Note: In such cases, it is convenient to split the table into two parts.

species, coded A–J in order of abundance, and recorded the number of each species present in a given weight of sediment (Table 7.27).

Although in theory you could run Fisher's exact test on these data, using *R* Commander for example, in reality, the programme will crash because it cannot cope with the large numbers. Chi-square is not an option because of the large number of cells that would produce expected values of less than five. Looking at the table, however, it is clear that the species fall into two groups: those that are common and those that are rare. It might make sense to treat these two groups differently and conduct the analysis in more than one step. You might begin by testing whether there is a significant difference in the common species, and this could be done by either ignoring the rarer species or by putting them into a single group marked 'other'. Taking the first option gives a table that contains only quite large numbers (Table 7.28), so the most appropriate test is chi-square, giving the results: $\chi^2 = 7.29$, $df = 2$, $p = 0.026$, Cramer's $V = 0.17$. We can conclude that in terms of the three common species, there is a small difference between the proportions (Cramer's V is <0.3) but it is statistically significant ($p = 0.026$). Now, we can look at the rare species using Fisher's exact test which will now run to conclusion giving a p-value of 0.28 suggesting that, in terms of the relative proportions of the rarer species, the two clay units are not significantly different.

Given these results, you would have to be cautious about how to interpret the difference between the two clay units. The rare species indicate no difference and although the common species show a statistically significant difference it is very small. It is statistically significant because the samples are large.

Plotting the percentage deviations for the three common species (Table 7.28) shows that the main differences are between species B and C. This might prompt a further layer of analysis, taking just those two species and comparing them using a 2×2 contingency table. That would yield a chi-square value of 7.02 with one degree of freedom, giving $p = 0.008$. The temptation is to leap to the conclusion that this is a strongly significant difference, but some caution is required when you dissect a contingency table in this way.

TABLE 7.28

Counts of the Three Most Common Species in Table 7.27 Together with the Percentage Deviations

	Counts						% Deviations		
	A	B	C	Totals			A	B	C
Clay 1	36	46	28	110		Clay 1	−4.8	25.9	−21.5
Clay 2	53	40	56	149		Clay 2	3.5	−19.2	15.9
Totals	89	86	84	259					

Note: It is clear that the largest differences are in the relative proportions of species B and C.

The probabilities that are produced by inferential statistical tests like chi-square are based on the assumption that you are only performing one test and that it is being applied to a hypothesis that was formulated before the data were collected. Collecting a lot of data, sifting through to find the biggest differences and then applying the tests is not a good practice and the probability values are not realistic when you do this. The only real justification for this approach would be as an exploratory tool, which in this case suggests that if there is a difference between these two clay units, it is likely to be reflected in the relative proportions of species B and C. To understand why this might be the case, you would need to investigate the literature on those two species. They might, for example, have different tolerances to temperature or salinity.

7.4 Two-Sample Tests for Counts in Ordered Categories

Where the aim is to compare two samples and the data have been collected as counts in categories that fall in some logical order, there are three options available (Figure 7.2). The first is to ignore the order of the categories, treat them as ordinal and use the chi-square test. That is rarely the best option, however, because chi-square (and Fisher's exact test) ignore the order of the categories. In most cases, the best option is to use the Kolmogorov–Smirnov two-sample test. The procedure is slightly different depending on whether your two samples are the same size or not. It is not a difficult test to compute and can be performed in a spreadsheet, using tables of critical values to determine the level of significance. Where the data sets are very large, the other option is to ignore the fact that the data were collected as counts in categories and treat them as if they are continuous. This is achieved by assigning a numerical value to each category and using it as a multiplier (coding). There is some controversy about this approach, as discussed below, but it seems to work very well as long as the sample sizes are very large.

7.5 Kolmogorov–Smirnov Two-Sample Test

The Kolmogorov–Smirnov two-sample test is designed to test the null hypothesis that two samples have been drawn from the same population. Unlike many other tests, it is sensitive not only to differences in where the middle of the two samples lie, it can be used to test for any differences at all. It is a non-parametric test, so the data do not have to be normally distributed. The important assumption is that the data represent continuous distributions. This does not mean that it cannot be used when the data are recorded as counts in categories, it means that the categories should represent something that is continuous in the population. The counts in categories are just a convenient way of measuring or displaying the data. For example, you might ask people to tick boxes to indicate their level of agreement, using a five-point 'Likert-type' scale from strongly agree to strongly disagree, but you can reasonably assume that underlying that classification levels of agreement are continuous. Similarly, you might choose to place people into a number of age classes, but of course everyone in one class is not exactly the same age. The test works by changing the raw counts into either cumulative counts, where samples are the same size, or cumulative

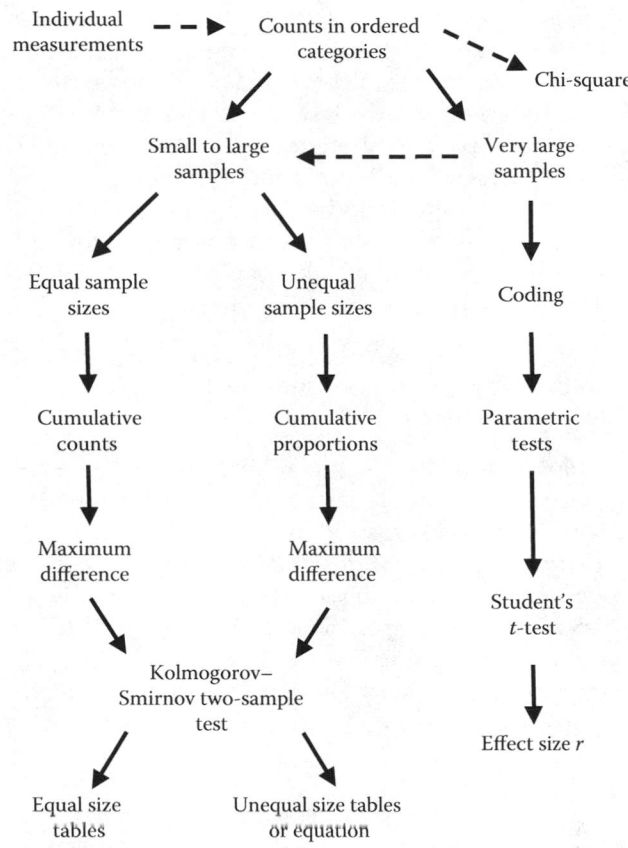

FIGURE 7.2
Logical series of steps for deciding how to compare two samples where the data are recorded as counts in categories that fall in a logical order.

proportions and finding the point of maximum difference between the two samples. It is convenient to perform by hand or in a spreadsheet. Significance values are obtained using tables of critical values (Section 12.5).

7.5.1 When It Is Useful

The Kolmogorov–Smirnov two-sample test is used when data are arranged in categories that fall in a logical order (ordinal scale). It must not be used where the classes are nominal (just names with no logical order e.g. religious or racial groups). Data may be collected as counts in categories or individual measurements can be degraded to represent the number in each of several categories. For example, the age of each person may have been recorded but for convenience of analysis, they can be arranged into a few age classes. There are no assumptions relating to sample size, so it can be used in cases where the assumptions of chi-square are violated and is a good alternative to combining categories. When data can be analysed using either the chi-square two-sample test or the Kolmogorov–Smirnov two-sample test, Siegel (1956) suggests that the Kolmogorov–Smirnov two-sample test is the better option because it uses more information and is therefore more powerful. However, if one of the samples is strongly bimodal, it seems to give very odd results (see example below) so that may not be universally true. If you have individual measurements, and are using either Student's *t*-test or the Mann–Whitney *U*-test to check for a difference in the position of the middle of the two samples (mean and mode for those two tests), you can use the Kolmogorov–Smirnov two-sample test to also check for any other difference, but with continuous data, it is applied in a different way, as described in Section 8.10.

Since the Kolmogorov–Smirnov two-sample test is looking for any difference rather than a difference in the position of the middle, it is often argued that it is inherently non-directional and therefore can only be used to test a two-tail hypothesis. However, it is sometimes used to test carefully devised one-tail hypotheses, and that option is discussed later. If in any doubt, stick to two-tail hypotheses with this test.

7.5.2 How to Perform the Kolmogorov–Smirnov Two-Sample Test with Counts in Categories

The Kolmogorov–Smirnov two-sample test can be calculated by hand or, more conveniently, in a spreadsheet and some simple calculators are provided on the companion website. The Real Statistics Resource Pack also supports Kolmogorov–Smirnov tests. There are a few online calculators but they are designed for continuous rather than categorical data and I would not recommend using them. Performing the Kolmogorov–Smirnov two-sample test in SPSS is only convenient if the data have been coded appropriately. I am not aware of an easy way of performing this test in *R* Commander. The way to apply the Kolmogorov–Smirnov two-sample test by hand, or in a spreadsheet, depends on whether your samples are the same size or not.

7.5.2.1 Equal Sample Sizes

As an example of the procedure imagine that you wish to compare the roundness of clasts (stones) in two glacial deposits. Roundness is recorded using the six-point Power's scale, but we can assume that there is an underlying continuum of roundness.

> *Step 1.* Arrange your data into the ordered categories (Table 7.29). You must use the same categories for both samples. You must use the actual number that you recorded, so that the total really is your true sample size. Do not use percentages.

> *Step 2.* Change the counts into cumulative counts (Table 7.30). This means that the first cell (left most) stays the same, and for each of the other cells, add the total of all cells to the left. The last cell will then contain the total sample size. Now, calculate the difference in cumulative counts for each category (ignore the sign – see discussion of one- and two-tail tests later). The test statistic '*D*' is the largest of these differences.

TABLE 7.29

Counts in Ordered Categories Suitable for Analysis Using the Kolmogorov–Smirnov Two-Sample Test

	Well-Rounded	Rounded	Sub-Rounded	Sub-Angular	Angular	Very Angular	Totals
Unit A	10	14	35	22	14	5	100
Unit B	2	5	24	40	20	9	100

Note: You must enter counts, not percentages. In this case, the sample size is 100 in each case.

TABLE 7.30

Where Sample Sizes Are Equal the Counts Are Translated into Cumulative Counts and the Test Statistic 'D' is the Maximum Difference between Them

	Well-Rounded	Rounded	Sub-Rounded	Sub-Angular	Angular	Very Angular
Unit A	10	24	59	81	95	100
Unit B	2	7	31	71	91	100
Difference	8	17	28	10	4	0

Note: Ignore the sign of the difference unless you are conducting a carefully designed one-tail test.

Step 3. Use tables of critical values (Section 12.5) to compare your largest difference with the critical values. In the example above (Table 7.30), with equal sample sizes of 100, the tabulated critical value at $p = 0.001$ is 28. Since the observed value of D (28) is 'equal to or larger than' the tabulated value, we can conclude that the difference between the two samples is strongly significant.

7.5.2.2 Unequal Sample Sizes

Older textbooks tend to state that when sample sizes are less than 40, they must be equal, but that limitation seems to have been removed and it is possible to calculate probabilities or critical values for any combination of sample sizes. However, using unequal sizes requires an extra step.

Step 1. Arrange the data into the ordered categories, using the same categories for both samples. Use the counts, so that the total really is the true sample size (Table 7.31). Do not use percentages.

Step 2. This is the extra step: change the counts into proportions (out of one) by dividing the count in each cell by the total for that row (Table 7.32). In the example above, the first cell on the first row gives 4/26 = 0.154. When performed correctly, the total equals one. Beware of rounding errors if calculating by hand (work to three decimal places).

TABLE 7.31

Counts Arranged in Ordered Categories, in This Case a 'Likert-Type' Five-Point Scale of Level of Agreement

	Strongly Disagree	Disagree	Neutral	Agree	Strongly Agree	Total
Group A	4	5	6	6	5	26
Group B	10	8	5	6	1	30

TABLE 7.32

Counts Are Translated into Proportions So That They Are Directly Comparable despite the Difference in Sample Size

	Strongly Disagree	Disagree	Neutral	Agree	Strongly Agree	Total
Group A	0.154	0.192	0.231	0.231	0.192	1.00
Group B	0.333	0.267	0.167	0.200	0.033	1.00

Note: The total should now be one.

TABLE 7.33

Proportions Are Translated into Cumulative Proportions and the Test Statistic is the Largest
Difference between Them; in This Case 0.25

	Strongly Disagree	Disagree	Neutral	Agree	Strongly Agree
Group A	0.154	0.346	0.577	0.808	1.00
Group B	0.333	0.600	0.767	0.967	1.00
Difference	0.18	0.25	0.19	0.16	0

Note: For a two-tail test ignore the sign of the difference.

Step 3. Change the proportions into cumulative proportions (Table 7.33). This means
that the first cell (left most) stays the same, and for each of the other cells add the
total of all cells to the left. The last cells in each row should contain the number
one. Now calculate the difference in cumulative proportions for each category
(ignore the sign – see discussion of one- and two-tail tests later). The test statistic
'D' is the largest of these differences.

Step 4. Use tables of critical values (Section 12.5) to compare the largest difference (D)
with the critical values. These tables are a little more complicated than those used
for equal sample sizes. The top right diagonal gives the critical values for $p = 0.05$
and the lower left diagonal gives the critical values for $p = 0.01$. To obtain the critical value, use the two-sample sizes. In this case, for sample sizes of 30 and 26, the
critical value at $p = 0.05$ is 36, but note that the tabulated values are multiplied by
100, to remove decimal places and save space, so the critical value is 0.36. Since the
value that we obtained (0.25) is lower than the critical value, we can accept the null
hypothesis that the two groups are drawn from the same population and conclude
that they are not significantly different.

If one sample is too large to use the tables of critical values (Section 12.5), there are two
options. If both samples are larger than 40, the critical values of D are obtained using the
equations explained in Box 7.1. These equations are already entered into the calculators on
the companion site, so the critical values are computed automatically. For cases where one
sample is less than 40 and the other is between 40 and 200 tables of critical values (two-
and one-tail, $p = 0.05$ and $p = 0.01$) are supplied on the companion website. The values in
the tables are based on the KSINV function in the Real Statistics Resource Pack for Excel
(Zaiontz 2015).

7.5.3 One-Tail Testing

Opinions seem to be divided on whether it is reasonable to use the Kolmogorov–Smirnov
two-sample test with one-tail hypotheses. The problem is that it is testing whether two
samples come from the same population, or populations with the same distribution, and is
therefore sensitive to any difference between two samples. It is not a test for the difference
in where the middles (mean or median) of the two samples lie.

With tests that are based on the difference in the mean or median, such as Student's *t*-test
or the Mann–Whitney *U*-test, the difference between a one- and two-tail hypotheses is
reasonably straightforward. If you are predicting the direction of the difference in advance
(e.g. sample *A* will have a higher mean value than sample *B*), then the hypothesis is one-tail.
With a test that is sensitive to any difference between two distributions, it is not so easy
to clearly define a directional hypothesis. However, the Kolmogorov–Smirnov two-sample

TABLE 7.34

For a One-Tail Test, It Is Necessary to Predict, in Advance, Which Series will Be Biased toward the Left and Which toward the Right of the Table

Raw Counts	Very Angular	Angular	Sub-Angular	Sub-Rounded	Rounded	Well-Rounded	Totals
Breccia	15	20	9	4	0	2	50
Glacial	4	6	12	20	5	3	50

Note: In this case, the hypothesis is that glacial sediments contain more rounded clasts and that is clearly true for these samples.

test has been used for one-tail testing and Siegel (1956), for example, provides tables of one-tail probabilities. He argues that for a one-tail test, you must predict that one of the samples is 'stochastically larger' than the other. Perhaps, the easiest way to understand how a one-tail test might be considered reasonable is to use an example and some diagrams.

Imagine that you are working on Quaternary (ice age) deposits on some coastal cliffs and you suspect that there is a change from locally derived periglacial deposits (breccia) to glacially transported (glacial) deposits. You might reasonably predict that a glacially transported deposit would comprise clasts (rocks) that are more rounded than those in a breccia. Your hypothesis, therefore, is not that there is a difference in roundness, which would be two-tail, but that the clasts in the glacial deposit are more rounded, which is one-tail. The appropriate null hypothesis is that the clasts in the glacial deposit are not more rounded. Tabulating the values, we would predict a greater proportion of the glacial clasts to lie towards the right (more rounded) side of the table and a greater proportion of the breccia clasts to lie towards the left. Since the sample sizes are the same, it is clear from Table 7.34 and Figure 7.3 that the prediction is correct (Table 7.35).

Converting the raw counts into cumulative proportions allows the same data to be presented as a line graph (Figure 7.4) in which the sample that is biased towards the left side of the table (breccia in this case) forms the upper line and the sample biased towards the right side forms the lower line. Predicting that the glacial deposit is the more rounded,

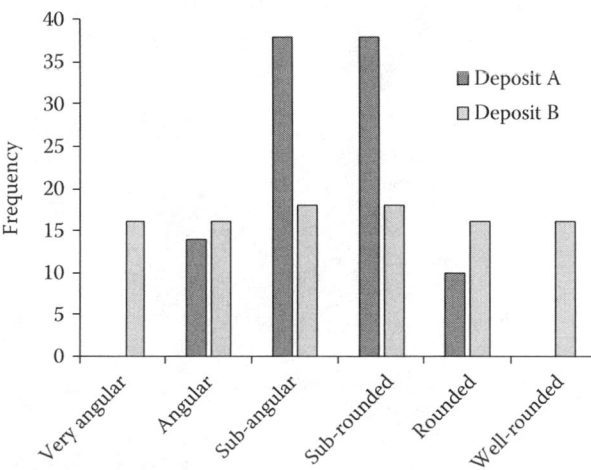

FIGURE 7.3
Clast roundness of two sediment units presented as histograms. This figure is part of the output from the companion site calculator for the Kolmogorov–Smirnov two-sample test with six categories.

TABLE 7.35

If a Series Is Biased towards the Left, It Will Produce Higher Cumulative Values in Most or All of the Categories, So the Direction of the Largest Difference Can Be Predicted in Advance

Cumulative Counts	Very Angular	Angular	Sub-Angular	Sub-Rounded	Rounded	Well-Rounded
Breccia	15	35	44	48	48	50
Glacial	4	10	22	42	47	50
Difference	11	25	20	6	1	0

Note: In this case, the one-tail hypothesis predicts that the breccia should give the higher of the two values that give the largest difference, which is true.

effectively predicts that the glacial deposit will form the lower line. When calculating D, the largest difference between the two lines, it is now possible to predict the sign of that difference, and in this case, we would predict that the cumulative sum for the breccia would be the larger number and the cumulative sum for the glacial deposit the smaller number. In this example, the largest difference ($D = 25$) is strongly significant ($p < 0.001$).

7.5.4 Effect Size

When the Kolmogorov–Smirnov two-sample test is applied to samples of unequal size, the value D is a proportion and so it already lies on a scale of zero to one, where zero represents no difference between the two samples and one means that there is no overlap. In this case, the value D is already an effect size. When the Kolmogorov–Smirnov two-sample test is used on samples of equal size, the value D can be translated into a proportion, and therefore an effect size, by dividing it by the sample size (not the sum of both samples).

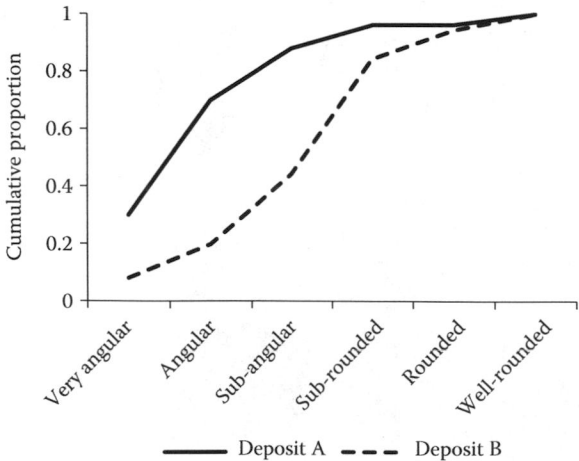

FIGURE 7.4

Same data as Figure 7.3 but presented as cumulative proportions. Note that the data set that is biased towards the left side of the range (breccia) forms the upper line. Predicting which series will form the upper line, and therefore the sign of the maximum difference between the cumulative counts or cumulative proportions is the basis for crafting a one-tail hypothesis for the Kolmogorov–Smirnov two-sample test. This figure is part of the output from the companion site calculator for the Kolmogorov–Smirnov two-sample test with six categories.

BOX 7.1   CRITICAL VALUES OF THE KOLMOGOROV–SMIRNOV TEST WITH LARGE SAMPLE SIZES

When both of the samples are larger than 40, the critical threshold for D at different levels of significance can be obtained by inserting the two-sample sizes (n_1 and n_2) into the formula below:

$$D_{crit} = KSM \times \sqrt{\frac{n_1 + n_2}{n_1 \times n_2}}$$

where KSM is the 'Kolmogorov–Smirnov multiplier' obtained from Table 7.36. For a two-tail test, for example, the critical value for $p = 0.05$ is 1.36, so if the two-sample sizes were 45 and 50, the critical value for D is

$$D_{crit} = 1.36 \times \sqrt{\frac{45 + 50}{45 \times 50}} = 1.36 \times \sqrt{\frac{95}{2250}} = 1.36 \times \sqrt{0.0422} = 1.36 \times 0.205 = 0.279$$

TABLE 7.36

Multipliers (KSM Values) Used for Estimating the Critical Value of D for Large Sample Sizes

Two-Tail Tests		One-Tail Tests	
$p = 0.05$	$p = 0.01$	$p = 0.05$	$p = 0.01$
1.3581	1.6276	1.2238	1.5174

7.5.5 How to Do It

The Kolmogorov–Smirnov two-sample test for counts in ordinal categories is not difficult to calculate by hand or in a spreadsheet, but the simple companion site calculators make it even easier. It is hardly worth using SPSS to perform this test and I am not aware of a simple way of conducting it using *R* Commander. The Real Statistics Resource Pack supports these tests.

7.5.5.1 Companion Site Calculators

Simple spreadsheet calculators for the Kolmogorov–Smirnov two-sample test are provided for cases where there are between 3 and 12 categories. Choose the appropriate calculator based on the number of categories and enter the raw counts (not percentages or cumulative counts). The names of the categories can be changed and these will appear on two simple graphs showing the counts as histograms and the cumulative proportions as line graphs (see Figures 7.3 and 7.4). The results include the maximum difference (D), which for equal sample sizes is an integer and for unequal sample sizes is a decimal fraction (which also serves as the effect size).

Unfortunately, there is no standard Excel function to allow probabilities to be obtained directly, but a close approximation can be obtained using the chi-square distribution. Both one- and two-tail probability values are returned but be aware that these approximations will be slightly conservative, meaning that that the *p*-values may be slightly higher than they should be. This means that if the estimate is $p = 0.049$, then the difference is certainly significant at $p < 0.05$ but if the estimate is $p = 0.054$, then it is sufficiently close to the boundary of significance to warrant a check against critical values. Where both samples are larger than 39, the critical values for $p = 0.05$, $p = 0.01$ and $p = 0.001$ are provided. Where either sample is less than 40, it is necessary to use the tables of critical values provided (Section 12.5).

These calculators also include the option to add scoring values to the categories to provide a mean and standard deviation for each sample and to apply Student's t-test or the F-test for equality of variance. This option is described later in the section on 'scoring categories for parametric tests' (Section 7.6).

7.5.5.2 Real Statistics Resource Pack

To use this function, the data have to be arranged in two columns, not rows. To change rows to columns in a spreadsheet, copy the rows, choose 'paste special' and tick 'transpose'. If you have loaded the Resource Pack, access a list of options using 'ctrl-m' and choose 'Non-parametric Tests'. Identify the cells that hold your table and choose 'Kolmogorov–Smirnov (freq)'. Also decide what critical threshold of probability you want to use; the default is $p = 0.05$. The table of outputs includes the test statistic 'D', a two-tail p-value, the critical value for your sample sizes at your chosen threshold level and the two-sample sizes. If you need the one-tail probability, you can half the two-tail probability.

7.5.5.3 In SPSS

Enter the data in the 'long format' so that one column identifies the two groups and the other identifies the choices as coding numbers. The number of rows equals the total sample size. Choose 'Analyze', 'Non-parametric Tests' and 'Independent Samples'. Under 'Objectives', choose 'Customize analysis' and under 'Fields', put the two groups in the lower box marked 'Groups' and the coded scores into 'Test Field'. Under 'Settings', check 'Customize analysis' and choose 'Kolmogorov–Smirnov (two-samples)' and click on 'Run'. The output box gives the two-tail probability and clicking on it takes you to some further information including the 'most extreme difference' which is D; the maximum difference between the cumulative proportions.

7.5.6 Examples

7.5.6.1 Example: Different Shaped Distributions

Two sediment deposits (A and B) are compared with respect to the roundness of their clasts. In each case, 100 clasts are chosen at random and allocated a roundness class using Power's scale. The results (Figure 7.5) reveal that although both deposits have modal counts in the sub-angular to sub-rounded categories, they differ with regard to the spread of results. The Kolmogorov–Smirnov two-sample test is sensitive to any difference between the samples and in this case, the maximum difference between the cumulative frequencies is 22, which exceeds the two-tail critical value for $p = 0.05$ (20), so we can reject the null hypothesis and conclude that the two deposits differ with regard to clast roundness.

Given the large sample size, it might be considered acceptable to code the categories from one to six and apply Student's t-test. In this case, the result would indicate no significant difference ($t = 0.32$, two-tail $p = 0.75$) because the t-test is only sensitive to a difference

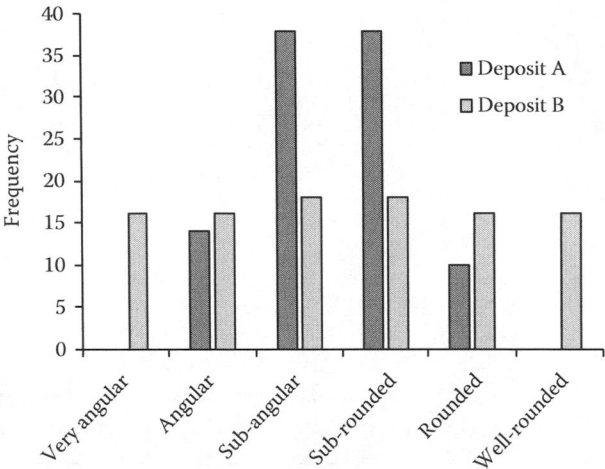

FIGURE 7.5
Kolmogorov–Smirnov two-sample test is unusual in that it tests for any difference between two samples, not just a difference in location (the position of the middle). In this case, there is virtually no difference in location but one sample is much more variable than the other. A Student's *t*-test would suggest no difference but the Kolmogorov–Smirnov two-sample test recognises the significant difference.

in the mean, and the mean values in this case are very similar. The *F*-test for equality of variance suggests a significant difference (*F*-ratio = 3.87, two-tail $p < 0.001$).

7.5.6.2 Example: Likert Scale, Small Samples

The Kolmogorov–Smirnov two-sample test is particularly useful for analysing the results of data collected using Likert-type scales where sample sizes are small or very unequal. In this case (Table 7.37), a survey of locals and tourists has resulted in a large disparity in sample size but a clear difference in the responses. It would not be appropriate to code these data and apply parametric tests, but the Kolmogorov–Smirnov two-sample test recognises the difference as statistically significant. The maximum difference in cumulative proportions is 0.55 and the tabulated critical values at $p = 0.05$ and $p = 0.01$ are 0.48 and 0.58 (comprehensive tables provided on the companion website). We can conclude that the difference is significant at $p < 0.05$. The chi-square approximation calculated in the companion site calculator gives $p = 0.016$.

7.5.6.3 Example: Likert Scale, Large Samples

In a comparison of opinions carried out in two villages, the sample sizes are both large (Table 7.38), so as well as applying the Kolmogorov–Smirnov two-sample test for any

TABLE 7.37

Kolmogorov–Smirnov Two-Sample Test is Well-Suited to Analysing Likert-Type Scales Where Sample Sizes are Small or Very Different

	Strongly Agree	**Agree**	**Neutral**	**Disagree**	**Strongly Disagree**	**Totals**
Local	10	16	8	2	4	40
Tourist	1	0	3	5	1	10

TABLE 7.38

Where Sample Sizes Are Large, It is a Common Practice to Code Each Answer Using a Number and Then Treat the Data as If They Are Continuous and Apply Parametric Statistics Such as Student's *t*-Test

	Strongly Agree	Agree	Neutral	Disagree	Strongly Disagree	Totals
Village A	24	32	14	2	4	76
Village B	14	20	27	10	3	74
Coding	2	1	0	−1	−2	

Note: The companion site calculators include this option.

difference, the data can also be coded and treated as continuous, allowing the parametric *t*-test for a difference in location (position of the mean) and *F*-test for equality of variance to be used. The data could be scaled from one to five in either direction, or symmetrically arranged around a zero value, as in this case. As long as the gaps between the steps are the same (one unit), it does not influence the results of the tests, only the values for the two means. In this case, the Kolmogorov–Smirnov two-sample test suggests a significant difference ($D_{max} = 0.28$, two-tail $p < 0.01$) and the Student's *t*-test gives a similar result ($t = 3.18$, $p < 0.01$). Both tests give low effect sizes ($D = 0.28$, $r = 0.26$) suggesting that although the difference in responses is statistically significant, it is small (effect sizes <0.3). You might conclude that there is a small but significant difference, with village A tending to be more strongly in agreement.

7.5.6.4 Example: Odd Case with a Bimodal Distribution

We are interested in perception of risk of natural disasters and on a fieldtrip to Vancouver ask locals and tourists what level of damage they think would occur if there was to be a movement of the tectonic plates at the subduction zone just offshore of Vancouver Island. A five-point scale of devastation has been devised and we are interested in any difference between the two groups, so this is a two-tail test.

Looking at the counts in Table 7.39 and the associated bar chart (Figure 7.6), it is clear that there is quite a big difference between the locals and tourists. The local opinions are centred in the middle of the options, so that the third category (widespread damage, some deaths) is the modal class and the numbers decline in both directions. The tourist opinions, however, are polarised, with almost equal numbers expecting minimal damage and total destruction and very few people opting for the central class. Since the sample sizes are reasonably large (40), I would intuitively expect a statistical test to demonstrate a strongly significant difference in this case. Since the Kolmogorov–Smirnov two-sample test is supposed to be sensitive to any difference, it seems appropriate. However, when the

TABLE 7.39

Fabricated Responses to a Question about the Likely Impacts of an Earthquake

	Minimal Damage	Some Property Damage	Widespread Damage, Some Deaths	Major Damage, Many Deaths	Total Destruction	Totals
Category	A	B	C	D	E	
Locals	2	8	22	7	1	40
Tourists	12	6	4	8	10	40

Note: The 'tourist' responses are polarised.

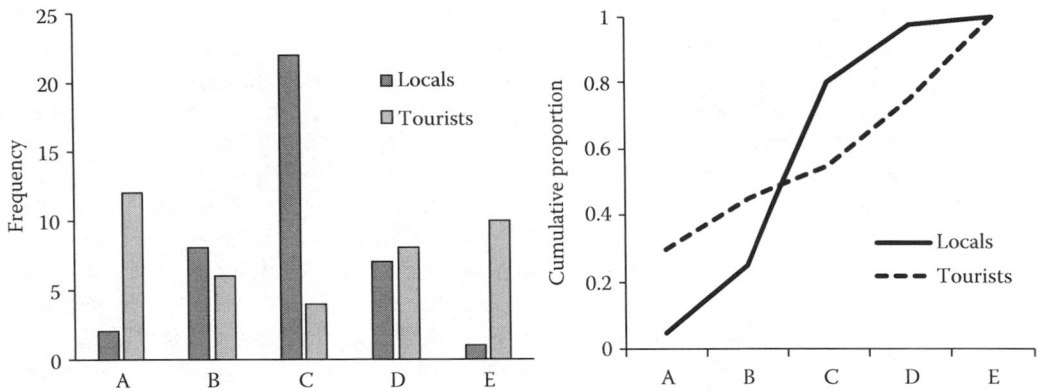

FIGURE 7.6

When one of the samples has a bimodal distribution, as for the tourist responses in this case, the Kolmogorov–Smirnov two-sample test gives very odd results. In this case, the maximum difference between the cumulative counts is 10, which is not significant (two-tail $p > 0.05$). However, a chi-square test on the same data suggests that the difference is strongly significant ($\chi^2 = 27.32$, $df = 4$, $p < 0.001$).

cumulative counts are compared the largest difference (D) is 10, but the number 10 occurs twice, once when the tourist curve is higher than the locals (category A) and once in the opposite direction (category C). The critical value at $p = 0.05$ for a two-tail test (Table 12.5) with equal sample sizes of 40 is 13, so according to the Kolmogorov–Smirnov two-sample test, the difference is not statistically significant and we have to accept the null hypothesis that the two samples have been drawn from the same or identical populations.

I would not be satisfied with this result. It seems the test cannot cope with one of the distributions being strongly bimodal. If the same data are re-analysed using the chi-square test, which ignores the order of the categories, the difference is very strongly significant ($\chi^2 = 27.32$, $df = 4$, $p < 0.001$). Cramer's V, a measure of the effect size, is 0.58, indicating that perception of risk is strongly contingent on the groups. In this case, I would ignore the results of the Kolmogorov–Smirnov two-sample test and use chi-square.

7.6 Scoring Categorical Data for Parametric Tests

In the social science and medical literatures, there has been a long lasting and at times ill-tempered discussion about whether it is reasonable to treat data that have been collected as counts in ordered categories as if they are continuous, and apply parametric statistical techniques (Jamieson 2004; Carifio and Perla 2008; Norman 2010). Likert-type scales (pronounced 'lick-urt'), with four, five or seven categories of 'level of agreement', are a particular bone of contention. On the one side, there are those who argue that such scales are strictly ordinal, which means that although the categories fall in rank order, we cannot assume that the distance between the categories is uniform (Jamieson 2004). They also point out that the results of surveys are often very strongly skewed, rather than near normally distributed. They typically argue that only non-parametric methods should be used. On the other hand, there are those who argue that the categories are just a convenient way of measuring something that is really continuous, and that parametric statistical methods are very robust to violations of continuity and non-normality (Norman 2010; Fagerland 2012).

There seems to be no end in sight to this debate. Computers have made it easier to apply non-parametric statistical tests, so their popularity has increased, but it is still very common practice to treat counts in ordered categories as continuous data and to apply parametric methods such as Student's *t*-test and ANOVA. The advantage of non-parametric methods is that they make very few assumptions and the ordinal nature of the data does not violate them. The advantage of parametric methods is that they are generally more powerful, in the sense that if there is a real difference in the population, you are more likely to detect it using parametric methods. Treating all of your data as continuous also makes it easier to compare results obtained using different types of measurement.

If you decide to treat categorical data as continuous, you should ensure that the categories are really ordinal, so that it is sensible to assume that underlying them there is a continuum. That is certainly true of Likert-type scales of strength of opinion, socioeconomic classes and of Power's roundness scale. Next you should consider whether the sample size is large enough to argue that parametric methods are justified. There is no simple answer to this question and it depends on the size of both samples and also on how skewed the distributions are. In the companion site calculators for the Kolmogorov–Smirnov two-sample test, I have suggested a minimum of 50 for the smaller of the two samples, but Student's *t*-test and *F*-ratio results are calculated if the smallest sample is 25.

In order to treat categorical data as continuous, numbers have to be allocated to each category. The important decision here is whether to use a simple numerical sequence, assuming that the distance between each category is uniform, or to use something more complicated. My advice is to keep it simple. Unless you have some very clear and defensible rationale for using a complicated scoring procedure just use numbers in a simple linear sequence. As long as the gap between the numbers stays the same, the way that you choose to order them does not influence the statistical results, but it does change the numbers that are obtained for the mean values of the two samples. For example, given a five-part Likert-type scale (Table 7.40), there are three simple options for how to score the categories. The categories can be scored from one for strongly disagree to five for strongly agree, in the opposite order, or they can be scored so that the neutral category is given a score of zero and the two categories of agreement are positive and the two for disagreement are negative. The last option has the advantage that 'no opinion' means zero, and the sign of a mean value is immediately informative about which side the sample falls on. A potential disadvantage is that mean values can fall below zero and there are some data treatments that do not work with negative values. The important point is that whichever of these options is chosen, it does not change the results of the *t*- or *F*-test (Table 7.41).

When you design a questionnaire, you should think carefully about how you intend to analyse the results. For example, you may ask several questions that are all linked in

TABLE 7.40

Three Simple Options for Scoring a Likert-Type Scale

	Strongly Agree	Agree	Neutral	Disagree	Strongly Disagree
Male	2	11	22	14	1
Female	12	16	12	8	2
Coding A	2	1	0	−1	−2
Coding B	1	2	3	4	5
Coding C	5	4	3	2	1

Note: The choice of which to use will change the mean values but not the results of a Student's *t*-test or *F*-test for equality of variance.

TABLE 7.41

Mean Values Resulting from the Three Different Coding Schemes Shown in Table 7.40

	Coding A	Coding B	Coding C
Male mean	−0.02	3.02	2.98
Female mean	0.56	2.44	3.56
Difference	0.58	0.58	0.58
Student's t	3.28	3.28	3.28
Two-tail p	0.001	0.001	0.001
F	1.03	1.03	1.03

Note: Although the mean values change, the difference between the two mean values remains constant, as do the results of the t- and F-tests.

some way, and when you have the results, it may be convenient to combine those answers, so that rather than dealing with each question individually, you can look at the average responses to a set of similar questions. When you do this, you must be careful to ensure that the coding allows you to produce a sensible average, and that similar questions are not actually cancelling each other out. For example, you may be interested in attitudes towards the environment and ask the following questions:

Q1: To what extent do you agree with the statement 'We must cut greenhouse gas emissions to protect future generations from extreme climate change'?

Q2: To what extent do you agree with the statement 'Cutting greenhouse gas emissions is too expensive and we need to keep using fossil fuels for the foreseeable future'?

Both of these questions are concerned with a similar theme, and it would make sense to combine them (perhaps with several other questions) to provide an overview of each respondent's attitudes. However, if these two questions are coded in exactly the same way, then they would cancel each other out. If the first one is coded so that 'strongly agree' is high and 'strongly disagree' is low, then it would make sense to code the second question in the opposite direction. Coding your data is not a simple mechanical procedure, you really have to think about what the numbers will mean in terms of the themes that you are interested in.

An exception to the general rule that the numbers used for coding should be consecutive, so that the distance between each category remains constant, is where Likert-type scales are produced without the central option of neutrality (Table 7.42). These are sometimes

TABLE 7.42

Three Logical Coding Schemes for a 'Forced Response' Question Where the Central Neutral Category Has Been Removed

	Strongly Agree	Agree	Disagree	Strongly Disagree
Male	18	25	5	2
Female	12	18	13	7
Coding A	2	1	−1	−2
Coding B	1	2	3	4
Coding C	4	3	2	1

TABLE 7.43

In This Case, the Choice of Coding is Critically Important Because It Changes Both the Distance between the Two Mean Values and Their Standard Deviations, and Hence the Results of the Statistical Tests

	Coding A	Coding B	Coding C
Male	1.04	1.82	3.18
Female	0.30	2.30	2.70
Difference	0.74	0.48	0.48
Student's *t*	3.29	3.13	3.13
Two-tail *p*	0.001	0.002	0.002
F	1.10	1.02	1.02

called 'forced response' questions. In this case, it is reasonable to assume that although the central 'neutral' category is removed, the underlying scale of opinion remains the same, so that the distance between the categories 'agree' and 'disagree' is still two units rather than one. On this basis, it is logical to assign the value zero to the missing category and +1 and −1 to the categories to either side. This ensures that mean values to the left of centre are positive and those to the right of centre are negative, which is a fair reflection of the level of agreement. Alternatively, the missing category can be ignored and the four remaining categories scored in simple numerical order from either end. Note that in this special case, the choice of coding does influence the statistics (Table 7.43). Choosing to ignore the missing zero category changes both the difference between the two mean values and the two standard deviations and hence the *t*-test and *F*-ratio.

7.6.1 How to Code Categorical Data

If you have a set of questionnaire data, you may choose to code everything so that all of the analysis can be conducted in a statistical package such as SPSS. In this case, there will be a row for each participant and each of their answers will be represented by a coding value in one of the columns. It is tedious to enter a big data set in this way, but it produces a column for each question and it is straightforward to treat the columns as if they contained continuous data. The same procedure can be followed in a spreadsheet and the usual functions can be used to compare pairs or multiples of columns.

If the data are already arranged as contingency tables, as is often the case, it is still possible to calculate the mean and standard deviation, which is all that you need to perform, for example, Student's *t*-test and the *F*-test for equality of variance. To calculate the mean, multiply the count in each cell by the coding value for that column and then sum the results for each row. That sum divided by the sample size gives the mean. To calculate the standard deviation, first calculate the sample variance. Take the sum of the squared differences between the values (coding numbers) and the mean and then divide by the sample size minus one. The sample standard deviation is the square root of the sample variance. The procedure is summarised in Table 7.44. The companion site calculators for the Kolmogorov–Smirnov two-sample test will make these calculations automatically.

TABLE 7.44

Example of How to Calculate Mean and Standard Deviations of a Sample of Data Collected as Counts in Categories

	Strongly Agree	Agree	Disagree	Strongly Disagree	Sum/Answer
Counts	6	9	5	5	= 25
Coding	4	3	2	1	
Calculate mean	$6 \times 4 = 24$	$9 \times 3 = 27$	$5 \times 2 = 10$	$5 \times 1 = 5$	= 66
				Mean =	$66 \div 25 = 2.64$
$(code - mean)^2$	$(4-2.64)^2 = 1.85$	$(3-2.64)^2 = 0.13$	$(2-2.64)^2 = 0.41$	$(1-2.64)^2 = 2.69$	
Times the count	$6 \times 1.85 = 11.10$	$9 \times 0.13 = 1.17$	$5 \times 0.41 = 2.05$	$5 \times 2.69 = 13.45$	= 27.76
	Sample variance = $27.76 \div n-1 = 27.76 \div 24$				= 1.157
	Sample standard deviation = square root of sample variance				= 1.075

Note: The companion site calculators for the Kolmogorov–Smirnov two-sample test will do this automatically and return the results of Student's *t*-test and the *F*-test for equality of variance.

References

Carifio, J. and Perla, R. 2008. Resolving the 50-year debate around using and misusing Likert scales. *Medical Education* 42: 1150–1152.

Fagerland, M.W. 2012. *t*-tests, non-parametric tests, and large studies – A paradox of statistical practice? *BMC Medical Research Methodology* 12: 78.

Freeman, G.H. and Halton, J.H. 1951. Note on an exact treatment of contingency goodness-of-fit and other problems of significance. *Biometrika* 38: 141–149.

Jamieson, S. 2004. Likert scales: How to (ab)use them. *Medical Education* 38: 1217–1218.

Norman, G. 2010. Likert scales, levels of measurement and the 'laws' of statistics. *Advances in Health Sciences Education* 15: 625–632.

Siegel, S. 1956. *Non-parametric statistics for the behavioural sciences.* New York: McGraw-Hill, 312pp.

Zaiontz, C. 2015. *Real Statistics Using Excel.* www.real-statistics.com.

8

Two-Sample Tests for Individual Measurements

8.1 Introduction

Where you have collected data as individual measurements, or as ranks, you have several options for comparing two samples. One is to put the numbers into categories and use the tests that have already been described for 'counts in categories' data. However, this is rarely the best option because it involves degrading your data and therefore losing information.

The first decision you must make is whether your data are paired or independent. Paired means that there is some logical reason for comparing one particular number in the first data set with one particular number in the second. Where a set of individuals sit two exams, for example, we would want to compare how each individual person performed in the first exam with how that same individual performed in the second exam. If we were comparing test results of one cohort with those of a different cohort then there would be no sense in pairing, we would want to compare the overall results.

For paired data a simple option is to use the sign test, which will tell you whether one sample has a strong tendency to give higher values than the other and is very easy to calculate. However, it does not use very much information and very large differences carry the same weight as very small differences. The Wilcoxon matched-pairs signed-ranks test is more powerful because it takes account of the size of the differences. If your samples are large (>30) and fit some assumptions, which you can check by just drawing histograms and seeing whether they are roughly bell-shaped, you can also use the paired-samples t-test, which is very easy to compute in a spreadsheet. All of these tests can be used with one-tail or two-tail hypotheses because it is possible to predict the direction of the difference in advance.

For independent data where you have small samples (<30), or where you have any doubts about whether your data meet the assumptions for Student's t-test, a very convenient alternative is the Mann–Whitney U-test. It tests whether there is a significant difference in the two medians, and can be used for two-tail or one-tail hypotheses. It is not difficult to compute, either using an online calculator or in a spreadsheet. If you have large samples that are near normally distributed then you can also use Student's t-test. If in doubt just apply both tests; they are both quick and easy. If they both agree on whether you should accept or reject the null hypothesis then you can be confident that your conclusions are robust, even if your samples are not huge and you are not absolutely sure that they are normally distributed.

Both the Mann–Whitney U-test and Student's t-test aim to test the null hypothesis that the two samples are drawn from either a single population or two populations that have the same mid-point. The U-test uses the median and the t-test uses the mean. However, sometimes we are not interested so much in the position of the middle, but in the amount

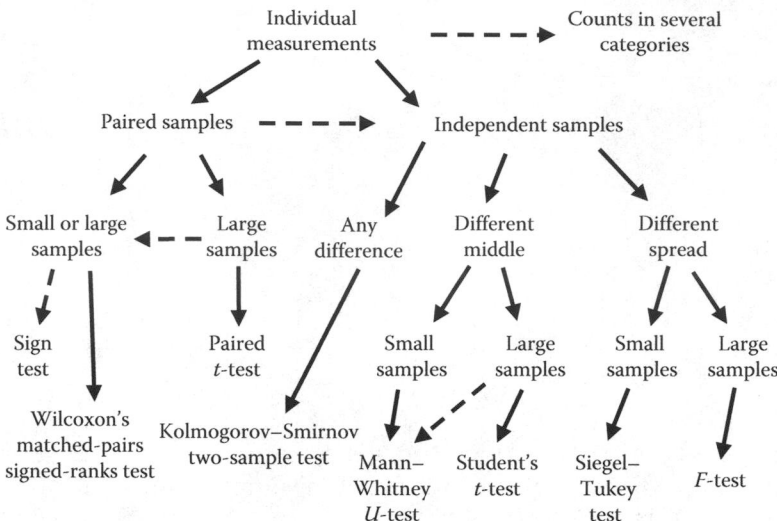

FIGURE 8.1
A logical series of steps for the comparison of two samples comprising individual measurements. Dashed lines indicate a loss of information.

of spread in the two samples. Given large samples the parametric F-test is appropriate, which tests the null hypothesis that the two samples have the same variance. When data are only available as ranks, or where the samples are too small or diverge strongly from the normal distribution, a clever non-parametric alternative is the Siegel–Tukey test. It is a little fiddly to compute, but not difficult if you use a spreadsheet. In addition it is possible to apply a version of the Kolmogorov–Smirnov two-sample test to two independent samples of individual measurements, providing a test for differences of any kind (Figure 8.1).

8.2 Wilcoxon's Matched-Pairs Signed-Ranks Test

8.2.1 When It Is Useful

This is a simple but powerful non-parametric test for paired data. Sample sizes can be small and there is no assumption that the two samples are symmetrically or normally distributed. It can be used whenever you have pairs of data and the differences between the pairs can be placed in rank order. Beware of confusing this test with 'Wilcoxon's rank-sum test' which is equivalent to the Mann–Whitney U-test used for independent rather than paired samples (Section 8.5).

8.2.2 What It Is Based On

The calculation procedure is perhaps best explained using a small example. Imagine that a group of 10 students sat a multiple choice examination prior to attending a course on statistics and a similar exam afterwards. You might wish to test the one-tail hypothesis that

TABLE 8.1

A Small Sample of Paired Data Showing the Workings Required to Calculate Wilcoxon's Matched-Pairs Signed-Ranks Test

Name	Before	After	Difference	Rank	Positive Ranks	Negative Ranks
Amy	52	52	0	Ignore		
Brian	26	50	+24	8	8	
Cath	35	45	+10	4	4	
Eve	50	70	+20	7	7	
Frank	45	50	+5	3	3	
Helen	58	57	−1	1		1
Joe	62	60	−2	2		2
Kevin	72	85	+13	6	6	
Nia	25	55	+30	9	9	
Pete	60	72	+12	5	5	
Sum of ranks					42	3

attending the course improves exam results. The exam results for each student, before and after the course, are presented in Table 8.1. Note that, of the 10 students, seven showed an improvement, two obtained a lower mark and one (Amy) got the same mark both times. We can immediately use these results to check for a significant improvement using the sign test (Section 8.6.1). Given seven successes from nine trials the one-tail probability is $p = 0.09$, so we would accept the null hypothesis that there is no improvement. However, from Table 8.1 it is clear that most of the improvements are rather large and the two differences in the opposite direction are very small. The sign test cannot use this information but Wilcoxon's matched-pairs signed-ranks test does. The differences are first ranked ignoring the direction of difference (i.e. ignore the sign) from smallest to largest (that is important). So in this case Helen is ranked first, with a difference of one, then Joe, then Frank, etc. Now we separate the ranks into two columns, depending on the sign of the difference. If there was no real difference between exam marks before and after sitting the course we might expect the ranks to be more or less evenly distributed between the two columns. If there was a large difference then we might expect most of the high ranks to be in one column and most of the low ranks in the other. In this example that is indeed what we see: the two differences that are in the opposite direction to our prediction (of improvement) are very small and therefore have low ranks. The 'test statistic', which for no obvious reason is called T is simply the smaller of the two sums of ranks, so in this case it is $T = 3$.

Tables are used to obtain critical values of T for one-tail and two-tail tests at various levels of significance (Section 12.6). In this case given a sample size of 9 (as in the sign test, cases where there is no difference are ignored) the critical value for a one-tail test at $p = 0.01$ is 3, and since T must be 'equal to or less than' the tabulated value we can confidently reject the null hypothesis that there is no improvement.

Note that in the example above the results of the sign test and Wilcoxon's matched-pairs signed-rank test conflict. The sign test gave $p > 0.05$, urging us to accept the null hypothesis whereas Wilcoxon's matched-pairs signed-ranks test gave $p < 0.01$, so that we can confidently reject it. The reason for the difference is that Wilcoxon's matched-pairs signed-ranks test is using more of the information, so it is a more powerful test. In such cases you should accept the result of the more powerful test, as long as any assumptions are met. Since Wilcoxon's matched-pairs signed-ranks test is non-parametric, and thus has very few assumptions, it is always more powerful than the sign test.

TABLE 8.2

Examination Results Reported Using a Ten-Point Scale So That Some of the Scores Are Tied

Name	Before	After	Difference	Rank	Positive Ranks	Negative Ranks
Adam	3	3	0	Ignore		
Bethany	3	5	+2	4.5	4.5	
Dave	4	5	+1	2	2	
Edith	5	7	+2	4.5	4.5	
Gayle	4	8	+4	6	6	
Harry	6	5	−1	2		2
Jane	8	7	−1	2		2
Kate	2	8	+6	8	8	
Liam	2	7	+5	7	7	
Robbie	1	8	+7	9	9	
Sum of ranks					41	4

Note: The tied scores are given the average of the tied ranks.

8.2.3 Tied Ranks

Where some of the differences are the same, the procedure is to give those cases the average of the tied ranks. In Table 8.2, for example, students are marked on a ten-point scale rather than using percentages, leading to several tied scores.

In this case there are three students with a difference of one (ignoring the sign) so they are each assigned the average of ranks 1, 2 and 3 ($6 \div 3 = 2$), so they all have a rank of 2. Since ranks 1–3 have been used up, the next available rank is 4, but again there are two differences that are the same (2). In this case the average of ranks 4 and 5 is given to both, so they both get a rank of 4.5 leaving 6 as the next available rank. When using non-parametric tests that involve ranks it is important that shared ranks are awarded in this way, so be careful to get it right (Box 8.1). A good check is to make sure that your highest rank is the same as the sample size (ignoring those with no change). In the example above $T = 4$ which for a one-tail test is significant at $p < 0.05$ (critical value is 8) but not at $p < 0.01$ (critical value is 3).

8.2.4 Effect Size

Most online calculators and software give a z-score, from the standard normal distribution, as well as a probability so it is relatively easy to calculate the effect size r, which is the z-score divided by the square root of the sample size. The sample size for Wilcoxon's matched-pairs signed-ranks test excludes any pairs that show no difference. If you are comparing 30 pairs, but two of them show no difference, then N is 28. However, for the effect size I think you should include the values that gave no difference as well, so the number of pairs is 30 in this case. The calculation uses the number of measurements, not the number of pairs (Field 2013, p. 234), so in this case the z-score is divided by the square root of 60 (not 28 or 30). This is the procedure used by the Real Statistics Resource Pack. If you are computing the test by hand, or in a spreadsheet, you will need to calculate the z-score. You can do this by converting the one-tail probability into a z-score using the NORMSINV function in Excel. The companion site calculator returns the effect size r.

8.2.5 How to Do It

8.2.5.1 *In a Spreadsheet*

The simplest way to conduct this test, even with large samples, is using a spreadsheet. Start by calculating the differences as above but add another column to give the absolute differences (Table 8.3). This is easy to do using the 'ABS' function which just removes the sign. So if cell E14 contains the number –1, and you want to ignore the sign, in cell F14 you can enter '=ABS(E14)' and the number 1 will appear. When you have a column with absolute differences you can sort your data using that column, so that assigning the ranks is easier.

If you have a very large data set, so that assigning the ranks is tedious, you can do it automatically in Excel using the 'RANK.AVG' function (Box 8.1). Beware because it is easy to make a mistake using this function, especially by forgetting to fix the reference array using dollar signs, so that it does not change for every cell, or by ranking from high to low instead of from low to high values. Note that you have to remove the cases where there is no change before you do any automatic ranking, otherwise they will be treated as zero values and assigned ranks.

8.2.5.2 *Real Statistics Resource Pack*

Use Ctrl-M to access a list of options and choose non-parametric tests and Wilcoxon signed ranks. A new pop-up appears, identify the two sets of data, with or without headers, and set your significance level (default is $p = 0.05$) and tails. For the 'options' choose two paired samples and 'test type' non-parametric. You can tick all of the non-parametric options boxes. The output includes T and both the one and two-tail probabilities as well as a z-score and effect size.

8.2.5.3 *Companion Site Calculator*

On the companion site a simple 'paired samples' calculator is provided that will perform the test for you. Enter the data in the two columns marked sample A and sample B and the test statistic and sample size are returned. For small samples look up the critical values

TABLE 8.3

How to Arrange Rank Data in a Spreadsheet in Order to Calculate Wilcoxon's Matched-Pairs Signed-Ranks Test

Name	Before	After	Difference	Absolute	Rank	Positive Ranks	Negative Ranks
Adam	3	3	0	0	Ignore		
Dave	4	5	1	1	2	2	
Harry	6	5	–1	1	2		2
Jane	8	7	–1	1	2		2
Bethany	3	5	2	2	4.5	4.5	
Edith	5	7	2	2	4.5	4.5	
Gayle	4	8	4	4	6	6	
Liam	2	7	5	5	7	7	
Kate	2	8	6	6	8	8	
Robbie	1	8	7	7	9	9	
Sum of ranks						41	4

TABLE 8.4

Selection of Free Online Calculators for Wilcoxon's Matched-Pairs Singed-Ranks Test

Provider	Website	Comments
SocSci Statistics	http://www.socscistatistics.com/tests/signedranks/Default2.aspx	Easy to use and gives a table with ranks and p-values based on critical values of T (called W in this case) and the z score for the normal approximation depending on sample size. Size limit is 200.
AITherapy	https://www.ai-therapy.com/psychology-statistics/hypothesis-testing/two-samples	Need to choose 'same subjects' and 'non-parametric' to reach this test. Easy to use but when there are samples with no difference it reports them as part of N, even though it uses the correct N in the calculation. Remove them first. Gives a z-score and r as effect size but they look wrong to me. Beware.
SciStat Calc	http://scistatcalc.blogspot.co.uk/2013/10/wilcoxon-signed-rank-test-calculator.html	Data must be comma delimited so cutting and pasting does not work. Put the two columns of data in a fresh spreadsheet and save as a csv file then use the 'open csv file' option. Works with huge samples.
Vassar stats	http://vassarstats.net/ Listed under 'ordinal data'	Cut and paste into the box on the right and ensure cursor is next to the last number not below it. Uses an odd calculation so does not return T but the probabilities are correct.
IFA services?	http://www.fon.hum.uva.nl/Service/Statistics/Wilcoxon_Test.html	Reset button does not seem to work so delete example data and replace. The W value seems wrong but the probability is correct. Use with caution!

using the tables provided (Section 12.6). When the sample size (number of pairs where there is a difference) reaches 25 the one and two-tail probabilities are calculated automatically using a normal approximation, together with an effect size. The calculator also produces the same results using the paired-samples t-test and the sign test.

8.2.5.4 Online Calculators

If you have access to the Internet you can also run Wilcoxon's matched-pairs signed-ranks test using free online calculators by simply cutting and pasting the data out of a spreadsheet. Some suitable sites, with comments, are included in Table 8.4.

8.2.5.5 In R Commander

Put the two columns of data in the top left corner of a new spreadsheet, including headers with no gaps, and save it as a comma-delimited text file. Open it in *R Commander* using 'data' and 'import from a text file'. Under the 'statistics' tab choose 'non-parametric statistics' and 'paired-samples Wilcoxon test'. Choose the two data sets and on the options tab either choose a two-sided (two-tail) test or predict the direction of difference. Choose the default option and it will choose how to calculate the probability based on the sample size. The output includes the text 'alternative hypothesis: true location difference is not equal to 0'. This does not mean that there is a significant difference, it is just telling you that the hypothesis is two tail. Unfortunately *R* Commander does not return a *z*-score, even when using a normal approximation.

8.2.5.6 *In SPSS*

Cut and paste the two columns of data into the 'data view' (spreadsheet) page then click on the 'variable view' table to assign names and define the 'measure' as 'scale'. Now use the menu at the top to go to 'analyze' then 'non-parametric tests' then 'related samples'. As always do not let SPSS do anything automatically, choose 'customize analysis' on the 'objective' page, then on the 'fields' page move both data sets into the 'test field' using the arrow. On the 'settings' page choose 'customize tests' and click on 'Wilcoxon matched-pair signed-rank (two-sample)' and click 'run'. The output box includes a 'Sig.'. value, which is the two-tail probability. If you are using a one-tail hypothesis then divide this by two. Double-clicking on the output box gives you more information, none of which is very interesting.

8.2.6 Examples

8.2.6.1 *Example: Attitudes to Recycling*

You are interested in attitudes to recycling and wish to determine whether a new kerb-side collection scheme changes the way that people feel. You choose an unbiased sample of local residents and ask them to mark their level of enthusiasm for recycling by placing a cross on a 20 cm line with very unenthusiastic (hate it) at one end and very enthusiastic (love it) at the other. A few weeks after the introduction of the new scheme you ask the same group the same question. By measuring the position of the cross you obtain individual measurements for each respondent before and after the introduction of the new scheme (Table 8.5). The measurements in mm are easily converted into % enthusiasm (e.g. 100 mm gives (100 ÷ 200) × 100 = 50), which can help with description of the results, though it makes no difference to the test because the rank order does not change. You are unsure whether the new scheme will increase enthusiasm or whether the extra sorting of waste into different coloured bags will annoy people and make them less enthusiastic, so a two-tail test, with the null hypothesis that there is no change in level of enthusiasm, is appropriate.

Checking the tables of critical values for a two-tail test with a sample size of 15 we find that the critical value for $p = 0.05$ is 25. To be significant the measured value of *T* must be equal to or lower than the tabulated value, so given that the value we obtained is higher (35.5) we accept the null hypothesis that the new scheme has not changed the level of enthusiasm for recycling amongst local residents. You could report the results in the following way. 'Of the 15 residents sampled five showed a decrease in enthusiasm and 10 showed an increase. However, most of the changes were small, with only two showing an increase of more than 10% on the scale of enthusiasm that was used. Results of the non-parametric Wilcoxon's matched-pairs signed-ranks test suggest that the change in enthusiasm is not statistically significant ($T = 34.5$, $N = 15$, two-tail $p > 0.05$)'.

8.2.6.2 *Example: Grazing and Plant Diversity*

You are interested in the effect of sheep grazing on plant diversity in Snowdonia and know of one mountain where sheep have been excluded from a large area using a fence. Rabbits

TABLE 8.5

Calculations for Wilcoxon's Matched-Pairs Signed-Ranks Test for Changes
in Attitudes to Recycling

Before	After	Difference	Absolute Difference	Ranks	Ranks Up	Ranks Down
58	56	−2	2	1	1	
90	93	3	3	2.5		2.5
42	45	3	3	2.5		2.5
52	56	4	4	5		5
32	28	−4	4	5	5	
8	12	4	4	5		5
51	45	−6	6	8	8	
31	25	−6	6	8	8	
78	84	6	6	8		8
43	50	7	7	10.5		10.5
68	75	7	7	10.5		10.5
20	10	−10	10	12.5	12.5	
48	58	10	10	12.5		12.5
43	56	13	13	14		14
43	75	32	32	15		15
Sum of ranks					34.5	85.5
				Counts	5	10
				$T=$	34.5	
				$N=$	15	

and feral goats are able to pass the fence. You use a paired-quadrat design by measuring
the number of plant species in 1 m² quadrats on either side of the fence at 30 regularly
spaced intervals on a transect up the mountain. Your hypothesis is that excluding sheep
increases plant diversity, so this is a one-tail test. Of the 30 quadrats 28 showed higher
diversity on the ungrazed side of the fence. In this case it is not really necessary to use
Wilcoxon's matched-pairs signed-ranks test because the difference is so large, in the predicted direction, that the sign test returns a one-tail probability of $p < 0.0001$.

BOX 8.1   RANKING DATA IN A SPREADSHEET

Many of the non-parametric tests in this book require the original measured values
to be replaced with ranks. With small data sets this procedure can be performed by
hand, but with larger samples it is very tedious and easy to make a mistake, especially in dealing with tied ranks. A spreadsheet makes it much easier.

Recent versions of Excel include the 'RANK.AVG' function which is very useful
indeed. It takes a target value and translates it into a rank relative to a reference
set. In Table 8.6, for example, the aim is to translate a single column of values into
ranks. The data are already arranged in rank order but there are some tied ranks
that need to be dealt with appropriately. The equation shown in cell B2 is identifying the target value (A2) and assigning it a rank relative to the set of values from

TABLE 8.6

Using the RANK.AVG Function in Excel Helps to Rank Data and Deal Properly with Tied Ranks

	A	B
	Values	Ranks
1		
2	16	=RANK.AVG(A2,A$2:A$19,1)
3	18	2
4	27	3
5	36	4.5
6	36	4.5
7	37	6
8	39	7
9	43	8
10	47	9
11	48	10
12	51	11
13	53	12
14	54	13
15	57	14
16	58	16
17	58	16
18	58	16
19	60	18

Note. However, it is easy to mess this up, so be sure to add the dollar signs in the right place in the formula before you copy it.

cells A2 to A19. Note that there are dollar signs before the 2 and the 19. These are critically important and you have to add them manually when you use this function. The dollar sign before a number means 'keep the row constant when this formula is copied'. If you forget to add the dollar sign, and copy this formula down column B, the reference set will move down as well as the target, producing nonsense. Note how the tied scores are dealt with by assigning the average of the ranks, so that the final rank is equal to the size of the sample. Do not use the RANK function because it deals with tied ranks in a different way that is not suitable for most of the non-parametric tests in this book. The number one at the end of the formula in cell B2 means 'rank from smallest to largest'. If you want to rank from largest to smallest enter a zero.

The simple case of ranking a single column of data is used for some tests, including Spearman's rank correlation (Section 8.10.5), but many tests require values to be ranked relative to two or more columns of data and it can become quite confusing. Before this function was available it required the data to be combined in a single column, ranked, and then separated. The RANK.AVG function is particularly useful because it allows you to leave the data in the original columns. The procedure for ranking values from three samples, so that the ranks represent the combined data set rather than the individual samples, is shown in Table 8.7. This is the procedure that is

TABLE 8.7

The RANK.AVG Function Can be Used to Rank Values Relative to Two or More
Columns of Data, Which Is Useful for Many Non-Parametric Tests

	A	B	C	D	E	F
9	One	Two	Three	Ranks 1	Ranks 2	Ranks 3
10	67	46	33	5	16.5	=RANK.AVG(C10,$A$10:$C$19,0)
11	69	60	62	4	9.5	8
12	66	26	50	6	26	14
13	48	15	78	15	28.5	1
14	76	60	43	2	9.5	18
15	39	46	37	20	16.5	21
16	53	15	56	13	28.5	12
17	72	8	58	3	30	11
18	18	34	36	27	23	22
19	65	32	40	7	25	19

Note: Be very careful with the dollar signs.

used in the Kruskal–Wallis *H*-test (Section 9.5). The Mann–Whitney *U*-test, described
in this chapter (Section 8.5) applies the same logic to two samples.

Note that in this case the formula shown in cell F10 now has dollar signs before
both the letters and the numbers for the reference set. This ensures that when that
formula is copied to populate the three columns where you want the ranks to appear,
the reference set remains the full set of three columns of original measurements. In
this case the final part of the equation is a zero, which means rank in descending
order, so the highest value (78) receives a rank of one.

Unfortunately, the LibreOffice Calc spreadsheet does not have the 'RANK.AVG'
function yet (check your version) and if you use just 'RANK' it will not share the joint
ranks properly. Better to do it by hand in that case. Place all of the data in a single
column and in the column next to it add something to indicate which group each
value belongs to. Also add a column of numbers that will allow you to put the data
back in the original order. Now pick up all three columns and sort them by value.
Now that they are in order you can carefully add a column of ranks, being sure to
share any joint ranks properly. If you get it right the last rank should be equal to the
sample size. When you are confident that it is all correct, pick up all four columns
and re-sort using the group identifiers. That will allow you to separate the groups
with their ranks. Take your time with this because if you mess it up the results of any
tests that use the ranks will be wrong.

If there are a lot of tied ranks a useful 'trick' is first to get the data into rank order,
and then add another column that automatically warns you if two values in the
sequence are the same. You can do this using the 'IF' function. For example if the
data were in column A of a spreadsheet, already in rank order, starting in row 10,
you could enter in cell B11 'IF(A11 = A10, TIE, 0)'. This means that if the cell to the left
ties with the cell above it enter TIE, otherwise enter a zero. Copy this equation down
the column and the ties are all identified, making it easier and faster to perform the
ranking properly.

8.3 Paired-Samples Student's *t*-Test

There are three varieties of Student's *t*-test, but they have nothing to do with university or school students, the name 'Student' was used by William Sealy Gosset when he published the tests in 1908. He worked for the Guinness brewing company and they did not want him to use his real name.

8.3.1 When It Is Useful

This is a parametric test, so only use it where you have individual measurements and your sample size is large (≥ 30 is a rule of thumb). This test assumes that the differences between the pairs of data (not the two samples) form a near normal distribution. To check this you can calculate the differences in a spreadsheet, calculate descriptive statistics and plot a histogram. If they are near normally distributed the mean and median should be similar and the skewness and kurtosis values near zero. The histogram should be roughly bell shaped (Section 3.3.1). Large samples will tend to be near normally distributed and a bit of deviation from normality is not disastrous. If in doubt just run both paired-samples *t*-test and Wilcoxon's matched-pairs signed-ranks test. You should not expect them to give exactly the same probability, but they should agree on whether the difference is statistically significant. The companion site calculator for paired samples returns the descriptive statistics for the differences, including mean, median, skewness and kurtosis.

8.3.2 Effect Size

The procedure for calculating the effect size *r* from a *t*-test is covered in Box 8.4. However, it is sometimes argued that the effect size calculated in this way for a paired test design is an overestimate of the real effect in the population (Dunlap et al. 1996, Field 2013). An alternative approach is to use Cohen's *d* as the effect size. I am not really convinced that this is useful but in the companion site calculator both results are given.

8.3.3 How to Do It

8.3.3.1 *In a Spreadsheet*

The most convenient way to run the paired-samples *t*-test is in a spreadsheet. Choose the function 't.test' and follow the instructions. Be careful because it is quite easy to mess this up, because there are three different tests included in this one function. You are first asked to identify the two 'arrays', which is your two samples. The test will compare the first value in array 1 with the first in array 2, and so on, so make sure that they are lined up properly in the spreadsheet. Put them in adjacent columns then you will not make a mistake. Then choose whether you are conducting a one-tail or a two-tail test. If your hypothesis predicts the dominant direction of the difference (i.e. which sample tends to give the higher of the two pairs of numbers) then use a one-tail test, if in any doubt use a two-tail. Now you have to choose the type of test, and this is where it can all go horribly wrong. You need to choose type one, which is the paired-samples test. A click gives you the probability.

8.3.3.2 *Companion Site Calculator*

The 'paired-sample' calculator on the companion site allows you to enter two columns of data and returns the results of Wilcoxon's matched-pairs signed-ranks test, the paired-samples *t*-test with effect size *r* and the sign test. It also returns descriptive statistics on the samples and, more usefully for this test, on the differences between the samples. The minimum sample size for the *t*-test is set at 20, but be wary of trusting the results from such small samples.

8.3.3.3 In R Commander

Place the two data sets side by side in the first two columns of a clean spreadsheet. Give each column a header with no gaps in it and save as a comma-delimited text file. Open this in *R* Commander using the 'data' and 'import data' options. No need to change any of the default settings. Check the data by clicking 'view data set' then choose 'statistics' and 'means'. If you have entered the data in the right format 'paired *t*-test' should appear as one of the options. Click and follow the simple instructions. Output includes the *t*-value, degrees of freedom and either the one-tail or two-tail probability, depending on your choice on the options tab.

8.3.3.4 In SPSS

Input the data as two columns in the 'data view' (spreadsheet) page then give them names and identify the 'measure' as 'scale' on the 'variable view' page. Now go to 'analyze' then 'compare means' and choose 'paired-samples *t*-test'. Move both data sets into the 'paired variables' box using the arrow and click 'run'. SPSS now generates a bewildering array of results that you did not ask for. The important numbers are right at the end of the last box; the *t*-value, degrees of freedom (number of pairs minus one) and the two-tail probability. To obtain a one-tail probability divide this by two.

8.3.3.5 Using a Calculator

I am not sure why anyone would want to conduct a paired-samples *t*-test by hand, but at least it is not as difficult as the independent samples *t*-test (see Box 8.4). The dependent *t*-test is not based on the means and standard deviations of the two samples, it is based on the differences between the pairs of samples. The equation is

$$t = \frac{\text{Mean difference}}{\text{Standard error of the differences}}$$

The standard error of the differences is the standard deviation of the differences divided by the square root of the sample size. The sample size is the number of pairs, not the number of individual measurements. Modern scientific calculators will return the mean and standard deviation.

8.3.3.6 Online Calculators

Several free online calculators are available for the paired-samples *t*-test (Table 8.8) and most are quick and easy to use. Make sure that you do not run the independent samples *t*-test by mistake. I could not find any sites that calculated an effect size but there are sites that will calculate an effect size from a *t*-test result.

TABLE 8.8

Selection of Free Online Calculators for the Paired-Samples Student's *t*-Test

Provider	Website	Comments
GraphPad	http://graphpad.com/quickcalcs/ttest1/	Easy to use. Choose option 'enter or paste up to 2000 rows' to allow cut and paste and check 'paired *t*-test'. No effect size.
Social Science Statistics	http://www.socscistatistics.com/tests/ttestdependent/	Easy to use. Cut and paste into the two boxes, check significance level and tails and it returns the probability. No effect size.
MathPortal.org	http://www.mathportal.org/calculators/statistics-calculator/t-test-calculator.php	Odd format but generates equations, so useful if you want to check your workings when forced to make the calculations by hand. Does not give a probability, just the critical threshold value. No effect size.
Vassar Stats	http://vassarstats.net/tu.html	Check the correlated samples option, cut and paste the data and receive one-tail and two-tail probabilities and other output, but no effect size. Easy.
SciStatCalc	http://scistatcalc.blogspot.co.uk/2013/10/paired-students-t-test.html	Odd format and data have to be separated by a comma. Draws histograms but they are not very nice.

Note: Be careful not to choose the independent samples *t*-test by mistake.

8.3.4 Examples

8.3.4.1 Example: Examination Marks

You wish to test the hypothesis that students tend to improve their exam scores in year 2 relative to year 1. You obtain an unbiased sample of results from 30 students (Table 8.9). Note that in this case each student 'owns' two exam marks, so a paired-sample design is appropriate. Some summary statistics for these two samples are provided in Table 8.10. Note that the difference in the mean exam scores for years 1 and 2 is very small (1.67). If you were to treat these as two independent samples, and just compare the averages using an independent samples *t*-test, the results would suggest that the difference is not statistically

TABLE 8.9

Paired Data Suitable for the Paired-Sample Student's *t*-Test

Name	Year 1	Year 2	Name	Year 1	Year 2	Name	Year 1	Year 2
Anne	60	58	Henry	48	52	Paul	58	60
Andy	66	65	Iain	55	57	Peter	48	48
Beth	67	71	Jane	57	59	Rita	54	56
Carol	60	65	Kate	46	47	Ryan	55	56
Debby	55	54	Len	53	55	Sandy	52	50
Dan	58	59	Laura	55	61	Steve	54	53
Dave	54	57	Mike	51	54	Tavi	58	60
Eve	59	62	Nia	62	63	Tony	51	54
Frank	59	60	Nora	46	48	Vera	52	55
Guy	49	52	Ollie	52	54	Will	51	50

Note: The data are paired, rather than independent, because each student 'owns' two values.

TABLE 8.10

Summary Statistics for the Example Using Examination Marks (Table 8.9)

	Year 1	Year 2
Mean	54.83	56.50
Standard deviation	5.27	5.58
Standard error	0.96	1.02
Difference in means		1.67
Improved		23
Worse		5
Same		2
Mean change		+1.67
Standard deviation of the differences		1.97
Standard error of the differences		0.36

Note: The standard deviation of the differences (1.97) is much smaller than the standard deviation of the full set of marks for either year. This suggests that the small improvement overall (1.67) is consistent, applying to most of the students, rather than being due to a few students making huge gains.

significant ($t = -1.19$, $df = 58$, one-tail $p = 0.12$). This is because the difference between the mean values is small relative to the standard error values of the two samples (Table 8.10).

If the data are treated as paired samples, the *t*-test is not based on the standard errors of the two samples but on the standard error of the differences between the two marks for each student. In this case although the overall difference is small, this is not because a few students made huge leaps forward, it is because most of the students made a small improvement. This means that the standard deviation and standard error of the differences is small. Since the paired-samples *t*-test simply divides the mean difference by the standard error of the differences, in this case t is large (paired $t = 4.63$, $df = 29$, one-tail $p = 0.00007$).

In this case we would reject the null hypothesis and conclude that there is a small but consistent improvement between years 1 and 2. If the same data are analysed using the non-parametric Wilcoxon's matched-pairs signed-ranks test the result is also strongly significant ($T = 49$, $n = 29$, one-tail $p < 0.0001$) and given that 23 of the 29 changes are in the predicted direction even a sign test produces a strongly significant result (one-tail $p < 0.001$). For the paired *t*-test the effect size r is 0.65, suggesting a strong effect whereas Cohen's d is 0.31 suggesting a rather weak effect. Given that there is such a consistent change in one direction the strong effect seems sensible, but the magnitude of the improvement is very small, so to be honest I am not sure what to believe.

8.4 Two-Sample Tests for Independent Data with Individual Measurements

Comparing two samples that are independent, in the sense that the samples do not fall into logical sets of pairs, is a common problem in geography. Generally we want to know whether the values in the first sample are significantly higher or lower than those in the second sample. The most commonly used test in such situations is 'Students *t*-test', which

---

BOX 8.2   CALCULATING THE EFFECT SIZE *r* FOR A *T*-TEST

The effect size *r* is obtained from the *t*-statistic (*t*) and the degrees of freedom (*df*). The equation is

$$r = \sqrt{\frac{t^2}{t^2 + df}}$$

If you have run the *t*-test in Excel then you may have a probability value but not the *t*-statistic. This can be obtained using the T.INV.2 T function for a two-tail probability and the T.INV function for a one-tail probability. The number of degrees of freedom depends on the size of the two samples and the type of test you are applying. For a paired-samples *t*-test the degrees of freedom are the number of pairs minus one. For an independent samples *t*-test it is the size of both samples combined, minus two. So for two samples of 30 the degrees of freedom for a paired-samples test is 29 and for an independent *t*-test it is 58.

For example, given two samples of 30 with mean values of 13.04 and 12.24 and standard deviations of 1.25 and 1.81 an independent samples *t*-test, run using the T.TEST function in Excel, would return a two-tail probability of 0.0498. To obtain the *t*-value use the T.INV.2T function with 58 degrees of freedom to give $t = 2.0035$. Now entering the *t*-value and degrees of freedom into the equation to obtain the effect size *r* gives

$$r = \sqrt{\frac{t^2}{t^2 + df}} = \sqrt{\frac{2.0035^2}{2.0035^2 + 58}} = \sqrt{\frac{4.014}{4.014 + 58}} = \sqrt{\frac{4.014}{62.014}} = \sqrt{0.065} = 0.254$$

It is sometimes argued that the effect size *r* results in an overestimate of the effect size in the population when applied to the paired-samples *t*-test (Dunlap et al. 1996). Field (2013), for example, recommends the use of Cohen's *d* instead. The calculation for a paired test, where samples sizes are always equal, is not difficult

$$d = \frac{\text{Mean } A - \text{Mean } B}{\text{Standard deviation}}$$

The only complication is deciding which standard deviation to use. If one of the samples is a control group of some sort then it would make sense to use the standard deviation of that sample, otherwise just take the average of the two values. To be honest I do not really see the point of calculating Cohen's *d* in this context, because it does not include any information about the variability of the pairwise comparisons, but it is included here for completeness and the companion site calculator for paired samples returns both *r* and Cohen's *d*. The scale for Cohen's *d* is not the same as that for *r* (Cohen 1988, 1992), and the rules of thumb are $d = 0.2$ is a small effect, $d = 0.5$ is a medium effect and $d = 0.8$ is a large effect.

tests the null hypothesis that the two samples have been drawn either from the same population or from populations that have the same mean (average) value. It uses the difference between the mean values and the standard deviations of the two samples, as well as the sample size, to make the calculation. It is not a difficult test to compute, but it is parametric, so you need to have a reasonably large sample (>30) and the samples should be near normally distributed. The appropriate non-parametric test is given a variety of names, the most common being the Mann–Whitney *U*-test. It is a very elegant little test that relies on first placing all of the values in rank order, ignoring which group they came from, and then putting those ranks back into the two groups and calculating the two 'sums of ranks'. Because it is based on ranks, rather than the measured values, it is sensitive to differences in the median of the two samples rather than the position of the mean. For the types of data and typical sample sizes that geography students use in project work the *U*-test is very useful indeed. Before spreadsheets and online calculators were available it was very tedious to calculate because of the need to rank the data, but now those steps are trivial and it is not difficult to compute.

Both Student's *t*-test and the Mann–Whitney *U*-test are concerned with differences in location (where the middle lies) but it is sometimes interesting to check whether the variability of two samples is different. The parametric test for this is the *F*-test for equality of variance and a non-parametric alternative is the Siegel–Tukey test. A non-parametric test that is sensitive to any difference between two samples is the Kolmogorov–Smirnov two-sample test.

8.5 Mann–Whitney *U*-Test

I was introduced to this test as the Mann–Whitney *U*-test, and that is the name that has generally been used in other geography-orientated texts. However, whilst writing this book I have found the same test, sometimes performed using a different calculation procedure, variously called the Wilcoxon two-sample test, Mann–Whitney–Wilcoxon test, Wilcoxon rank-sum test, or Wilcoxon–Mann–Whitney test, so there is clearly some room for confusion! If you are using online calculators beware of confusing it with Wilcoxon's matched-pairs signed-ranks test, which only works with paired data. There are also some variants of the test that have been given different names but are effectively simplified versions that allow you to skip some of the steps and avoid mathematical calculations. I am always in favour of avoiding maths and equations wherever possible so I have incorporated one version into the description below as the option where you have small samples of equal size.

8.5.1 When It Is Useful

This test is used when you want to know whether the values in one sample are significantly higher or lower than those in another sample. If you can predict the direction of the difference in advance, on the basis of theory rather than by looking at the results, then use a one-tail test, otherwise, or if you are not sure, use a two-tail test. It is generally regarded as testing the null hypothesis that the medians of the two samples are the same, so it is used as a test for differences in location (where the middle lies). That is a slight oversimplification, because differences other than the position of the median can influence the results (Lehmann 1975, Fagerland and Sandvik 2009). It is a non-parametric test so there are no restrictions of sample size and the samples do not have to be normally distributed or the same size. It can be regarded as the non-parametric equivalent of the independent samples Student's *t*-test. When samples are small or not exactly normally distributed it is more powerful than the *t*-test and when samples are large it is almost as powerful, so I would recommend this as the default test when you want to compare two independent (unpaired) samples. If your result falls very close to a significance boundary, and your sample is large and near normally distributed, you can always perform both tests and present both results.

8.5.2 What It Is Based On

One of the reasons I like this test is that it is easy to demonstrate how it works in a lecture room without using any equations. Every year I use it to test the null hypothesis that there is no difference in the height of male and female geography students. It works by taking all of the values and placing them in rank order, ignoring which sample they come from. In the lecture theatre this means taking equal unbiased samples of male and female students and arranging them in order of height, irrespective of gender. Everyone is then given a number that represents their rank order of height, from shortest to tallest. Then the male and female students disengage and form two groups, and we simply add up the ranks for the males and females. Every year (so far) the sum of ranks for the females is less than the sum of ranks for the males because the null hypothesis is wrong. Male geography students, in Swansea at least, tend to be on average (in the case of this test 'on median' would be more correct), a bit taller than the females.

This is a simple case because I always use samples of equal size, and if there was no gender effect then you would clearly expect the sum of ranks to be about equal. If you use unequal sample sizes then just comparing two 'sum of rank' values does not make sense, and a few more mathematical steps are required, but the basic logic is the same and you end up with the same result, which is a probability value that tells you the chances that a difference at least as large as that observed could have occurred by chance if the two samples had been drawn from identical populations. The important assumption for this test is that the data are drawn from populations that are 'continuous', which in practice means that they do not fall into groups and can be placed in rank order. Where more than one individual has the same value (e.g. in the example used above they are exactly the same height) then the ranks have to be shared (Box 8.1).

To add a bit of variety to my statistics lectures I have tried this practical exercise using level of intelligence rather than height, but the students complained that it was difficult to sort themselves into rank order. I explored using less quantifiable metrics such as 'attractiveness' or 'weirdness' but my boss was not happy with that suggestion and made some regrettable comments about where I would lie on those scales.

8.5.3 Dealing with Tied Ranks

If ranks are tied within a single sample it has no effect on the calculation of the *U* statistic and so can be ignored. However, ranks that are tied between the two samples do have some influence on the results of *U* and they also influence the variability of the results and strictly speaking should be included in the calculation of the standard deviation part of the equation used to calculate the *z*-score and therefore the probability (Box 8.3). Most online calculators, and the Real Statistics Resource Pack make a correction for tied ranks when reporting the probability. In practice, however, the effect of tied ranks is very small indeed, even when the majority of the ranks are shared in some way. Also the effect of the correction is always to very slightly increase the value of z, which means that the difference becomes very slightly more significant. Ignoring the correction for ties just means that you are being very slightly more conservative in your testing, which means you are more likely to accept the null hypothesis. That is actually not a bad thing, and it means that if you have significant differences without making the correction they would certainly remain significant with the correction. If you are performing the Mann–Whitney U-test by hand, or in a spreadsheet, my advice is to ignore the correction for ties and accept that your test is slightly on the conservative side.

8.5.4 Effect Size

The most appropriate effect size for the Mann–Whitney U-test is r (as usual). Since the significance of the U-statistic is generally determined by first calculating a z-score, it is easy to convert this to an r-value by dividing it by the square root of the sample size, which in this case is the size of the two samples combined. If you are using an online calculator that gives a probability but no z-score you can obtain one by entering the one-tail probability into a spreadsheet and using the NORMSINV function to translate it into a z-score.

8.5.5 How to Do It

The most convenient way to perform this test is using an online calculator. In most cases this involves simply cutting and pasting the two data sets. Doing it by hand is tedious rather than difficult, because of the need to place the data in rank order, so a spreadsheet is much better. When the two samples are the same size the calculations are a bit easier. It is very easy to run this test using the Real Statistics Resource Pack. In SPSS or R Commander the data need to be arranged in the 'long' format, which seems odd if you are more used to using spreadsheets.

8.5.5.1 Online Calculators

A selection of online calculators is provided in Table 8.11. In most cases the data can be entered by cutting and pasting from a spreadsheet and the output is generally the probability value for your chosen one-tail or two-tail test. As always, be careful to ensure that you do not copy and paste any spaces that might be interpreted as zero values. Also be sure to include 2 or 3 decimal places when you copy and paste the data because many sites will not see decimal places that are hidden by the spreadsheet, resulting in spurious shared ranks. Some sites will only return a two-tail significance value, so if you want the one-tail probability (p-value), and the difference in the medians is in the predicted direction, just halve the two-tail p-value. If the difference in the medians is in the wrong direction then you accept the null hypothesis without applying the test. Few of the calculators will directly provide a measure of effect size.

8.5.5.2 Real Statistics Resource Pack

When you have downloaded the add-in following the clear instructions provided you will be able to access the resources using Ctrl-M (hold them both down) and from the list choose 'T-tests and non-parametric equivalents'. You now identify the two data sets that you wish to compare by first placing the cursor in the box marked 'input range 1',

TABLE 8.11

Selection of Free Online Calculators for the Mann–Whitney *U*-Test

Source	Website	Comments
Social Science Statistics	http://www.socscistatistics.com/tests/mannwhitney/Default2.aspx	Easy to use, gives detailed results including *z*-score. Maximum sample size is 200.
Free statistics and forecasting software (Holiday 2012)	http://www.wessa.net/rwasp_Reddy-Moores%20Wilcoxon%20Mann-Witney%20Test.wasp	Cut and paste both data sets together as two columns. Uses R-code so copes with huge data sets. Basic output. Easy to use.
StatPages.net	http://www.fon.hum.uva.nl/Service/Statistics/Wilcoxon_Test.html	Easy to use. Calls it 'Wilcoxon two-sample test', only gives two-tail *p*-value. Copes with huge data sets.
Saarland university	http://www.ccb.uni-saarland.de/?page_id=812	Seems to require manual input of data, no cut and paste option.
Vassar stats	http://vassarstats.net/utest.html	Rather awkward to use. Accepts large data sets then crashes.
SciStatCalc	http://scistatcalc.blogspot.co.uk/2013/10/mann-whitney-u-test-calculator.html	Data must be separated by commas, so cut and paste then manually replace gaps with commas. Plots simple scatter graph and histogram.

highlighting the part of the spreadsheet that contains those data and clicking the 'fill' button. Repeat for the second set. You have to decide whether there is a header included in your selection. The 'alpha' value, is the level of statistical significance that you are interested in, so the default value of 0.05 is usually appropriate. In the options choose 'two independent samples' and for the test choose 'non-parametric'. For the non-parametric test option tick both the 'use ties correction' and 'include exact test' options. The output is quite comprehensive as shown in Table 8.12.

TABLE 8.12

Output for the Mann–Whitney *U*-Test Conducted Using the Free 'Real Statistics Resource Pack for Excel'

Mann–Whitney Test for Two Independent Samples		
	Sample *A*	**Sample *B***
Count	23	20
Median	25.78	22.72
Rank sum	584	362
U	152	308
	one tail	two tail
Alpha	0.05	
U	152	
Mean	230	
Std dev	41.06905	ties
z-score	−1.89924	
Effect *r*	0.289631	
U-crit	161.9474	149.0061
p-value	0.028766	0.057533
Sig	Yes	No

Check the sample sizes are correct using the 'count' values. If you are one short you have probably ticked the 'column heads included with data' without picking up the header. If you are using a one-tail test check that the difference in medians is in the direction that you predicted. The important numbers to report are the test statistic U, the appropriate probability (p-value) and the effect size r. The 'mean' and 'std dev' values are those used to calculate the z-score, which is the basis for the effect size and p-value calculation, they are not the mean and standard deviation of either of the samples. U-crit is the critical value of the test statistic U at the chosen level of significance.

In the case of Table 8.12, if we had predicted in advance that the values in sample A would be higher than those in sample B we might report the results as: 'The median value of sample A (25.78, $n = 23$) was, as expected, higher than that of sample B (22.72, $n = 20$) and using a one-tail Mann–Whitney U-test the difference is statistically significant ($U = 152$, $p = 0.03$) but with a small effect size ($r = 0.29$)'. If we had not predicted the direction of difference in advance we would use a two-tail test and conclude that the difference is not statistically significant ($p > 0.05$).

8.5.5.3 In a Spreadsheet

If you are using a university or public computer, where you do not have administrative rights, you will not be able to download the Real Statistics Resource Pack, and if you do not have access to Excel it will not work, in which case you can still perform the Mann–Whitney U-test, it just takes a little more effort.

The first step is to change the raw measurements into ranks, ignoring which sample they come from. With a small sample this can be done by hand, as in the example here (Table 8.13). For large samples follow the instructions in Box 8.1. If there are any identical values then the ranks are shared. If there was no difference between these two samples we might expect the ranks to be approximately evenly distributed between the two samples. If there was no overlap at all then one sample would be given ranks 1–10 and the other ranks 11–20. Clearly the difference between the two sums of ranks gives us some indication of how different the two samples are. In a simple case like this, where the samples

TABLE 8.13

With Small Samples It Is Not Difficult to Assign the Ranks by Hand

Sample A	Sample B	Ranks A	Ranks B
10.69	12.89	1	2
19.22	13.32	8	3
19.31	16.40	9	4
19.68	17.44	11	5
19.73	17.81	12	6
20.16	18.81	13	7
21.27	19.60	16	10
23.22	20.57	18	14
24.15	21.02	19	15
24.39	22.55	20	17
Sum of ranks		127	83
Sample size		10	10

are small and of equal size it is convenient to use tables of critical values to determine whether the difference is statistically significant. Such tables are supplied in Section 12.8.1 for samples up to 20.

If your sample sizes are unequal, or larger than 20 you cannot use the tables of critical values based on the sum of ranks and will need to calculate the 'U-statistic'. The formula for this (Box 8.3) looks rather complicated, but don't panic, if you break it into several steps it is not too bad. The good news is that since U is a function of the two sample sizes and the sum of ranks, it can be calculated in a spreadsheet by just entering those values. The companion site includes a simple calculator that allows you to input your two sample sizes and the sums of the ranks so that you can obtain the U value without doing any calculations, or to check your results. That calculator will also give you both one and two-tail p-values and an effect size.

Once you have a U value you have two choices: you can check to see if it is lower than a critical value at your chosen level of significance (e.g. $p = 0.05$) using tables, or you can convert it into a probability value. Tables (Section 12.8) are supplied for samples of equal size in the range 4–50 and for unequal size up to 20. If you have larger samples of unequal size then the p-values are obtained by translating the U value into a z-score, because even with quite small sample sizes U is near normally distributed. If you want to have a go at making the calculation, or the evil professor is forcing you to do so, then follow the instructions in Box 8.3.

8.5.5.4 Companion Site Calculator

A simple calculator is provided that allows you to enter two independent samples of up to 100 and receive the results for the Mann–Whitney U-test and several others. Where the larger of the two samples is less than 20 it is still necessary to use tables of critical values, but for larger samples both one tail and two-tail probabilities are given together with the effect size r. Tied ranks are ignored in this calculator, so it may be slightly conservative if you have a lot of ties.

8.5.5.5 Using R Commander

It is not difficult to perform the Mann–Whitney U-test in R Commander as long as you input the data in the correct format, which is not the same format that you would typically use in a spreadsheet. When working in a spreadsheet I typically put each set of data in a separate column. If you enter data like that into R Commander and click on non-parametric tests the U-test appears in faint type and clicking on it has no effect. You have to enter all of the data in a single column and in the column next to it you identify which group each sample belongs to. The groups can be identified using a number, a letter or a name, it does not matter. Give each column a header such as 'groups' and 'values' with no gaps. It is easiest to do this in the top left corner of a clean spreadsheet and then save it as a tab-delimited text file. In R Commander use the 'data' then 'import data' buttons to import the data and then click the 'view data set' button to check it looks okay. Now you can go to 'statistics' and 'non-parametric tests'. The one you need is called the 'two-sample Wilcoxon test' (which is basically the same as the Mann–Whitney U-test but uses a different calculation procedure). Using the 'data' button you can choose which column header contains the groups and which contains the 'response variable', which is the values. The 'options' button allows you to choose a two-tail test, which is the default, or to define the direction of difference that you expect using a one-tail test. If the difference in the median values is the opposite of your prediction it will return a probability of 1. You can also choose to calculate the probability using the 'exact' method, the normal approximation or the normal approximation including a continuity correction. The easiest option is to choose 'default' and let R make the decision based on the sample size.

8.5.5.6 In SPSS

The data have to be entered in the 'long' format, which means put all of the data in a single column and use the column next to it to identify the two groups (e.g. using A and B). It is easiest to do this in a spreadsheet and then cut and paste without the headers. On the 'variable view' page give them names (such as 'data' and 'groups') and under 'measure' choose 'scale' for the data and 'nominal' for the groups. Go to 'analyze', 'non-parametric tests' and 'independent samples'. Under 'objective' choose 'customize analysis', and on the 'fields' page put the data in the 'test fields' box and the groups in the 'groups' box using the arrows. Under 'settings' choose 'customize tests' and check the option for 'Mann–Whitney U (two samples)' and press 'run'. The output summary box only gives the two-tail significance but double-clicking on it reveals a table with more information including the value for U and a 'standardized test statistic' which is the z-score. For a one-tail test you need to check that the medians are different in the predicted direction, which is easier in a spreadsheet but in SPSS the medians can be obtained using 'analyze', 'compare means', and choosing 'means' where one of the options is median. SPSS does not give an effect size but the effect size r can be obtained from the z-score (z divided by the square root of the sample size, which is n_1 plus n_2).

8.5.6 Examples

8.5.6.1 Example: Exam Performance and Gender

We wish to test whether there is a gender difference in examination performance (Table 8.14). Average test scores of 15 male and 15 female students are examined. The null hypothesis is that there is no difference (two tail).

Given small samples and equal sizes we do not have to calculate U, the significance can be obtained from the table in Section 12.8.1 using just the lower of the two sums of ranks. Since 232 is much higher than the tabulated value for $n = 15$ and two-tail probability of 0.05 (critical value is 211), we accept the null hypothesis and conclude that there is no gender difference. Note that in this case the sums of ranks for the two samples are almost identical. The larger the difference the more likely it is to be significant. Using the Real Statistics Resource Pack for Excel gives the two-tail probability of $p = 0.98$, which means there is a 98% chance that the two samples have been drawn from populations with the same median.

8.5.6.2 Example: Biochar

We wish to test claims that adding biochar to soil stimulates growth and therefore draws down more carbon dioxide. Ten tomato plants were grown in pots of compost with biochar added and another 10 were grown in compost without biochar. All 20 were grown under identical conditions and at the end of the season the weight (g) of tomatoes produced by each plant was recorded (Table 8.15). Since we predict in advance that the plants with biochar added should be more productive, a one-tail test is appropriate.

TABLE 8.14

Examination Scores from Male and Female Students

Male	Female		Male	Female
Scores	Scores		Ranks	Ranks
81	82		29	30
76	75		28	27
74	74		25.5	25.5
73	70		24	23
69	68		22	21
67	67		19.5	19.5
66	65		18	17
61	60		16	15
58	59		13	14
55	57		9.5	12
55	56		9.5	11
54	53		8	7
52	51		6	5
48	50		3	4
46	42		2	1
$n = 15$	$n = 15$	Sum	233	232

Note: In this case there is no sense in treating the data as pairs because every value belongs to a different person; they are independent. Sample size is too small for the Student's *t*-test but the Mann–Whitney *U*-test is appropriate

Given small samples like this, with results that are strongly skewed because a few plants clearly did not grow well at all, the Mann–Whitney *U*-test is suitable. Note, however that in this case the median weight for the plants treated with biochar is lower than the median weight of those that were not treated. Our hypothesis predicted the opposite, and since we have already decided that a one-tail test is appropriate we must now accept the null hypothesis without performing the test. We can conclude that there is no evidence that biochar increases productivity.

8.5.6.3 Example: Schmidt Hammer and Glacial Moraines

In the introduction I gave the example of using the Schmidt hammer to test whether two glacial moraines differ in age. It is a simple hand-held instrument used to measure the hardness of concrete, but it also works on boulders. Weathering tends to lead to a reduction in surface hardness so it provides a measure of degree of rock weathering and therefore the amount of time that the rock has been exposed to weathering, or exposure age. A typical sampling strategy is to take one reading from each of 30 boulders. Histograms showing the results from two moraines, A and B, are shown in Figure 1.2. They are clearly a bit skewed, rather than normally distributed, so the *U*-test is more appropriate than Student's *t*-test. Since we do not know in advance which moraine is likely to be older the null hypothesis is that there is no difference in age or degree of weathering, which is a two-tail hypothesis.

TABLE 8.15

Results of a Plant Growth Experiment

	With Biochar	Without Biochar
	984	1025
	754	946
	738	845
	693	723
	578	613
	534	524
	25	327
	15	117
	0	26
	0	10
Median	556	568.5

Note: The medians are not in the order that was predicted by the one-tail hypothesis so the null hypothesis is accepted and the test is not conducted.

Entering the data into the companion site calculator yields a U-value of 169.5 and a two-tail probability of $p < 0.001$ with a large effect size ($r = 0.54$). There is clearly a large and significant difference in surface hardness of the boulders on these two moraines, so if we are confident that it cannot be explained by some other factor, such as a difference in rock type, then we can interpret it as a substantial difference in exposure age.

BOX 8.3   CALCULATING THE MANN–WHITNEY U-VALUE AND A z-SCORE

The test statistic U is calculated twice, using the two sums of ranks (one for each sample) and the sample sizes. It is the smaller of the two U values that is used to test for significance

$$U1 = n_1 \times n_2 + \frac{n_1(n_1 + 1)}{2} - R_1$$

$$U2 = n_1 \times n_2 + \frac{n_2(n_2 + 1)}{2} - R_2$$

where
R_1 is the sum of ranks of the first sample, with sample size of n_1.
R_2 is the sum of ranks of the second sample with sample size of n_2.

Using the example in Table 8.13 where: $n_1 = 10$, $n_2 = 10$, $R_1 = 127$, $R_2 = 83$

$$U1 = 10 \times 10 + \frac{10(10 + 1)}{2} - 127 = 100 + \frac{10(11)}{2} - 127 = 100 + \frac{110}{2} - 127$$

$$U1 = 100 + 55 - 127 = 155 - 127 = 28$$

and

$$U2 = 10 \times 10 + \frac{10(10+1)}{2} - 83 = 100 + \frac{10(11)}{2} - 83 = 100 + \frac{110}{2} - 83$$

$$U2 = 100 + 55 - 83 = 155 - 83 = 72$$

In this case the U value that we need is 28.
A useful shortcut is available here because it is always true that

$$U1 + U2 = n_1 \times n_2$$

which means that having calculated one of the U values the other can be obtained because

$$U1 = (n_1 \times n_2) - U2$$

and

$$U2 = (n_1 \times n_2) - U1$$

So using the examples above

$$U1 = (10 \times 10) - 72 = 100 - 72 = 28$$

To obtain a z-score mathematically use

$$z = \frac{U - \frac{n_1 \times n_2}{2}}{\sqrt{\frac{(n_1) \times (n_2) \times (n_1 + n_2 + 1)}{12}}}$$

giving

$$z = \frac{28 - \frac{10 \times 10}{2}}{\sqrt{\frac{(10) \times (10) \times (10 + 10 + 1)}{12}}} = \frac{28 - \frac{100}{2}}{\sqrt{\frac{(10) \times (10) \times (21)}{12}}} = \frac{28 - 50}{\sqrt{\frac{100 \times 21}{12}}}$$

$$z = \frac{28 - 50}{\sqrt{\frac{100 \times 21}{12}}} = \frac{-22}{\sqrt{\frac{2100}{12}}} = \frac{-22}{\sqrt{175}} = \frac{-22}{13.23} = -1.66$$

To obtain a probability from z use tables of the normal distribution (Section 12.14) or enter the value into a spreadsheet and use the NORMSDIST function which translates a z-score into a one-tail probability. If you want the two-tail probability, double the one-tail probability. If doubling the one-tail probability gives a value higher than one, just report it as one. There are also plenty of online calculators that will translate z-scores into probability values (e.g. http://www.socscistatistics.com/pvalues/normaldistribution.aspx).

In this case the probabilities are

One-tail $p = 0.048$

Two-tail $p = 0.096$

To obtain the effect size r use

$$r = \frac{z}{\sqrt{n_1 + n_2}} = \frac{-1.66}{\sqrt{10 + 10}} = \frac{-1.66}{\sqrt{20}} = \frac{-1.66}{4.47} = -0.37$$

The sign does not matter so report it as $r = 0.37$

8.6 Student's t-Test for Two Independent Samples

8.6.1 When It Is Useful

This is a test for the difference in the means of two independent samples. It is very easy to perform, even using a calculator, and has become perhaps a bit too popular with geographers, who often ignore the assumptions. It is a parametric test, so samples have to be reasonably large (>30 is a good rule of thumb) and they need to be near normally distributed. See Section 3.3.1 on how to decide whether your data are near normally distributed. If in doubt use the Mann–Whitney U-test instead.

8.6.2 What It Is Based On

Student's t-test works by looking at the difference in the two mean values and the amount of overlap between the two distributions, using the standard error values and the two sample sizes (which do not need to be the same). Where there is a lot of variability in the samples (a wide spread of results), and sample sizes are small, then we might expect quite a large difference in the two mean (average) values to occur just by chance. With large samples with a narrow spread then even small differences in the position of the mean are unlikely to be due to chance. There are actually three variants of the t-test, so although it is not difficult to perform it is also quite easy to get completely wrong. The most common error is to confuse the paired-sample t-test, which is only used when data fall naturally into linked pairs (see Section 8.3) with one of the independent samples t-tests. There are two test variants for use on two independent samples: one which assumes equal variance and one that does not make that assumption.

8.6.3 Effect Size

As usual the preferred effect size is r and it is calculated directly from the t-value and sample sizes using the equation described in Box 8.2. There are some online calculators that will produce an effect size for a t-test result but most return Cohen's d rather than r. An exception, that is very easy to use, is provided by Dr. Lee Becker from the University of Colorado (http://www.uccs.edu/~lbecker/). The companion site calculator includes this calculation.

8.6.4 How to Do It

The standard t-test, assuming equal variance, can be conducted by hand as long as you can calculate the mean and standard deviations of the two samples, which is possible using modern scientific calculators. However, I doubt anyone actually does this in practice so I will not go into the equation here. If you are being tortured into performing the t-test by hand then follow the instructions in Box 8.4. The simplest way to conduct the independent samples t-test is using a spreadsheet.

8.6.4.1 *In a Spreadsheet*

Both Excel and Calc have 'T.TEST' functions that have the same syntax and return the probability according to your choice of test and tails. Making those choices is where it can all go horribly wrong, so make sure that you understand this test and the decisions you have to make before you use it.

When you open the 'T.TEST' function in Excel there are four boxes to complete. The first two are 'Array 1' and 'Array 2', which are your two samples. Now choose whether you want to conduct a one-tail or two-tail test. If you are predicting in advance the direction of difference in the mean values of the two samples then you can use a one-tail test, otherwise or if you are not sure use a two-tail. The final decision is which test to use, and there are three possibilities

1. Paired: This test is only used where your data are in pairs; it is not suitable for two independent samples. If you use this version by mistake your results will be complete nonsense.
2. Equal variance: This is the most common test variant. It assumes that your two samples have about the same variability, which you can check by looking at the standard error values or more formally by conducting an F-test (Section 8.8).
3. Unequal variance: This variant of the t-test, sometimes called Welch's test, is used when you expect the two samples to have a large difference in variance.

In practice choosing either of options 2 and 3 will not make much difference and many people always use the equal variance option without checking whether there is a difference in variance. In most cases that is actually quite a reasonable procedure because we are usually interested in testing whether two samples have been drawn from the same population. If there is a very big difference in the variance of the samples then it seems unlikely that they are from the same population, irrespective of the magnitude of the difference in the mean values. Also, although the normal t-test assumes equal variance the results are apparently not very sensitive to that assumption, and it is not disastrous if your samples have different standard errors as long as the sample sizes are not hugely different. If in doubt I think it is reasonable to use the equal variance option for the *t*-test. The important thing is to be clear about which test you have used.

If you have some reason to expect your samples have come from populations with different variance, or (more likely) you have been told to check for equality of variance you can do that using the *F*-test, also in a spreadsheet. The function 'F.TEST' just requires you to identify the two samples and it will return the probability that they are drawn from populations with the same variance. If your result is greater than 0.05 (more than a 5% chance) then you can use the equal variance *t*-test. If it is lower than 0.05 then you might choose to use the unequal variance version. You should really think carefully about why there are big differences in variance however, because that alone suggests the two samples are not drawn from the same population.

When you perform the *t*-test in a spreadsheet just be aware that it is your responsibility to check the assumptions, the spreadsheet will not do that for you. Excel will perform the *t*-test with just two or three values, or on two data sets that are very strongly skewed or have extreme outliers without issuing any warnings. The results will, however, be complete nonsense. Do not use this test for small samples (<30 rule of thumb) or where the data are clearly not normally distributed (see Section 3.3.1). If in doubt use the Mann–Whitney *U*-test.

8.6.4.2 Online Calculators

There are plenty of online calculators for Student's *t*-test, though it is probably easier to perform it in a spreadsheet than to cut and paste the data out of the spreadsheet and into the calculator. If you are determined to use one a selection is provided in Table 8.16.

8.6.4.3 Companion Site Calculator

The companion site calculator for independent samples will return the results of both the equal variance and unequal variance versions of the *t*-test together with those of the

TABLE 8.16

Selection of Free Online Calculators for Student's *t*-Test

Provider	Website	Comments
Graphpad Quickcalcs	http://graphpad.com/quickcalcs/ttest1/?Format=C	Use second option to paste data from a spreadsheet. All three test variants. Easy.
Social Science Statistics	http://www.socscistatistics.com/tests/studentttest/	Choose tails and significance level. Gives *t*-value and *p*-value. Easy to use.
Kirkman, T.W. (1996)	http://www.physics.csbsju.edu/stats/t-test_bulk_form.html	Gives *t* and *p*-values but also lots of other descriptive stats. Will draw a box plot for you.
Vassarstats	http://vassarstats.net/tu.html	Choose independent samples. Gives results for equal and unequal variance and for the *F*-test for equal variance. Good site.
MathPortal	http://www.mathportal.org/calculators/statistics-calculator/t-test-calculator.php	Easy to use. Gives *t* and *p*-values but also shows the equations used in the calculations
Useablestats	https://www.usablestats.com/calcs/2samplet	Gives results for equal and unequal variance tests.

Mann–Whitney U-test, F-test for equality of variance and the Kolmogorov–Smirnov two-sample test for continuous data. Probability values are returned if the smaller sample is at least 15, even though that is really a bit too small for a t-test.

8.6.4.4 In R Commander

It is hardly worth the bother of importing the data into R Commander to conduct a t-test, but if your evil professor is forcing you to use R then it is not too difficult. The important thing is to put the data in the right format, which is not a very familiar format if you usually work with spreadsheets. Rather than putting each sample in a separate column, you have to put them in the same column and then identify which group they belong to in a second column. You can identify them using a number, letter or text. Give each column a header, such as 'data' and 'group'. It is easiest to do this in a clean spreadsheet then save it as a text file (tab delimited, .txt format). Import the data into R Commander, check it using 'view data' then choose 'statistics' and 'means'. If your data are in the right format you should see the option 'independent samples t-test'. On the 'data' tab you choose which column has the data (response variable) and which identifies the groups. On the 'options' tab choose a one-tail or two-tail test and choose either the equal variance or unequal variance test. The output includes the two mean values, the test statistic t, degrees of freedom based on the sample size and the p-value. If you are not sure whether to use the equal or unequal variance options it is easy to run an F-test for equality of variance on the same data. In the 'statistics' menu choose 'variances' and 'two variances F-test'. If the result is greater than 0.05 use the equal variance option and if it is lower you might choose to use the unequal variance option (the output calls this the 'Welch two-sample t-test').

8.6.4.5 In SPSS

In SPSS it is not difficult to run Student's t-test but the output is quite confusing. To run the test you have to input your data in the 'long' format which means that you cannot put your two samples in adjacent columns, they have to be in a single column and in the next column you identify which group they belong to using names, letters or numbers. It is easiest to arrange your data in a spreadsheet then copy and paste it over to SPSS, (without headers). Now go to the 'variable view' tab at the bottom and in the first column give your data a name (I use 'data') and your groups a name (use groups). Now go to the 'analyze' button and choose 'compare means' where you will find several options including 'independent samples T test'. Be careful to choose this one. Now you are prompted to place one set of data in the 'test variables' box, that is where your data set goes, and then put groups into the 'grouping variable' box. Now click the 'define groups' button and tell it which group is which, so if you used A and B in the groups column enter these letters. Now click 'ok'.

As usual SPSS gives you much more output than you really need or want and this can be a real source of confusion. The first box marked 'group statistics' gives you the count for each group, the mean values, standard deviations and standard errors. If you are using a one-tail hypothesis, where you predict the direction of the difference, then use these results to check that your mean values fall in the correct order (i.e. when you predicted that the mean of one sample would be higher, the prediction is correct). The main box starts with the results of a test for equality of variance called Levene's test. The idea is that this guides your choice of which kind of *t*-test to use (see above). Next you have the results for two *t*-tests, the top row is the one that assumes equal variance and the lower row is the one that does not assume equal variance (it is what R calls the Welch two-sample test). If the *p*-value for Levene's test is higher than 0.05 then you should use the *t*-test that assumes equality of variance. If it is less than 0.05 you might choose to use the one that does not assume equal variance. They are normally very similar. The numbers you need are the *t*-value and the probability which is in the column marked 'sig.(two-tail)'. If you are using a one-tail hypothesis you divide this probability value in half to get the one-tail value.

8.6.5 Examples

8.6.5.1 Example: Male Underperformance

Research has indicated that there is a tendency for male students to underperform in examinations relative to their female colleagues. We wish to test this by comparing unbiased samples of male and female examination results. Sample sizes are reasonably large (Table 8.17) and near normally distributed so an independent samples *t*-test is appropriate. A one-tail test is called for because we are predicting in advance the direction of the difference in the mean values. Our prediction is based on theory, it is not based on looking at the results. The null hypothesis is that male students do not underperform.

Since the mean male score is, as predicted, lower than the female score we do not immediately accept the null hypothesis and continue with the test. Using the independent samples calculator on the companion site yields the output presented in Table 8.17. The one-tail probability is greater than 0.05 so we would accept the null hypothesis.

Note that we do not conclude that the male students have not underperformed; that is not true because their marks are lower, it is just that given samples of this size, and with this amount of variability, there is more than a 5% chance that such a large difference could occur just by luck. In a study like this it is useful to quote an effect size as well, even though the result is not statistically significant. Others may wish to combine the results of this small study with the results of many other studies (meta-analysis), providing a much larger overall sample size and therefore more power to recognise a small but real effect in the population.

TABLE 8.17

Example of the Output for Student's *t*-Test Generated by the Companion Site Calculator for Independent Samples

Student's *t*-Test Independent Samples		
	Male	**Female**
Count *n*	33	45
Mean	56.88	61.96
Standard deviation	12.78	14.01
Student's *t*	1.64	
Degrees of freedom	76	
Two-tail *p* =	0.105	
One-tail *p* =	0.053	
Effect size *r*	0.18	

BOX 8.4   PERFORMING AN INDEPENDENT STUDENT'S *t*-TEST BY HAND

Performing the *t*-test by hand is terribly dull rather than terribly difficult. You need to know six numbers (Table 8.18) and the equation you use depends on whether your sample sizes are equal or not:

TABLE 8.18

The Six Numbers Required to Calculate Student's *t*-Test

Name	Description	Example 1	Example 2
Mean of *A*	Average of sample *A*	35.4	35.4
Mean of *B*	Average of sample *B*	29.0	29.0
N of *A*	Size of sample one	30	20
N of *B*	Size of sample two	30	40
stdev*A*	Standard deviation of sample *A*	12.57	12.57
stdev*B*	Standard deviation of sample *B*	11.70	11.70

Note: The calculation procedure depends on whether the two sample sizes are the same (example 1) or different (example 2).

For equal sample sizes the full equation is

$$t = (\text{Mean of } A - \text{Mean of } B) \div \sqrt{\frac{(\text{Stdev}A)^2}{N \text{ of } A} + \frac{(\text{Stdev}B)^2}{N \text{ of } B}}$$

So using the numbers in the first example above and taking it in small steps

$$t = (35.4 - 29.0) \div \sqrt{\frac{(12.57)^2}{30} + \frac{(11.70)^2}{30}}$$

$$t = 6.4 \div \sqrt{\frac{158}{30} + \frac{136.89}{30}} = 6.4 \div \sqrt{5.27 + 4.56} = 6.4 \div \sqrt{9.83}$$

$$t = 6.4 \div 3.13 = 2.04$$

For unequal sample sizes the two standard deviations are replaced by a 'pooled' version (here I have called it stdev*P*) that is effectively a weighted average of the two standard deviations with the relevant weight being determined by the sample size.

$$\text{Stdev}P = \frac{(N \text{ of } A - 1)\text{stdev}A + (N \text{ of } B - 1)\text{stdev}B}{N \text{ of } A + N \text{ of } B - 2}$$

Using sample sizes from example 2 above gives:

$$\text{Stdev}P = \frac{(20 - 1)12.57 + (40 - 1)11.70}{20 + 40 - 2}$$

$$\text{Stdev}P = \frac{(19 \times 12.57) + (39 \times 11.70)}{58}$$

$$\text{Stdev}P = \frac{(238.83) + (456.3)}{58}$$

$$\text{Stdev}P = \frac{695.13}{58} = 11.985$$

Note that the new weighted standard deviation lies between the two original values but is not quite in the middle (that would be 12.1), it is weighted a little bit towards the lower value from the larger sample.

Now you replace the two standard deviation values in the *t*-test equation to give:

$$t = (\text{Mean of } A - \text{Mean of } B) \div \sqrt{\frac{(\text{Stdev}P)^2}{N \text{ of } A} + \frac{(\text{Stdev}P)^2}{N \text{ of } B}}$$

$$t - (35.4 - 29.0) \div \sqrt{\frac{(11.985)^2}{20} + \frac{(11.985)^2}{40}} = 6.4 \div \sqrt{\frac{143.64}{20} + \frac{143.64}{40}}$$

$$t = 6.4 \div \sqrt{7.182 + 3.591} = 6.4 \div \sqrt{10.77} = 6.4 \div 3.282 = 1.95$$

Having calculated the *t*-statistic you can use tables (Section 12.7) to check against critical values for a chosen level of significance for a one-tail or two-tail test. Assuming the two examples above are testing a two-tail hypothesis, where the null hypothesis is that the two samples are drawn from the same population or from populations with the same mean, then the critical value at $p = 0.05$ and for 58 degrees of freedom (total sample size minus 2) is 2.00, so the result for the first example (2.04) just passes the critical threshold whereas that from the second example does not. In the first case we would reject the null hypothesis and in the second case accept it.

An alternative approach it to use your *t*-value and sample sizes to obtain a probability using either a spreadsheet or online calculator (in which case why are you doing it by hand?!). In Excel the function 'T.DIST.2T' returns the two-tail probability when you enter the *t*-value (called X for some reason) and the degrees of freedom (total sample size minus 2). In this case the two results are

Two-tail *p*-value = T.DIST.2 T(2.04) = 0.046

Two-tail *p*-value = T.DIST.2 T(1.95) = 0.056

The one-tail probabilities are half the two-tail values, so 0.023 and 0.028 for examples 1 and 2. There are also plenty of online calculators that will translate a *t*-value into a probability (Table 8.19). Alternatively, calculate a *z*-score, using the instructions in Box 8.2, which allows you to check the significance using tables of the normal distribution (Section 12.14) and to calculate the effect size *r*.

TABLE 8.19

Selection of Free Websites that Will Translate a *t*-Value into a Probability

Provider	Website
Daniel Soper	http://www.danielsoper.com/statcalc/calculator.aspx?id=8
Easycalculation.com	https://www.easycalculation.com/statistics/p-value-t-test.php
Graphpad Quickcalcs	http://graphpad.com/quickcalcs/pValue1/
Vassar stats	http://vassarstats.net/tabs.html#t
SocSciStatistics.com	http://www.socscistatistics.com/pvalues/tdistribution.aspx
SurfStat Australia	http://surfstat.anu.edu.au/surfstat-home/tables/t.php

8.7 Two Independent Samples: Tests for Difference in Variability

When comparing two independent samples it is often interesting to ask whether they differ with respect to variability as well as the position of the middle. Tests that do this are sometimes called 'tests for homogeneity of variance'. One of the assumptions of most parametric tests is that the samples that are being compared have equal variance, and the most common use of tests for homogeneity of variance is to test that assumption. If you have two reasonably large samples that are reasonably close to being normally distributed then the simplest option is the '*F*-test'. This is a parametric test, and should not really be used on small or very skewed samples, although it often is. There are several non-parametric alternatives that are discussed below.

8.8 The *F*-Test for Equality of Variance

The '*F*' refers, I think, to Ronald Fisher and the test is sometimes called 'Fisher's *F*' or the 'variance ratio test'. It is a very easy test to apply because it is just the ratio of the two variances.

8.8.1 When It Is Useful

The variance of a sample is the average squared difference between each value and the mean, so it is a measure of variability. We tend not to use variance very often because the squaring step makes the scale a bit confusing. If you have a sample of measurements in mm, for example, the units of variance are actually mm^2. If we take the square root of the variance then it reverts to the original units of measurement, which is much easier to interpret. The square root of the variance is the standard deviation. If you have standard deviation values you can translate them into variance by squaring them.

8.8.2 What It Is Based On

If two samples have been drawn from the same population, or from populations with identical variance, then we would expect their variances to be very similar. With huge samples they should be nearly identical, but as always when samples are smaller there could be a difference in variance just due to chance. The F-ratio is the larger variance divided by the smaller variance. Clearly if they are identical then the result is one, and the further it gets from one, the bigger the difference between them. If the samples are near normally distributed we can estimate the probability of a given ratio value occurring just by chance, which is what this test does.

8.8.3 How to Do It

The equation for the F-ratio is

$$F = \frac{\text{Larger variance}}{\text{Smaller variance}}$$

In the example used to calculate the independent samples t-test in Box 8.4, for example, there were two samples of 30 and the standard deviation values were 12.57 and 11.70. Squaring those values gives you two variances (158 and 137) so the F-ratio is

$$F = \frac{158}{137} = 1.15$$

To obtain the probability you can use tables of the probabilities of the 'F-distribution'. These are widely available, but not particularly user friendly. Since in geography we often use samples of the same size I have compiled a simpler version (Section 12.9). For the example above we can see that given two samples of 30 the critical value for a two-tail test at $p = 0.05$ is 2.10, which is much higher than our ratio. We can accept the null hypothesis that the two samples are drawn from populations with equal variance. Looking at the table in Section 12.9 it is clear that a variance ratio of 1.15 is actually very small indeed, and the two samples would have to contain more than 700 values each before a ratio that small would be considered as statistically significant.

8.8.3.1 *In a Spreadsheet or Companion Site Calculator*

You can also run the F-test for equality of variance in a spreadsheet. In Excel use the function 'F.TEST' and simply identify the two data sets and you will obtain the two-tail probability. The companion site calculator for independent samples also returns the two-tail probability for an F-test.

8.8.3.2 *In R Commander*

Under the menu tabs for 'statistics' and 'variance' you will find 'two-variances F-test'. To run this you need to have your data arranged in a single column with another column indicating which group each value belongs to. *R* Commander will recognise which is which and in the options you can choose a two-tail or directional test (where you predict in advance which variance will be higher). Output includes the F-ratio, the degrees of freedom based on the two sample sizes ($df = n - 1$), and the p-value that you need. If p is more than 0.05 you can accept the null hypothesis and conclude that any difference in variance between your two samples is not statistically significant. You are also given 95% confidence limits for the F-ratio.

8.8.3.3 In SPSS

When you run a Student's t-test in SPSS the output includes the results of another test for equality of variance called Levene's test. It is more complicated to calculate and with big samples it can suggest that there is a significant difference in variance even where the difference is rather small. I prefer the F-ratio, which is easier to interpret. There is no point in using SPSS to calculate the F-ratio because it is just one number divided by another.

8.8.3.4 Online Calculators

There are plenty of free online calculators that will perform the F-test for equality of variance (Table 8.20).

8.8.4 Examples

8.8.4.1 Example: Organic Strawberries

You are interested in the difference between organically and conventionally grown fruit and vegetables and have taken unbiased samples of ripe strawberries grown using both methods. You measure the size of each strawberry in mm and wish to compare the two samples (Table 8.21). You are interested in differences in average size but also in the variability of sizes because that affects the market price.

The F-ratio is $42.25 \div 23.04 = 1.83$

Using the table for equal sample sizes of 50 (Section 12.9) we can see that the critical values for a two-tail test at $p = 0.05$ and 0.01 are 1.76 and 2.11. Since our measured ratio falls between these two values we can conclude that the difference in variance is significant at $p < 0.05$ but not at $p < 0.01$. To obtain a more precise probability we can enter the values

TABLE 8.20

Selection of Free Online Calculators for the F-Test for Equality of Variance

Provider	Website	Comments
Daniel Soper	http://www.danielsoper.com/statcalc3/calc.aspx?id=7	Enter the larger of the two possible ratios (larger variance divided by smaller variance).
Stat Trek	http://stattrek.com/online-calculator/f-distribution.aspx	Beware – you must enter the *smaller* of the two possible ratios (smaller variance divided by larger variance).
GraphPad.com	http://graphpad.com/quickcalcs/pValue1/	Enter the larger of the two possible ratios (larger variance divided by smaller variance).
Vassarstats	http://vassarstats.net/tabs.html#f	Enter the larger of the two possible ratios (larger variance divided by smaller variance).
EasyCalculation.com	https://www.easycalculation.com/statistics/f-test-p-value.php	Enter the larger of the two possible ratios (larger variance divided by smaller variance).
Social science statistics	http://www.socscistatistics.com/pvalues/fdistribution.aspx	Enter the larger of the two possible ratios (larger variance divided by smaller variance).

TABLE 8.21

Summary Statistics Describing the Size of Strawberries Grown Using Conventional and Organic Methods

	Sample Size	Mean Size (mm)	Standard Deviation	Variance
Organic	50	19	6.5	42.25
Conventional	50	23	4.8	23.04

into an online calculator or use the F.DIST.RT function in Excel to find that the probability is 0.018. We can conclude that there is less than a 2% chance that the strawberries grown organically and conventionally are equally variable in terms of size.

8.9 Non-Parametric Tests for Equality of Variance

If you are dealing with very small samples, or have samples that are clearly very far from being normally distributed, by virtue of being very skewed or including extreme values, for example, there are some non-parametric tests for equality of variance available, though they are rarely used in geography. The Siegel–Tukey test has the advantage that the results are comparable to the Mann–Whitney U-test and are also near normally distributed, which means that you do not need special tables of probabilities. It is the easiest option if you are using small samples and need to make the calculations by hand or in a spreadsheet. I have not found any online calculators for this or similar tests. If you use R Commander you can use the similar but perhaps slightly better 'Ansari–Bradley' test.

8.10 Siegel–Tukey Test

8.10.1 When It Is Useful

This test tests the null hypothesis that your samples have come from populations that have the same variance. It is a non-parametric test so makes no assumptions about the underlying distribution and it can be applied to small samples.

8.10.2 What It Is Based On

To perform this test you have to assign scores to the data in a very odd way! First you place all of the values in rank order, ignoring which group they come from. The lowest value is given a score of 1, but scores 2 and 3 go to the two highest values, then 4 and 5 go at the bottom, 6 and 7 at the top and so on until you get to the middle. Where two or more values are identical, use shared scores, in exactly the same way as for the Mann–Whitney U-test. The effect is that values towards the extremes receive low scores and values near the middle receive high scores. If there is no difference in terms of the spread of values between your two samples, and they are really drawn randomly from the same population, then they will be randomly arranged when placed in rank order and will have about

TABLE 8.22

Example of the Strange Scoring Scheme Used by the Siegel–Tukey Test

Group	A	A	B	A	B	B	A	B	B	B	A	A	B	A
Value	2.1	2.3	2.8	4.3	7.4	7.7	8.5	12.4	23.0	28.3	35.0	37.5	45.2	71.8
Score	1	4	5	8	9	12	13	14	11	10	7	6	3	2

Note: The result is that values towards the extremes receive low scores and those near the middle high scores.

the same average scores. If one of the samples has a much lower spread than the other then it will tend to occur more often towards the middle and less often at the extremes and so will tend to have higher scores.

8.10.3 How to Do It

Look at the example in Table 8.22 to see how the strange scoring scheme works in practice. The beauty of this scoring scheme is that the sums of scores for each group have the same properties as the sum of ranks used in the Mann–Whitney U-test, and so the same tables of critical values can be used (Section 12.8). Having obtained the sum of scores for each sample these can be treated as if they were the sum of ranks and the U-value can be calculated as for the Mann–Whitney U-test (Box 8.3). Since the distribution of the Mann–Whitney U-statistic quickly becomes near normally distributed it is possible to covert U into a z-score to obtain both a probability and an effect size.

The example shown in Table 8.22 would result in sum of scores values of 41 and 64, giving U-values of 36 and 13. The lower of these values is compared with tables of critical values (Section 12.8). The critical value for a two-tail test at $p = 0.05$ is 8 and since to be significant the measured value of U must be 'equal to or smaller than the tabulated value' we can accept the null hypothesis and conclude that the difference in spread of the two samples is not statistically significant.

If you have some good reason to predict the direction of the difference in variability in advance then you can use a one-tail test, but be careful to check that the difference is really in the predicted direction. If your sample sizes are equal it is easy because the more variable sample will have the lower sum of scores. If your samples are uneven in size then you will need to calculate the two U values, in which case the more variable sample will give the larger value of U. As usual, if the result is the opposite to your prediction then you accept the null hypothesis and do not carry on with the test.

8.10.3.1 Online Calculators and Spreadsheets

I have not been able to find any online calculators that will directly perform the Siegel–Tukey test, however there is a cunning trick that you can use and that is to perform the scoring in a spreadsheet and then treat the scores as if they are the raw data for a Mann–Whitney U-test. Because you have already allocated scores to each group using a ranking procedure, the software will allocate exactly the same ranks to the two groups and so the output will refer to the difference in variability, rather than the difference in medians. The same logic applies when you use spreadsheet add-ins. The Real Statistics Resource Pack for excel, for example, includes the Mann–Whitney U-test and if we were to apply it to the raw values of the example above we would obtain sum of rank values of 51 and 54, giving U-values of 26 and 23. However, if we 'trick' the software by entering the scores instead of the raw data then we obtain sum of rank values of 41 and 64 and U values of 36 and 13, which is exactly what we found when calculating by hand. If your sample sizes are not terribly small (>10) then having obtained a probability value using one of these methods you can obtain an effect size r using the usual procedure of converting the one-tail probability into a z-score using the NORMSINV function and then dividing that value by the square root of the total sample size.

8.10.3.2 In R Commander

The simplest option here is to perform the scoring by hand and then enter the scores (not the raw values) into R Commander in the usual way and perform the Mann–Whitney U-test. However, if you have large samples performing the scoring by hand is very tedious. Alternatively you can load a plug-in to R Commander called Coin (Leucuta 2011). You have to use 'packages', 'load packages', choose a 'Cran mirror' somewhere near where you live and then find RcmdrPlugin.coin. When you open R Commander, use the 'tools' menu to load the plugin. A new button will appear called 'coin' that includes 'independent location tests' and in that list you will find the 'Ansari–Bradley test'. It runs in much the same way as the Mann–Whitney U-test using data arranged in a single column with a group identifier in the next column. The pop-up box which allows you to choose the data and a one-tail or two-tail test is a bit confusing because it includes an option called 'block' that needs to be unchecked (hold ctrl and click it) or it will not run. This test is not identical to the Siegel–Tukey test because it assigns scores in a different and probably superior way (more symmetrical), with the penalty that they are no longer equivalent to the ranks used in a U-test. However, since R is doing the calculations it does not matter and you obtain a probability value and a z-score.

8.10.3.3 In SPSS

I cannot find any way to run the Siegel–Tukey or Ansari–Bradley tests automatically in SPSS. If they are there, they are well hidden. The only option is to 'trick' the software by allocating the scores in a spreadsheet and then entering those instead of the raw values. Now if you run the Mann–Whitney U-test the results will actually refer to the Siegel–Tukey test.

8.10.4 Examples

8.10.4.1 Example: Extremity of Opinion

You are interested in attitudes towards the building of a wind farm and decide to compare the attitudes of people who live in a village close to the proposed site (A) and a similar sample from a village that is much further away (B). You ask the question: 'Are you in favour of wind-farm developments?' Responses are recorded by marking a cross on a bar marked 'strongly in favour' at one end and 'strongly against' at the other. By measuring the distance along the bar in mm you obtain individual measurements of level of support. The results are presented in rank order in Table 8.23 and scored according to the procedure described above.

   If the measured values are entered into the companion site calculator for independent samples it is clear that there is little difference in location between these two samples (means 45.2 and 49.6) so it is not surprising that the Mann–Whitney U-test suggests that they are not significantly different ($U = 100$, $p > 0.05$). However, entering the scores, rather than the values or ranks, into the companion site calculator, returns the values $U = 54.5$,

TABLE 8.23

Values Arranged in Rank Order, Ignoring the Two Groups, and Then Scored for the Siegel–Tukey Test

Value	3	7	9	15	16	18	20	23	26	28	30	36	38	40	44
Rank order	1	2	3	4	5	6	7	8	9	10	11	12	13	14	15
Group	A	B	A	A	A	B	A	A	A	A	B	B	B	B	B
Score	1	4	5	8	9	12	13	16	17	20	20	24	25	28	29
Value	49	51	56	59	63	65	69	74	75	77	80	82	85	90	95
Rank order	16	17	18	19	20	21	22	23	24	25	26	27	28	29	30
Group	B	B	B	A	A	B	B	A	B	B	A	A	A	B	A
Score	30	27	26	23	22	19	18	15	14	11	10	7	6	3	2

Note: Group A has more values towards the two extremes and group B has more near the middle.

and this now represents the value for the Siegel–Tukey test. The tables for the Mann–Whitney U-test can be used (Section 12.8) and for two samples of 15 the critical value for a two-tail test at $p = 0.05$ is 64. Since 54.5 is 'less than or equal to' the critical value we can conclude that the difference in variability is significant.

8.10.5 'Measuring from the Middle' Approach

An alternative approach that avoids the strange scoring procedure and is much easier with large samples is to measure not the distance along the line but the distance from the middle. The values now become a measure of strength of opinion rather than level of agreement, with zero being awarded to values in the middle of the line and 50 to those at either extreme. Of course this can be achieved mathematically by taking the absolute difference between each of the original measurements and the value for the middle of the line (50 in this case). Applying this methodology to the example used above, and using the Mann–Whitney U-test again, returns a U-value of 54, which is almost identical to the result obtained from the Siegel–Tukey test. If the data are large enough then the t-test could also be used (in this case $t = 2.7$, $df = 28$, two-tail $p = 0.012$). I have never actually seen this method applied, but given a fixed scale where the middle is clearly defined in advance, I see no reason why it is not valid.

8.11 Kolmogorov–Smirnov Two-Sample Test for Continuous Data

This test has already been described in Chapter 7 where it provides a useful way to compare two samples collected as counts in numbered categories. It is an unusual test in that it is sensitive to any difference between two samples. It is based on the maximum vertical difference between two cumulative frequency distributions (expressed as cumulative proportions).

8.11.1 When It Is Useful

The cost of the generality of the test is a loss of power, so if you are really interested in a difference in location (the middle) use either the Mann–Whitney U-test or Student's t-test

TABLE 8.24

Examination Marks Arranged in Rank Order Ignoring the Groups from Which They Are Taken

Mark	92	85	84	77	76	68	64	64	60	58	56
Rank	1	2	3	4	5	6	7	7	9	10	11
Group	A	A	A	B	A	B	B	A	B	B	B
Mark	53	52	50	48	47	45	45	35	15	12	
Rank	12	13	14	15	16	17	17	19	20	21	
Group	B	B	B	B	B	B	B	B	B	B	

Note: Tied ranks are not given the average as for other non-parametric tests.

and if you want to test specifically for a difference in spread, use either the *F*-test or the Siegel–Tukey test. The Kolmogorov–Smirnov test is most useful for comparing very small samples, and in this case it may be more powerful than the *U*-test even in recognising a difference in location.

8.11.2 How to Do It

One way to use the test with individual measurements is to cast them into classes, but this involves a loss of information and the fewer categories used, the more information is lost. A simple solution is to place the values in rank order and use every value that occurs as a single class. In fact there is no need to use the values themselves, because Neave and Worthington (1988) explain that using the ranks produces exactly the same results. All we really need to know to apply the test is the rank order of the values, ignoring the group from which they were obtained. It is sometimes argued that the test can only be used for two-tail hypotheses, since the direction of 'any difference' cannot be predicted, but Siegel (1956), for example, demonstrates that it can be used to test a carefully designed one-tail hypothesis. The procedure for one-tail testing is described in Chapter 7 (Section 7.5.3).

As an example (Table 8.24), imagine that you wish to test whether the examination results of a small group of exchange students (group A) are significantly different from those of the other students (group B). These data are not suitable for analysis using parametric methods because of the small and unequal sample sizes (5 and 16). The data could be re-arranged into 5% or 10% classes, but that would ignore some of the information. A better solution is to allocate a class to every number, or every rank, since it makes no difference. Note that in this case the ranking is not performed using the usual averaging procedure, instead all cases with the same number are given the same rank and a suitable number of ranks are then ignored. In this case there are six values higher than 64 then the value 64 occurs twice, so they are both given a rank of 7 and the next case is ranked 9. This is the procedure used by the 'RANK' function in Excel.

Using average ranks would not change the result of the test but the advantage of the simple ranking procedure is that it makes it easy to produce a pair of cumulative frequency distributions and the horizontal scale is simply the ranks. In the case of the example in Table 8.24 the result is effectively a tally sheet listing the number of occurrences for each rank (Table 8.25). This is easily converted into cumulative counts and then into cumulative proportions (by dividing each value by the total count for that column). The largest difference between the cumulative proportions is the test statistic *D* (0.81). Tables of critical values (Section 12.5) for a two-tail test using sample sizes of 5 and 16 show that to be significant at $p = 0.05$ and $p = 0.01$ the value of *D* must be equal to or larger than 0.70 and

TABLE 8.25

Example of Data Arranged in a Spreadsheet in Order to Perform the Kolmogorov–Smirnov Two-Sample Test for Continuous Data

	Tally Sheet		Cumulative Count		Cumulative Proportions		Difference
Ranks	A	B	A	B	A	B	
1	1	0	1	0	0.20	0.00	0.20
2	1	0	2	0	0.40	0.00	0.40
3	1	0	3	0	0.60	0.00	0.60
4	0	1	3	1	0.60	0.06	0.54
5	1	0	4	1	0.80	0.06	0.74
6	0	1	4	2	0.80	0.13	0.68
7	1	1	5	3	1.00	0.19	0.81
8	0	0	5	3	1.00	0.19	0.81
9	0	1	5	4	1.00	0.25	0.75
10	0	1	5	5	1.00	0.31	0.69
11	0	1	5	6	1.00	0.38	0.63
12	0	1	5	7	1.00	0.44	0.56
13	0	1	5	8	1.00	0.50	0.50
14	0	1	5	9	1.00	0.56	0.44
15	0	1	5	10	1.00	0.63	0.38
16	0	1	5	11	1.00	0.69	0.31
17	0	2	5	13	1.00	0.81	0.19
18	0	0	5	13	1.00	0.81	0.19
19	0	1	5	14	1.00	0.88	0.13
20	0	1	5	15	1.00	0.94	0.06
21	0	1	5	16	1.00	1.00	0.00

0.83, respectively, so in this case the difference is statistically significant at $p < 0.05$. The test statistic D serves as an effect size.

8.11.2.1 Companion Site Calculator

The independent samples calculator includes the results of the Kolmogorov–Smirnov two-sample test for continuous data, returning the test statistic D and one and two-tail probabilities based on a chi-square approximation. These values will be on the conservative side, so if they suggest a significant difference there is no need to use tables. If the probability is slightly above the chosen level of significance then it is worth checking against the tables of critical values (Section 12.5). In the example above the estimated two-tail probability is $p = 0.013$. The same calculator gives the results of the Mann–Whitney U-test, which in this case are also strongly significant at ($U = 3.5$, two-tail $p < 0.01$) and of the Student's t-test, which with these very small and skewed samples would be nonsense.

8.11.2.2 In SPSS and R Commander

Follow the instructions for the Mann–Whitney U-test but on the 'settings' page choose 'Kolmogorov–Smirnov (two-samples)'. The summary results only give the two-tail probability but double-clicking it opens a table that includes the 'most extreme difference', which is the test statistic. The probability value given is an estimate, probably based on the chi-square approximation, and so is likely to be on the conservative side. If your result is just above a boundary, use tables of critical values. I could not find a simple way to run the test in *R* Commander, though there is R-code available (https://stat.ethz.ch/R-manual/R-devel/library/stats/html/ks.test.html).

References

Cohen, J. 1988. *Statistical Power Analysis for the Behavioural Sciences* (2nd ed.). Academic Press, New York.

Cohen, J. 1992. A power primer. *Psychological Bulletin* 112, 155–159.

Dunlap, W.P., Cortinal, J.M., Vaslow, J.B., Burke, M.J. 1996. Meta-analysis of experiments with matched groups or repeated measures designs. *Psychological Methods* 1, 170–177.

Fagerland, M.W. and Sandvik, L. 2009. Performance of five two-sample location tests for skewed distributions with unequal variances. *Contemporary Clinical Trials* 30, 490–496.

Field, A. 2013. *Discovering Statistics Using IBM SPSS Statistics* (4th ed). Sage, London. 916pp.

Leucuta, D.C. 2011. Coin's package permutation tests graphic user interface extension for *R* commander. *Applied Medical Informatics* 29, 3.

Lehmann, E.L. 1975. *Nonparametrics-Statistical Methods Based on Ranks*. Prentice Hall, San Francisco.

Neave, H.R. and Worthington P.L. 1988. *Distribution-Free Tests*. Unwin-Hyman Ltd, London. 430pp.

Siegel, S. 1956. *Nonparametric Statistics for the Behavioural Sciences*. McGraw Hill, New York. 312pp.

9

Comparing More Than Two Samples

9.1 Introduction

In geography, we often wish to compare not just two groups but several. A common approach to this problem is to use one of the two-sample tests but to run it for all possible pairs. For example, you might be interested in whether different farming practices result in differences in soil properties and wish to compare four fields treated in different ways: we can call the fields A, B, C and D. You could now use a two-sample test, such as the independent samples Student's t-test to make the following comparisons:

A–B   A–C   A–D   B–C   B–D   C–D

The problem with this approach is that the probabilities are not the same as when you run a single test.

When we use statistical tests, we often take $p = 0.05$ as our critical threshold. That is the same as 5%. If we are using a test for the difference between two sets of values, such as Student's t-test, the 5% is based on the chances that a difference as large as the one that you observe could have arisen just by chance. A good analogy is to take a bag with 100 numbered balls in it, five of which are 'winning numbers'. If you pay once, and put your hand in the bag and pull out one ball, you have a 5 in 100 or 5% chance of winning. However, if you pay six times, and have six dips at the bag, clearly your chances of finding a winning number are much better. In the example above, where you are making six comparisons rather than just one, there is a similar problem in that your chances of one of the six crossing the usual 'significance' threshold is no longer 5%. A statistician would tell us that the degrees of freedom have changed.

In geography, in published papers as well as in student work, it is quite common to see multiple tests used without making any correction to the degrees of freedom. The result is that spurious positive results occur more often than they should. You should either make a correction, such as that described in Section 9.2 or, much better, use statistical tests that are specifically designed for dealing with more than two samples. These are called k-sample tests and are the main subject of this chapter.

9.2 'Family-Wise' Error and the Bonferroni Correction

There is a way to deal with the change in the degrees of freedom, or your chances of crossing the usual significance threshold just by chance, and that is to adjust the significance values according to the number of tests that you are conducting. The most common approach is the 'Bonferroni correction' (Dunn 1959, 1961). It is not difficult to apply, you just divide the critical significance level that you are using by the number of tests conducted. This means that if your critical threshold for a single test is $p = 0.05$, then with two tests, it is halved to $p = 0.025$ and with 10 tests, it is divided by 10 to give $p = 0.005$ (Figure 9.1).

In practice, there are two ways to use the Bonferroni correction, depending on whether you are using tables of critical values, and so only have the values for some test statistic, such as t-values from the Student's t-test, or already have probability values. If you have a probability value for each of your tests, then it is quite easy to change these to 'corrected probabilities' by simply multiplying them by the number of tests that you are conducting. For example, if you were running a 'family' of five tests, and one of them individually gave a probability value of $p = 0.01$, then that would be multiplied by 5 to give $p = 0.05$. If you were running a family of six tests, the same probability for that individual test would be corrected to $p = 0.06$ and would be considered not statistically significant. When you use this method, you sometimes produce a probability value that is greater than one, which makes no sense. A simple solution is to report any numbers above $p = 0.1$ (10%) as $p > 0.1$.

When you are dealing with tables of critical values, using the Bonferroni correction is more difficult because the critical values that you need are unlikely to be tabulated. For example, with a family of six tests, an original significance threshold of $p = 0.05$ becomes $p = 0.0083$ $(0.05 \div 6)$ and of course there are no tables of critical values for $p = 0.0083$.

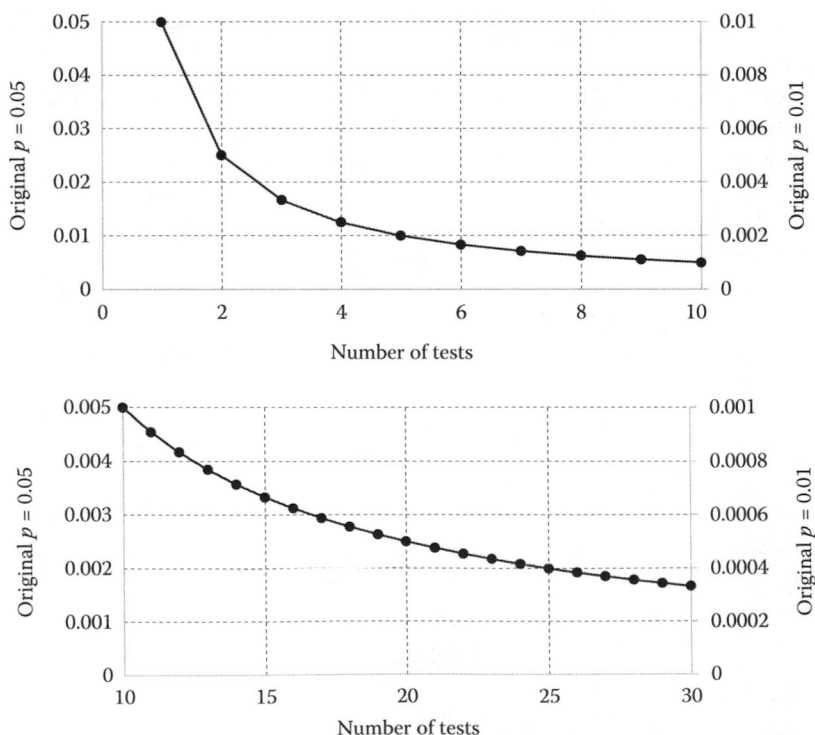

FIGURE 9.1
Effect of the Bonferroni correction on probability values. Note the change of vertical scales in the lower graph. Given an original critical threshold of $p = 0.05$, the corrected value after five tests is $p = 0.01$, after 10 tests it is $p = 0.005$ and after 25 tests it is $p = 0.002$.

A simple option is to check whether your result is significant at the next level that is tabulated. If, for example, you were conducting six independent Student's t-tests using two samples of 30 (58 degrees of freedom) and all of them gave t-values of more than 3.5, you can conclude that they are all significant at $p < 0.0083$ because the critical value for $p < 0.001$ is slightly less than 3.5. Similarly, if some are not significant at $p < 0.05$, then they cannot be significant at the more stringent level. Where values fall between the tabulated critical values, however, you can often find the critical value for any level of significance using the appropriate probability distribution for that test. The t-test, for example, uses the t-distribution and you can find the critical values for any level of significance and degrees of freedom using the 'TINV2T' function in Excel. However, it is usually easier to just find some way to obtain the significance level for each test and then correct those significance levels. If your tests result in z-scores (from the standard normal distribution), you can compare them to a table that includes the Bonferroni correction (Section 12.14).

If you use the Bonferroni correction, or similar but more complicated methods, a critical decision is how many tests are in the 'family' that you are trying to correct. This may sound simple, in that it is just the number of tests that you have completed, but that is not actually true. For example, you might want to compare the examination results of the four groups of students in Table 9.1 using pairs of t-tests. In this case, the aim would be to determine whether there are any significant differences between all possible pairs. To compare all possible pairs, you would need to conduct six individual t-tests. The pairs being tested would be

A–B, A–C, A–D, B–C, B–D, C–D

By just looking at the mean and standard deviation values, however, it is clear that there cannot possibly be a statistically significant difference between groups A and B because the difference in the mean is only 0.2%. Similarly, no test is required to conclude that there is no significant difference between C and D, again because the mean values are too similar, with a difference of only 0.05%. You might choose, therefore, to conduct only four comparisons, ignoring A–B and C–D and only computing A–C, A–D, B–C and B–D. Even if you only conducted the four tests, however, you would still have to multiply the resulting significance values by 6 rather than 4. The reason is that you only decided not to bother comparing A–B and C–D after looking at the data and noticing how similar they are. In statistics that is effectively regarded as cheating because the decision is not made 'a priori' (in advance), as part of the experimental design, it is an ad hoc decision based on looking at the results and minimising the number of tests by cherry picking just those pairs that are most likely to produce a significant difference.

Another way to deal with the four sets of examination results in Table 9.1 is to pose a rather different question. Rather than asking whether there are any significant differences

TABLE 9.1

Summary Statistics for Four Sets of Examination Results

	Group	Number of Students	Mean as %	Standard Deviation
Physical geography	A	54	66.40	12.3
Human geography	B	55	66.60	12.1
Earth science	C	72	58.15	12.02
Geography	D	75	58.2	12.10

between pairs, you could ask how many 'homogeneous groups' the results fall into. To use this approach, you place the four groups in rank order according to the mean values and only compare the adjacent pairs. Now, you only need three tests, rather than six, and so the Bonferroni correction is much less harsh. Assuming a starting value of $p = 0.05$, the critical threshold for three tests becomes $p = 0.017$ rather than $p = 0.008$ with six tests. In the example above, you would conclude that A and B are not significantly different from each other, so they must be in the same group and C and D are not significantly different from each other, so they must also be in the same group. If B and C are significantly different, then you can conclude that there are two homogeneous groups, if B and C are not significantly different, using the corrected critical value of $p = 0.017$, then you conclude that they all fall in the same group. A similar approach is used as one of the options when you use SPSS to perform 'post hoc' tests after conducting k-sample tests such as the Kruskal–Wallis one-way ANOVA by ranks test (Kruskal–Wallis H-test), and there are some examples below. Whatever approach you use to define your 'family' of tests, and therefore the magnitude of the Bonferroni correction, the important thing is to support your decision with a clear argument. Often it is a matter of judgement, and you are unlikely to be penalised if you clearly explain your reasoning.

The advantage of the Bonferroni correction is that it reduces your chances of 'false positive' results. The disadvantage, however, is that the correction is so harsh that there is an increased chance that you will miss real effects because you have too many 'false negatives'. More importantly, this approach is not really testing the overall question that we usually want to answer. In the introduction to this chapter, for example, the main question was whether farming practice influenced soil properties and using multiple two-sample tests does not really answer that question. What we really need is a test specifically designed for dealing with more than two samples. That is what the k-sample tests do.

9.3 *K*-Sample Tests

9.3.1 Introduction

K-sample tests are those designed specifically to compare more than two samples. I have no idea why the letter k is used but in statistics, it seems generally to indicate more than two. A range of tests is available, and as with the two-sample tests, you need to look first at your data, or experimental design, to decide which tests would be appropriate (Figure 9.2). First decide if you have (or will collect) individual measurements, this means that the data are continuous enough to be placed in rank order without too many ties, or if they are really counts in categories.

For counts in categories, you can use complex chi-square. If you have individual measurements, you need to decide whether your data are independent or linked. If you have three samples, for example, you need to decide whether a particular value in sample A is connected in some way to particular values in samples B and C. If you were interested in comparing exam results from years 1, 2 and 3 of a university degree, for example, there are two quite different approaches you could use. If you were to take an unbiased sample of students from each year, and compare the results, those data would be independent of each other. There would be no logical reason for linking together any particular student's

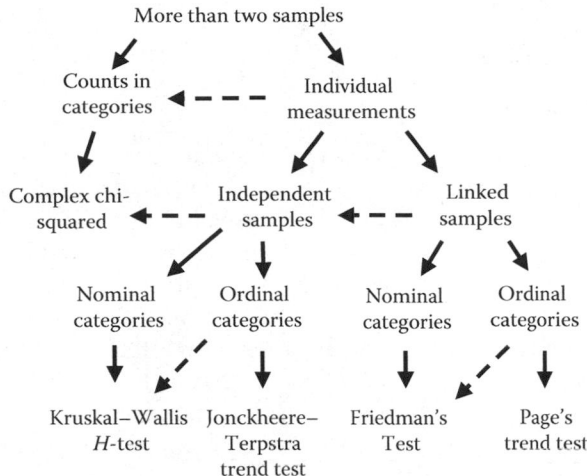

FIGURE 9.2
Logical series of steps for deciding which test to use when there are more than two samples to compare. Dashed lines indicate a loss of information.

result in year 1 with some other student's result in year 2 or yet another student's result from year 3. The alternative approach is to look back at the results obtained by a single cohort of students, so that for each individual person, you have three results; one from each year of study. In this case, the data sets are not independent, they are linked because each student 'owns' one mark in each year.

The next step it to decide whether you are interested in the order of the groups. In the example above, for example, are you interested to know if there is a difference in exam results between years, or do you want to know whether there is a general increase or general decrease from years 1 to 3? These are different questions and they require different tests. Finally, you need to decide whether to use parametric or non-parametric tests. Most statistical text books will advise you to use the parametric methods, which are various forms of 'analysis of variance' or ANOVA. The non-parametric alternatives are generally regarded as only suitable when there is some terrible 'problem' with your data. I would advise the complete opposite. I have seen a lot of student projects over the years that have tried to compare more than two samples and I can say that very few of them actually meet the requirements for a parametric ANOVA. In this book, I will focus mainly on the non-parametric methods. If you want to use the parametric methods, there are plenty of other books to choose from.

I suspect the real reason that the non-parametric methods have not been very popular is that before computers and the Web, they were tiresome to compute. That is no longer the case, and it is as easy to perform, for example, the Kruskal–Wallis H-test, which is a non-parametric ANOVA method for comparing three or more independent samples, as it is to perform the equivalent parametric ANOVA. Under conditions where the assumptions of parametric ANOVA are perfectly met, the H-test is almost as powerful (Siegel 1956 estimates it as 95.5%) and when the conditions are not perfectly met, which in my experience for geography projects is almost always, the H-test is likely to be more powerful. For linked samples, the non-parametric 'Friedman test' is similarly powerful. Where your hypothesis is that there is a trend across the groups, and they can be placed in a logical order without looking at the results, the appropriate tests are the Jonckheere–Terpstra trend test for independent data and Page's trend test for linked data.

9.4 Complex Chi-Square

9.4.1 When It Is Useful

To be honest, when a student comes to me with the intention of using complex chi-square, where there are several columns and several rows with counts in categories, my heart just sinks. There are cases where it is a useful approach, but normally it is just the last resort for students who have collected a load of questionnaires and have no idea what to do with the numbers. The fundamental problem is that if you have a very complicated design, with several kinds of 'subject' and several different 'treatments', it becomes very difficult to narrow down the question that you are actually asking. In effect, there is no clear hypothesis being tested. In many cases, the method is being used as a bit of a fishing expedition, in the desperate hope that something interesting will leap out. I am including the method here for completeness, but my advice is wherever possible to avoid complex chi-square, where everything is lumped in together, and focus on more specific questions using two-sample chi-square or other methods. In the examples below, I will give some ideas for how this might be achieved.

9.4.2 What It Is Based On

Complex chi-square is not conceptually different to the much more useful two-sample test described in Chapter 7, so read that section first. In essence, it is a way of analysing a large contingency table, where you want to know whether the 'responses' in the columns are 'contingent upon' the categories in the rows. 'Contingent' means something similar to dependent but with less certainty. The columns and rows are treated as nominal, in the sense that they cannot be placed in any logical order. For each cell of the table, you have an observed value and you can also calculate an 'expected' value. The expected values are the frequencies that you would expect if there was no contingency.

   Although chi-square is a non-parametric method, there are some strict assumptions and if you violate them, you will just produce nonsense. It is surprisingly easy to do that and I have seen some disastrous dissertation marks as a result of misuse of chi-square, and particularly of complex chi-square. The three critical assumptions are: data are counts, not percentages or proportions, data are independent and sample size is adequate.

9.4.2.1 The Data are Counts, Not Percentages or Proportions

The probabilities for this test are based on the chances of producing an uneven distribution of values across the contingency table just by luck, when there is really no relationship between the classes (often subjects) in the rows and the classes (often responses) in the columns. If you have small samples, it is easier to produce such an imbalance and as sample size increases the probability declines. The probabilities depend on the totals for the rows and columns and if they do not reflect the true sample size, then the results are just nonsense.

9.4.2.2 The Data Must Be Independent

This means that there is no link between any of the data points. So, if the rows represent people, and the columns represent their response to a question, a single individual can only occur in one row and they can only occur in one column. For example, you might ask someone to say if they prefer whisky, gin or vodka but you cannot let them tick more than one option, otherwise your row and column totals do not represent the true sample size.

9.4.2.3 Adequate Sample Size

The probabilities are not calculated mathematically (i.e. exactly) as in Fisher's exact test, they are based on the fact that the chi-square values produced by chance alone have a distribution that has a characteristic shape with known properties. However, that is only true when the sample size is big enough. For complex chi-square that means you need to check two things:

1. No cell has an expected frequency of less than 1
2. Fewer than 20% of cells have expected frequencies less than 5

9.4.3 Effect Size

The appropriate effect size for chi-square tests is Cramer's *V*, which is described in detail in Section 7.2.5. SPS and some online calculators return this value. *R* Commander does not but it is not difficult to calculate, although a little more complicated than the two-sample case. The equation is

$$V = \sqrt{\frac{\chi^2}{n \times k - 1}}$$

where *n* is the sample size and *k* is the number of either rows or columns, whichever is the smaller.

9.4.4 How to Do It

The chi-square equation and the procedure for calculating the expected values are explained in Section 7.2, so I will not repeat them here. It is not difficult to complete the test by hand, but it is tedious and easy to make a mistake. You can do it in a spreadsheet if you are careful. If you do calculate the chi-square value by hand, you can use tables of the chi-square distribution (Section 12.2) to find the probability. The degrees of freedom for a complex chi-square are

$$df = (nrows - 1) \times (ncolumns - 1)$$

where 'nrows' is the number of rows and 'ncolumns' is the number of columns. There are some online calculators that you can use but beware, most of them will not check the assumptions for you and if they are violated, your results will be nonsense. One of the easiest and best options is to use *R* Commander, which returns the expected counts but not Cramer's *V*. Using SPSS is a bit more complicated but the output includes Cramer's *V*. There are a few online calculators. To report the results give the chi-square (χ^2) value, degrees of freedom, probability and effect size.

9.4.4.1 In R Commander

No need to enter any data initially. Click on 'Statistics' and 'Contingency tables' then 'Enter and analyze two-way contingency tables'. You can produce a table with up to nine rows and nine columns and on the 'statistics' tab, choose various options, including 'expected frequencies'. When you run the test, a warning will appear in green at the bottom if you have violated the sample size assumptions. There is also an option for 'Fisher's exact test' (see Section 9.5) which is very helpful because if your samples are small, and the assumptions of chi-square are violated, this method will often (if no sample is very large) give you a probability.

9.4.4.2 In SPSS

To run complex chi-square in SPSS, you need to make sure the data are entered in the right format. You can use the 'long' format, where every individual case has a row, so if you were producing a 3 × 3 contingency table with red, green and blue as the columns and A, B and C as the rows, you would have a long column of colours and in the adjacent column, you would enter the letters. Unless you are deliberately coding data in this way, to perform a lot of analyses in SPSS, it is very tedious. The alternative is to enter the data as shown in Table 9.2, which is much faster. However, you must remember to go to 'Data' and 'Weight cases' and place the counts data into the 'Frequency variable' box. Now go to 'Analyze', 'Descriptive statistics' and 'Crosstabs'. Choose the data for the rows and columns and on the 'Statistics' tab, choose chi-square and 'Phi and Cramer's V'. The output includes a contingency table, which unfortunately uses an alphabetical order, and a table that gives the Pearson chi-square value, the degrees of freedom and the probability. Probabilities are always two-tail because you cannot predict a direction when the categories are nominal. Cramer's V provides an effect size. Beneath the table, there is some text that tells you how many cells have expected counts of less than five and the minimum expected count. Use these to check that the sample size assumptions are met.

TABLE 9.2

Short Method of Entering Data into SPSS to Run Complex Chi-Square

Columns	Rows	Count
Red	A	10
Red	B	12
Red	C	15
Green	A	12
Green	B	10
Green	C	8
Blue	A	5
Blue	B	2
Blue	C	10

Note: Each row represents a single cell of the contingency table. Use 'Data' and 'Weight cases' to inform SPSS that the final column represents counts.

TABLE 9.3

Selection of Free Online Calculators for the Complex Chi-Square Test

Provider	Website	Summary
Vassar stats	http://vassarstats.net/index.html Listed under frequency data	Up to 5×5 table. Easy to use. Warns if expected values too small but does not plot them. Gives Cramer's V effect size, standardised residuals and percentage deviations
Social Science Statistics	http://www.socscistatistics.com/tests/chisquare2/default2.aspx	Up to 5×5 table. Easy to use but no warnings if assumptions violated
Lawrence Turner	http://turner.faculty.swau.edu/mathematics/math241/materials/contablecalc/entry.php	Up to 6×6 table. Plots expected values and warns if some values are too low
Quantpsy.org	http://www.quantpsy.org/chisq/chisq.htm	Up to 10×10 table. Warns if any expected values are too low but does not plot them
T.W. Kirkman	http://www.physics.csbsju.edu/stats/contingency_NROW_NCOLUMN_form.html	Up to 9×9 table. Plots expected values. No warning when samples are too small

9.4.4.3 Online Calculators

There are a few online calculators that will run complex chi-square and if you already have your data in the form of a contingency table, they are very convenient. Be careful to check the assumptions though (Table 9.3).

9.4.5 Examples

9.4.5.1 Example: Typical Act of Desperation

Imagine that you have conducted a questionnaire survey to investigate local residents' attitude towards the building of a wind farm nearby. You have information on their age and gender and asked them to tick a box to record their opinions, using the usual five-point scale from strongly in favour to strongly against. You could place all of the information in a single contingency table (Table 9.4) and use complex chi-square in the vain hope that it will give you the ultimate answer to life, the universe and everything. It is a big data set, no cells have expected frequencies of less than 5, so running the analysis would return a strongly significant result ($\chi^2 = 65.591$, $df = 20$, p-value = 9.4E-07, Cramer's $V = 0.2$). The probability value is in scientific notation because it is such a small number. To translate it into a decimal fraction, move the decimal point 7 places to the left.

Cleary, the result is very strongly significant, so we can conclude that opinions are contingent on the 'type of person' and in this study, 'type of person' is defined according to age and gender. That is a good start, but if that is the only statistical result that you get for all the effort involved in conducting more than 400 questionnaires, it would be a bit sad. The general conclusion that 'the results were strongly statistically significant' is unlikely to satisfy either you or the person who marks the work.

There are some complicated methods available to deal with a big data set like this, such as log-linear analysis, and I am reliably informed that they do not require a brain the size of a planet. However, they do require a brain slightly larger than a walnut that has been soaked in whisky and then pecked by chickens, which is how mine feels at the moment. If you want to try any of that fancy stuff, read the book by Andy Field (2013). A simpler alternative is to break the data up a bit and ask some more focussed questions.

TABLE 9.4

Counts in Categories of Answer to a Question Where the Respondents are Divided According to Age and Gender

	Strongly Against	Against	Not Sure	In Favour	Strongly in Favour	Sum
Young females	12	8	21	24	16	81
Middle-aged females	21	15	8	12	10	66
Older females	25	12	21	8	4	70
Young males	8	6	14	16	9	53
Middle-aged males	26	18	12	10	12	78
Older males	28	5	23	2	6	64
Sum	120	64	99	72	57	412

An obvious question here is whether it is age or gender that is driving the overall difference. A simple way to investigate this is to check whether there is a significant gender difference within the same age groups using a series of much more targeted tests. These two-sample tests can be run using the methods described in Chapter 7, including the companion site calculators which give an effect size, Cramer's V, as well as a probability. The results are shown in Table 9.5.

When you use multiple tests like this, there are some complications about how to interpret the significance levels (see Section 9.2 on family-wise error and the Bonferroni correction) and if the differences were significant at $p < 0.05$, it would be a good practice to make a correction for multiple testing. However, the correction always raises the p-value, so when they are already higher than 0.05, there is no point. In this case, all comparisons give results that are very far from significant. This simple procedure clearly shows that, in terms of attitudes to the wind farm, gender is irrelevant and can just be ignored. Clearly, the differences in opinion must be driven by differences in age.

If we are to ignore gender, then we can just add together the male and female counts in the three age classes to give a more focussed comparison where the only classification of 'type of person' is the age groups that we have so inelegantly placed them in (Table 9.6).

Running this contingency table as a complex chi-square gives an even more strongly significant result, with nine zeros after the decimal point ($\chi^2 = 59.833$, $df = 8$, p-value = 5.03E-10, Cramer's $V = 0.27$). The result is clearly very strongly significant, so we can safely conclude that opinion is contingent on age. Notice, however, that Cramer's V is actually quite low. This is a very interesting result, because the very strong significance values might lead you to conclude that opinions are very strongly contingent on age, but actually this is not true. This is a very large data set, so the chances of producing these differences between

TABLE 9.5

Results of More Focussed Two-Sample Chi-Square Tests to Investigate the Importance of Gender

Compare	With	χ^2	Cramer's V	p-Value
Young females	Young males	0.20	0.04	0.995
Middle-aged females	Middle-aged males	0.98	0.08	0.914
Older females	Older males	6.89	0.23	0.142
All females	All males	3.25	0.09	0.516

Note: Since all of the probabilities are higher than 0.05, there is no need to make a correction for family-wise error. The results suggest that gender is irrelevant.

TABLE 9.6

Since Gender Has Been Shown to be Irrelevant, It Can Be Ignored and the Data Re-Tabulated by Age Alone

	Strongly Against	Against	Not Sure	In Favour	Strongly in Favour	Sum
Young	20	14	35	40	25	134
Middle	47	33	20	22	22	144
Older	53	17	44	10	10	134
Sum	120	64	99	72	57	412

the observed and expected values just by chance is very remote indeed, but that is not a measure of the strength of the association. The effect size is actually quite small and since it falls below 0.3 would generally be regarded as a 'weak' effect.

Calculating the percentage deviations (explained in Section 7.2.5) reveals where the biggest differences between observed and expected values lie, and we can see that young people tend to be strongly overrepresented in the categories 'in favour' and 'strongly in favour' and the older people are strongly underrepresented in those categories (Table 9.7). The middle-aged people seem to be more split in their opinions, with the strongest positive deviation in the 'against' category but also a positive value in 'strongly in favour'. It is also notable that middle-aged people are the most likely to actually have an opinion, because they are underrepresented in the 'not sure' category. This sort of interrogation of the results is useful, not least because it gives you something to discuss!

You could also go on to make focussed comparisons between the individual pairs. The results show that all comparisons are strongly statistically significant (Table 9.8). In this case, I have applied a Bonferroni correction assuming a family of three tests, which are all possible pairs (Section 9.2). The Cramer's V results show that middle-aged people have opinions that are more similar to older than to younger people and that the largest difference is between the two age extremes.

TABLE 9.7

Percentage Deviations Provide an Indication of Where the Largest Differences between Observed and Expected Values Occur

	Strongly Against	Against	Not Sure	In Favour	Strongly in Favour
Young	−49	−33	+9	+71	+35
Middle	+12	+47	−42	−13	+10
Older	+36	−18	+37	−57	−46

TABLE 9.8

Pairwise Comparisons Using Two-Sample Chi-Square Tests

Compare	With	χ^2	Cramer's V	p-Value	Corrected p-Value
Young	Middle	27.75	0.32	0.00001	0.00003
Middle	Older	23.15	0.29	0.0001	0.0003
Young	Older	40.66	0.39	3.20×10^{-8}	9.6×10^{-8}

Note: The corrected probabilities take account of the family-wise error introduced by making multiple comparisons.

TABLE 9.9

Mixture of Lithologies Obtained from Four Sedimentary Units Forming a Stratigraphic Sequence

	Granitic	Basaltic	Sandstone	Limestone	Other	Sum
Upper gravel	17	30	12	25	16	100
Upper diamicton	22	25	16	20	17	100
Lower gravel	18	23	14	23	22	100
Lower diamicton	10	15	23	4	48	100
Sum	67	93	65	72	103	400

9.4.5.2 Example: Glacial Deposits

Imagine you are interested in glacial sediments and have decided to work on an exposure where there are two gravels and two diamictons (Table 9.9), all of which contain a mixture of different lithologies (rock types). You would like to use the differences in lithology to determine whether all four units are similar, and could therefore represent a single glaciation, or whether there are differences, which might indicate separate glaciations or at least a big change in the direction if ice movement. You sample and identify 100 clasts from each unit. Note that these are actual counts, they are not percentages.

Running this 5×4 contingency table as a complex chi-square gives a strongly significant result ($\chi^2 = 55.4$, $df = 12$, $p = 1.5E-07$, Cramer's $V = 0.215$), from which we can glean that the mixture of lithologies is in some way contingent on the units from which they were sampled. Note that Cramer's V is low (<0.3), so we do not conclude that the degree of contingency is strong. The contingency is actually weak, but the large sample size means that the differences between the observed and expected counts are very unlikely to have arisen purely by chance. This tells us that the samples from the four units are very unlikely to have been drawn from the same population, but it does not really get us very far.

In this case, it would be less useful to produce a table of percentage deviations because that would just tell us the magnitude of the differences between the observed and expected values. That helps a lot when your samples differ in size, but since all of these samples are the same size, we can see the differences between them just as well by looking at the counts.

You could now split the table up and compare all possible pairs using two-sample chi-square tests. However, there is a clear stratigraphic order to these sediments, so comparing them in sequence seems more sensible and means that the 'family' of tests is restricted to three. The results are shown in Table 9.10. Note that the first two probabilities are not given a Bonferroni correction because they are already greater than 0.05 and this is all we need to know. It is now clear that the odd one out here is the lower diamicton. The other three units are statistically indistinguishable on the basis of their lithologies, so they form one homogeneous group and the lower diamicton lies outside it.

TABLE 9.10

Focussed Comparisons of Adjacent Pairs of Sediment Units

Compare	With	χ^2	Cramer's V	p-Value	Corrected p-Value
Upper gravel	Upper diamicton	2.25	0.11	0.69	>0.05
Upper diamicton	Lower gravel	1.47	0.09	0.83	>0.05
Lower gravel	Lower diamicton	29.19	0.38	7.20×10^{-6} (<0.0001)	2.16×10^{-5} (<0.0001)

Note: In this case, the pairings are based on the stratigraphic sequence.

9.5 Fisher's Exact Calculation for Small Samples

9.5.1 When It Is Useful

Sometimes, in research, large samples are not a realistic option. In such cases, it may be possible to analyse a contingency table using an exact calculation rather than relying on the assumption that the data set fits the chi-square distribution. This is an extension of Fisher's exact test (Section 6.10) and is very computationally demanding, but given modern computing power it is now feasible, at least with small samples. As with the more common two-sample case, Fisher's exact tests tend to be rather conservative, so the probability values tend to be slightly on the high side. For this reason, irrespective of the computational problems, it should only be used when the sample size assumptions of chi-square are violated.

9.5.2 How to Do It

It is not feasible to make the calculations by hand or in a spreadsheet (at least for me!). There are a few online calculators that will cope with small sample sizes. The best approach is to use *R* Commander or SPSS.

9.5.2.1 In SPSS

Follow the instructions for complex chi-square and check the warnings under the chi-square test results. If more than 5% of the cells have expected counts of less than five or the minimum expected value is less than one, then go back to 'Analyze', 'Descriptive statistics' and 'Crosstabs' and click on the 'Exact' tab and choose the option 'Exact'. On the statistics tab, keep 'Chi-square'. Now, when you click 'Run', the results include the exact probability produced by Fisher's exact test.

9.5.2.2 In R Commander

Click on 'Statistics' and 'Contingency tables' then 'Enter and analyse two-way contingency tables'. Enter counts in the contingency table (up to 5×5). On the 'Statistics' tab, choose 'Chi-square test of independence' and 'Print expected frequencies'. Run the test and check the expected frequencies. If they violate the sample size assumptions go back and replace chi-square with the Fisher's exact test option. The output is simply the *p*-value, which is always two-tail for this test.

9.5.2.3 Online Calculators

There are a few online calculators that will deal with small contingency tables with small counts in them (Table 9.11). Some include a 'mid-*p*' option for the probability, which reduces the conservative bias of the test. Most of them crash very easily.

9.5.3 Examples

9.5.3.1 Example: Snails

Imagine that you are comparing three deposits that contain snail shells and you want to know whether there are significant differences in the types of snail. This is a quick pilot

TABLE 9.11

Selection of Free Online Calculators for Fisher's Exact Calculation on a Contingency Table with More Than Two Rows and/or Columns

Provider	Website	Comments
Vassar stats	http://vassarstats.net/index.html	Separate pages for 2×3, 2×4 and 3×3 contingency tables. Two p-value options, I think PA is equivalent to R Commander
Daniel Soper	http://www.danielsoper.com/statcalc3/calc.aspx?id = 59	Separate pages for 2×3 and 3×3. Odd format for entering the values
In Silico project support for the life sciences	http://in-silico.net/tools/statistics/fisher_exact_test	Supports 3×2 and 4×2 tables. Odd format with groups in the columns. Choose 'both' for 'tail/side'
Cog-Genomics	https://www.cog-genomics.org/software/stats	Only for 3×2 tables. Allows a 'mid-p adjustment', which gives a less conservative probability. Good site
Simple Interactive Statistical Analysis	http://www.quantitativeskills.com/sisa/statistics/fiveby2.htm	Up to 2×5 tables. Remove zero values from empty cells. Use the mid-p option from the output

TABLE 9.12

Small Sample of Snail Shell Types from Three Sedimentary Units

	Small Circular	Big Coiled	Small Conical
Unit A	3	1	7
Unit B	8	2	2
Unit C	2	10	3

Note: Fisher's exact calculation is a good choice when the sample sizes are too low to satisfy the assumptions of complex chi-square

study and you just want to know if there are clear differences or not. You have not identified the snails to species yet, but it is clear that there are three different shell shapes present (Table 9.12).

It looks from the raw counts as if there is a difference between the three units in terms of the types of snail, but the samples are too small to use the complex chi-square test because six of the nine cells would have expected counts of less than five. Fortunately, however, the numbers are small enough to allow Fisher's exact calculation to be used and R Commander returns a probability of $p = 0.002$. You can safely conclude that these three units are not the same in terms of snail content.

9.5.3.2 Example: Use of Space

You are interested in why people visit the city centre and have asked small samples of people in three different age categories their main reason for visiting (Table 9.13). Most people gave reasons that fall into four categories (shopping, leisure, socialising and eating) and for this exercise, you have ignored any that do not fit. The sample sizes are too small to use chi-square because six of the 12 cells would have expected frequencies of less than five. Using Fisher's exact test, the result is strongly significant ($p = 0.001$), so you can conclude that the different age groups do differ in their main reasons for visiting.

TABLE 9.13

In This Contingency Table, the Counts Are Too Low to Allow a
Chi-Square Test So Fisher's Exact Test Is Used

	Shopping	Leisure	Social	Eating
Young	2	4	12	2
Middle-aged	12	2	2	0
Older	6	4	6	5

Note: The result is strongly significant ($p = 0.001$).

TABLE 9.14

Probability Values Obtained Using Fisher's Exact Test to Compare
All Possible Pairs

Groups Compared	*p*-Value	Corrected *p*
Young/middle-aged	0.0003	0.0008
Young/older	0.18	>0.1
Middle-aged/older	0.026	0.08

Note: Performing multiple comparisons increases the chances of a significant
result, so significant *p*-values are corrected using the Bonferroni procedure described in Section 9.2

As with the chi-square test, it is possible to take the analysis a bit further by comparing the three possible pairs (Table 9.14), using Fisher's exact test, but some caution is necessary because running multiple tests increases the chances of a significant result. To deal with this 'family-wise error', the probability values can be corrected using the Bonferroni procedure described in Section 9.2. Where there are three tests, it simply involves multiplying the significant probability values by three. This is equivalent to changing the critical value from $p = 0.05$ to $p = 0.0167$. The results indicate that the largest difference is between the 'young' and 'middle-aged' groups and that difference is very strongly significant even after the correction ($p = 0.0008$). The 'young' and 'older' groups, in contrast, are not significantly different even without a correction ($p = 0.18$). The difference between the 'middle-aged' and 'older' groups is not so easy to interpret. Without a correction, it is significant ($p = 0.03$) but after the correction, it is not ($p = 0.08$). Given that Fisher's exact test tends to be conservative, and the Bonferroni correction rather harsh, I would not just ignore this result, but conclude that more research, and a bigger sample, would be required to explore this difference in behaviour.

9.6 Kruskal–Wallis *H*-Test

9.6.1 When It Is Useful

This test works in the same way as the Mann–Whitney *U*-test and can be seen as an extension of that procedure (Kruskal and Wallis 1952). It is often described as a 'non-parametric one-way ANOVA by ranks'. It tests the null hypothesis that three or more samples have been drawn from the same population. It is a non-parametric test, so the data do not have to be normally distributed and there is no need to assume that each of the groups has equal variance. The important assumption is that your data can sensibly be placed in rank order, albeit with some tied values, rather than forming counts in categories.

9.6.2 What It Is Based On

To perform this test, you place all of the values in rank order irrespective of which group they come from, assign ranks from lowest to highest, then separate the groups and count the sum of the ranks for each group. In the simple case where you have equal numbers in each group, it is apparent that if the values were really drawn randomly from the same population, then we would expect the groups to be approximately randomly arranged when placed in rank order, so the sum or ranks for each group should be about the same. If one of the groups is dominated by relatively low values, then it will have a low sum of ranks and if one group has relatively high values, the sum of ranks will be high. When the group sizes are unequal, you need to take account of the differences in sample size.

By way of example (Table 9.15), take three samples each with four values. If there is no overlap at all between the three groups, then when they are arranged in rank order one group will receive ranks 1–4, another ranks 5–8 and the last ranks 9–12. This is the maximum possible difference for these samples and would give sum or ranks values of 10, 26 and 42. If, however, three samples are really drawn randomly from the same distribution, then we would expect them to be arranged more like the second example, where the sum of ranks for all three groups is 26. With very small sample sizes, we might expect differences in the sum of ranks to occur just by chance, but as the sample sizes increase the chances of a substantial difference arising by chance declines.

9.6.3 Obtaining a Probability Value

I have not included special tables of the critical values of the H-test here because H very quickly becomes distributed in a predictable way and so we can use much more familiar tables, of the chi-square distribution (Section 12.2), to obtain the probability values. All you need to know is the value of H and the degrees of freedom, which is the number of groups minus one (k–1). This approximation works well even if you only have sample sizes of five in three or more groups, and I doubt you will be using samples that are smaller than that. When you have your H value, you can either look up the critical values in tables (Section 12.2), or you can enter the H value and degrees of freedom into the 'CHISQ.DIST.RT' or 'CHIDIST' function in Excel to get the probability value.

Note that the probability values for the H-test are always two-tail. It assumes that the three or more groups are nominal in character, which means that they have names but cannot be placed in any sensible rank order. In that case, the concept of direction of prediction makes no sense and you cannot perform a one-tail test. If there is some logical order to

TABLE 9.15

Three Small Groups Arranged to Show the Two Most Extreme Outcomes of the Ranking Procedure Used in the Kruskal–Wallis H-Test

Group	A	A	A	A	B	B	B	B	C	C	C	C
Rank	1	2	3	4	5	6	7	8	9	10	11	12
Group	C	B	A	C	B	A	A	B	C	A	B	C
Rank	1	2	3	4	5	6	7	8	9	10	11	12

your groups and you want to know if there is a rising or falling trend in the values across the groups, then you need the Jonckheere–Terpstra trend test instead.

9.6.4 Effect Size

Andy Field (2013) argues that it is not useful to produce an effect size for the Kruskal–Wallis *H*-test because it is not a focussed comparison and he prefers to only use effect sizes for the post hoc pairwise comparisons.

9.6.5 Post Hoc Tests

Having run the Kruskal–Wallis *H*-test and found a significant difference between the groups overall it can be interesting to look at individual pairs to see where the big differences lie. You can do this qualitatively by comparing the sum of rank values, if the samples are of equal size, otherwise by using the average ranks (sum of ranks divided by the size of that sample). If you want to know the statistical significance of the differences, then there are several different methods that can be used. A simple strategy is to use the Mann–Whitney *U*-test to compare individual pairs, but this is not the best approach because it involves re-ranking the pairs and ignores the overall variability between all of the other samples. The appropriate tests use the sum of ranks that you have already given to each group based on the total sample. There are a number of options but the easiest and most flexible is Dunn's test, which is explained in Section 9.7.

9.6.6 How to Do It

With very small samples, the Kruskal–Wallis *H*-test can be computed by hand or in a spreadsheet using equations (Box 9.1), but it is a bit tedious. The easiest options are the tree Real Statistics Resource Pack for Excel, SPSS or the 'Coin' add-in to *R* Commander. SPSS is probably the best option for this test because it allows 'post hoc' tests and provides graphs with the output. The companion site includes a spreadsheet-based calculator that can deal with up to six groups with sample sizes up to 100 and also conducts the post hoc tests.

9.6.6.1 *Real Statistics Resource Pack for Excel*

When you have uploaded the Resource Pack (Zaiontz 2015), you can access the list of options using Ctrl-M and from that list choose 'Non-parametric Tests' and then 'Kruskal–Wallis Test'. Your data can be arranged in adjacent columns with headers and there is no need to do any ranking, it is all automatic. Take this simple example (Table 9.16) where we have three groups with unequal sample sizes. I have used small sample sizes so that you can count the ranks easily.

In this case, we can see that the two samples of equal size have quite a different sum of rank values and the highest sum of ranks comes from the smallest sample. Without doing any analysis, we might suspect that these three samples have not been drawn from a single population.

To run the *H*-test, we need only identify the three columns with the raw data and note whether or not we have included headers. There is no need to assign ranks. The output (Table 9.17) includes the three median values, the sum of ranks for each group and the count.

The values for r^2/n can be ignored, they are used in the calculation (Box 9.1). The important numbers are the test statistic *H*, the degrees of freedom which are equal to the number of groups minus one, and the probability value. In this case, the probability

TABLE 9.16

Kruskal–Wallis Test Involves Translating the Original
Values into Ranks, Ignoring the Groups, and Then
Counting the Sum of Ranks for Each Group

	Raw Values			Ranks	
A	B	C	A	B	C
96	82	115	4	2	7
123	124	149	8	9	13
83	132	166	3	10	14
61	135	147	1	11	12
101	109		5	6	
		Sum:	21	38	46

is less than 0.05 and so the result is considered statistically significant and we reject the
null hypothesis. We can conclude that the three samples are unlikely (3% chance) to have
been drawn from the same population. You can also make post hoc comparisons one at
a time and there is a good explanation on the website (http://www.real-statistics.com/
one-way-analysis-of-variance-anova/kruskal-wallis-test/follow-up-tests-kruskal-wallis/).

9.6.6.2 Companion Site Calculator

The companion site includes a simple calculator that will run the Kruskal–Wallis *H*-test
with up to six categories and with sample sizes up to 100. The results include the average
ranks and sum of ranks for each group as well as the *H*-statistic and a probability value
derived using the chi-square approximation with $k-1$ degrees of freedom. It also produces
a matrix that shows the difference between all possible pairs (in terms of mean ranks), and
calculates the Dunn's test (Section 9.7) scores (which are effectively *z*-scores). To interpret
these, you need to decide on the size of your 'family' of tests. To help, there is a table of
critical values for three levels of significance ($p = 0.05$, $p = 0.01$ and $p = 0.001$) and values
are provided assuming that you wish to make no correction, only compare adjacent pairs
when placed in rank order (of mean sum of ranks), or if you intend to compare all possible

TABLE 9.17

Output for the Kruskal–Wallis *H*-Test Performed Using the
Real Statistics Resource Pack for Excel

	Kruskal–Wallis Test			
	A	B	C	
Median	96	124	148	
Rank sum	21	38	46	
Count	5	5	4	14
r^2/n	88.2	288.8	529	906
H				6.77
df				2
p-value				0.03
Alpha				0.05
Sig				yes

TABLE 9.18

Selection of Free Online Calculators for the Kruskal–Wallis *H*-Test

Provider	Website	Comments
Vassar stats	http://vassarstats.net/kw3.html Look under 'ordinal data'	Available for three, four or five groups. Easy to use, gives *H* and *p*-values
Mathcracker	http://www.mathcracker.com/ kruskal-wallis.php	Allows up to five groups. Gives lots of output including the worked equations. Good site
Scistatcalc	http://scistatcalc.blogspot. co.uk/2013/11/kruskal-wallis-test-calculator.html#	Allows lots of groups. Odd horizontal format and data must be separated by commas, so paste data then add the commas and remove the spaces

pairs. The critical values are obtained by applying a Bonferroni correction (Section 9.2). If the Dunn's test score is equal to or higher than the tabulated value for your chosen level of significance, and level of correction, it is significant.

9.6.6.3 Online Calculators

There are a few online calculators that will perform the *H*-test (Table 9.18).

9.6.6.4 In R Commander

The *H*-test is included under the 'Statistics' and 'Non-parametric Tests' option in *R* Commander. To use it, you have to import your data in the correct format, which is not the format you would typically use in a spreadsheet. Place all of your data in a single column and use the column next to it to identify the groups. You can use names, letters or numbers, it does not matter as long as there are no spaces. When you choose the test, you are given no options other than to choose which column has the data and which has the group names. The output includes the 'Kruskal–Wallis Chi-square', which you can quote as the test statistic *H*, the degrees of freedom, which is the number of groups minus one, and the probability. Unfortunately, *R* Commander does not automatically perform any 'post hoc' comparisons between your samples. However, there is a plug-in to *R* Commander called coin that includes a more complete treatment of the data. You have to use 'packages', 'Load packages', choose a 'Cran Mirror' somewhere near where you live and then find RcmdrPlugin.coin. When you open *R* Commander, use the 'Tools' menu to load the plugin. A new button will appear called 'coin' that includes 'Independent location tests' and in that list you will find the 'Kruskal–Wallis Test'. When the pop-up appears, you choose the data and group columns as usual but there is also a 'block' box that will be automatically checked. You have to uncheck that using ctrl-click to make it work. There is now an option for 'pairwise comparisons of groups' and if you click this, you will obtain post hoc probabilities for all pairwise comparisons. See examples (Section 9.7.2) for how to interpret the output.

9.6.6.5 In SPSS

You cannot run this test if your data are arranged in separate columns, you have to use the 'long' format where you place all of the data in one column and use the column next to it to identify the groups. It is easiest to do this in a spreadsheet then cut and paste the two columns, without headers, into the 'Data view' (spreadsheet) page of SPSS. Now go to the 'Variable view' page and assign names in the first column (I always use 'data' and 'groups') and in the column marked 'Measure' define your data as 'scale' and your groups

BOX 9.1   CALCULATIONS FOR THE KRUSKAL–WALLIS TEST

The equations for the Kruskal–Wallis test look pretty terrifying to me, and without a worked example I would not get very far, so I will take you through it step by step using a simple example with small sample sizes (Table 9.19).

First, you have to assign the ranks, ignoring which group the samples come from. In this case, the lowest number (61) is in group A and is ranked 1 and the second lowest is in group B and is ranked 2, etc. If there are any identical numbers, then the ranks are shared using the usual procedure. For large samples, it is convenient to use the RANK.AVG function in Excel to perform the ranking (Box 8.1).

The numbers that we need for the equation are shown in Table 9.19. They include the sum of ranks for each of the groups (r_1, r_2 and r_3) and the number of samples in each group (n_1, n_2 and n_3). The overall sample size is 14 and we call that N.

The full equation is

$$H = \frac{12}{N(N+1)} \sum \frac{r^2}{n} - 3(N+1)$$

But we can split it into three parts

$$H = \frac{12}{N(N+1)} \times \sum \frac{r^2}{n} - 3(N+1)$$

$$H = (Part\ one) \times (Part\ two) - (Part\ three)$$

Part one is the easiest to solve

$$Part\ one = \frac{12}{N(N+1)} = \frac{12}{14(14+1)} = \frac{12}{14 \times 15} = \frac{12}{210} = 0.05714$$

TABLE 9.19

Very Small Data Set Used to Demonstrate How to Calculate the Kruskal–Wallis Test

Raw Values			Ranks		
A (1)	**B** (2)	**C** (3)	**A** (1)	**B** (2)	**C** (3)
96	82	115	4	2	7
123	124	149	8	9	13
83	132	166	3	10	14
61	135	147	1	11	12
101	109		5	6	
N		Count	n_1	n_2	n_3
14			5	5	4
		Sum	r_1	r_2	r_3
			21	38	46

Note that it is a good idea to keep several decimal places here because this number is used as a multiplier.

Part three is also not too bad

$$Part\ three = 3(N+1) = 3 \times (14+1) = 3 \times 15 = 45$$

So far so good, but part two looks more daunting. However, it is actually not as bad as it looks! What it means is take each sum of ranks, square it and divide by the sample size, then add them up. So to rephrase it in a way that a normal human might actually understand

$$\sum \frac{r^2}{n} = \frac{(r_1)^2}{n_1} + \frac{(r_2)^2}{n_2} + \frac{(r_3)^2}{n_3}$$

So part two is

$$Part\ two = \frac{(21)^2}{5} + \frac{(38)^2}{5} + \frac{(46)^2}{4} = \frac{441}{5} = \frac{1444}{5} + \frac{2116}{4} = 88.2 + 288.8 + 529 = 906$$

So going back to the original equation

$$H = (Part\ one) \times (Part\ two) - (Part\ three)$$

$$H = 0.05714 \times 906 - 45$$

Now follow the BODMAS rule (Section 4.2) and multiply before you subtract so that

$$H = 51.77 - 45 = 6.77$$

If you look back at the example earlier, which was solved using the Real Statistics Resource Pack in Excel (Section 9.6.6), you will see that this is the correct answer.

For very small sample sizes, there are special tables of critical values for the H-test. The critical values for $k = 3$ and sample sizes of 5, 5, 4 are $p = 0.05 = 5.666$ and $p = 0.01 = 7.823$. Since our H-value is 'greater than or equal to' the tabulated critical value for $p = 0.05$, we can reject the null hypothesis and conclude that there is less than a 5% chance that the differences between the groups could have arisen by chance if the three groups were drawn from a single population. Note that our result is lower than the tabulated critical value for $p = 0.01$, so we know that the probability lies somewhere between 0.05 and 0.01. The Real Statistics result was $p = 0.03$.

as 'nominal'. When the data are ready go to 'Analyze' then 'Non-parametric Tests' and choose 'Independent Samples'. On the 'Objective' tab, choose 'Customise analysis'. On the 'Fields' tab, use the arrows to put your 'data' in the top box marked 'Test fields' and your 'groups' in the bottom box marked 'Groups'. On the 'Settings' tab, check the 'Customize tests' button at the top and then choose the 'Kruskal–Wallis one-way ANOVA (*k*-samples)'.

Press 'Run' and the answer will appear, which is a probability value and a decision that tells you to accept or reject the null hypothesis at the default level of $p < 0.05$. If the probability value is less than 0.001, you will just see a value of 0.000. That does not mean that the probability is really zero, it just means $p < 0.001$ and that is how you should report it, not as zero.

To perform post hoc tests go to 'Analyze', 'Non-parametric tests' and 'Independent samples' and on the 'Settings' page, you can choose to perform 'Multiple comparisons'. The options are 'all pairwise' and 'stepwise step-down'. For the all pairwise option, you will obtain two significance levels ('sig' and 'adj.sig') and it is the second one that you should report because it has been corrected to take account of the number of tests you are doing. If you use the step-down procedure, you do not get significance values, the test statistic is just the difference.

9.7 Dunn's Test (Post Hoc Tests for Kruskal–Wallis Test)

Post hoc tests are so-called because you only apply them if you have already performed a k-sample test and obtained a result that is statistically significant. They allow you to test whether the difference between individual pairs is statistically significant. Several options are available but Dunn's test (Dunn 1964) has the advantage that it is not too complicated to calculate and the groups do not have to be the same size. For each comparison, it uses the absolute difference between the average ranks of the two samples and divides that by the standard deviation of the difference. It is effectively a z-score, so probabilities are obtained from the standard normal distribution (Section 12.14).

The full equation is

$$T = \frac{\left| mean\ R_A - mean\ R_B \right|}{\sqrt{\frac{N(N+1)}{12} \times \left(\frac{1}{n_A} + \frac{1}{n_B} \right)}}$$

where T is Dunn's test statistic, 'mean R_A' and 'mean R_B' are the average ranks for groups A and B and n_A and n_B are the two-sample sizes. In this case, N is not the two-sample sizes combined, it is the sum of all of the samples included in the Kruskal–Wallis test. The straight lines around the top of the equation mean take the absolute value (ignore the sign of the difference).

Taking the same example used in Box 9.1 to calculate the Kruskal–Wallis test and using Dunn's test to compare groups A and B we obtain $N = 14$ and

$$mean\ R_A = \frac{21}{5} = 4.2$$

$$mean\ R_B = \frac{38}{5} = 7.6$$

So the equation for Dunn's test becomes

$$T = \frac{\left| 4.2 - 7.6 \right|}{\sqrt{\frac{14(14+1)}{12} \times \left(\frac{1}{5} + \frac{1}{5} \right)}} = \frac{3.4}{\sqrt{\frac{210}{12} \times 0.4}} = \frac{3.4}{\sqrt{17.5 \times 0.4}} = \frac{3.4}{\sqrt{7}} = \frac{3.4}{2.646} = 1.29$$

TABLE 9.20

Example of the Output from the Companion Site
Calculator for the Kruskal–Wallis *H*-Test

	A	B	C
A		3.40	7.30
B	1.29		3.90
C	2.60	1.39	

Note: The cells in the upper right represent the difference in mean ranks and the values in the lower left are the results of Dunn's test (effectively z-scores).

Since Dunn's *T*-value is effectively a z-score, a probability can be derived directly from the standard normal distribution, but that is not a good strategy because it ignores the family-wise error that accompanies multiple testing of this kind. It is better to use critical values and adjust them to take account of the number of tests that are being conducted. Taking a threshold of $p = 0.05$, for example, the critical threshold for *T*, ignoring any family-wise error would be 1.96, but if you were to perform a family of three tests (all possible pairs in this case), then the Bonferroni correction requires the probability value be divided by the number of tests, so $p = 0.05$ becomes $p = 0.0167$ and the critical threshold for *T* rises to 2.39 (these are all two-tail values). Bonferroni-corrected critical thresholds for z-scores are included in Section 12.14.

Entering the example used in Box 9.1 into the companion site calculator for the Kruskal–Wallis test produces a matrix of output (Table 9.20) that shows the difference in mean ranks in the upper right cells and Dunn's statistic in the lower left cells. It also produces (separately) the critical values for three different significance levels adjusted to take account of the number of tests being conducted.

9.7.1 Effect Size

Since the Dunn's test values are effectively z-scores, they can be translated into the effect size *r* by dividing z by the square root of the combined sample size. For example, the comparison of samples A and B from Table 9.19 resulted in a Dunn's test score of 1.29 (Table 1.20) based on two samples of 5. The calculation of the effect size *r* is therefore

$$r = \frac{1.29}{\sqrt{10}} = \frac{1.29}{3.162} = 0.408$$

9.7.2 Examples

9.7.2.1 Example: Soil Compaction

You are interested in the impact of different types of grazing on soil compaction and have made 15 randomly located measurements in each of four very similar flat fields on the same farm. The results are shown in Table 9.21 as both original measurements and as ranks. When values are identical, the ranks are shared. Note that the sum of ranks values are very different.

Entering these data into the companion site calculator gives a strongly significant result ($H = 49.1$, $p = 1.2E{-}10$). We can conclude that the four samples are not drawn from identical populations, there are clear differences in soil hardness under different grazing regimes.

TABLE 9.21

Soil Hardness Data (Fabricated) for Fields with Different Grazing Regimes and the Same Data Transformed to Ranks

	Original Measurements					Changed to Ranks			
	None	Cattle	Sheep	Pigs		None	Cattle	Sheep	Pigs
1	70.61	78.16	70.40	84.90		17	40	16	58
2	70.94	74.06	71.77	82.00		18	29	19	53
3	61.45	74.23	72.80	80.68		1	31	22	47
4	74.48	78.61	76.00	81.35		32	42	34	49.5
5	69.68	79.03	73.67	85.84		13	43	25.5	60
6	67.19	78.19	77.80	80.90		8	41	37	48
7	61.90	80.42	71.90	81.39		3	46	20	51
8	64.87	76.84	73.10	79.61		6	35	24	45
9	68.29	78.00	72.63	83.19		11	38	21	55
10	61.97	79.52	67.80	83.65		4	44	9	57
11	61.58	75.61	69.37	82.97		2	33	12	54
12	69.74	74.16	72.97	78.13		14	30	23	39
13	64.94	73.90	73.67	81.35		7	27	25.5	49.5
14	63.23	77.26	68.03	83.42		5	36	10	56
15	69.94	81.52	73.97	85.52		15	52	28	59
Mean	66.72	77.30	72.39	82.33		10.40	37.80	21.73	52.07
					Sum	156	567	326	781

Note: When values are identical the ranks are shared. The difference in the average ranks indicates the magnitude of the difference between the groups.

To take the analysis a little further, we can examine the difference in mean ranks and the results of Dunn's test (Table 9.22). The smallest difference in mean ranks is between no grazing and sheep grazing and the largest difference is between no grazing and pigs. Since all of the samples are the same size, these differences are directly reflected in the values for Dunn's test.

The results of Dunn's test are compared to critical values (Section 12.14), also generated by the companion site calculator (Table 9.23). If we were to ignore the family-wise error that accompanies multiple testing, and treat every result as independent, then the critical value for $p = 0.05$ would be 1.96 and five of the six possible comparisons would be considered statistically significant. Soil hardness under sheep grazing in this case is not significantly different from that under no grazing. To take the other extreme, we might decide to compare all possible pairs and make a correction for multiple testing. In this case, the critical

TABLE 9.22

Difference in Mean Ranks for All Possible Pairs (Upper Right Cells) and Dunn's Test Results (Effectively z-Scores)

	None	Cattle	Sheep	Pigs
None		27.40	11.33	41.67
Cattle	4.30		16.07	14.27
Sheep	1.78	2.52		30.33
Pigs	6.53	2.24	4.76	

TABLE 9.23

Critical Values for Dunn's Test (Effectively z-Scores) Adjusted According to
the Number of Tests Being Conducted

Critical Values of T	No Correction	Adjacent Pairs	All Pairs
No. of tests		3	6
$p = 0.05$	1.96	2.39	2.64
$p = 0.01$	2.58	2.94	3.14
$p = 0.001$	3.29	3.59	3.76

Note: The Bonferroni correction on which these are based is explained in Section 9.2

TABLE 9.24

Effect Size *r* Results for the Post Hoc Comparisons Using
Dunn's test

	None	Cattle	Sheep
Cattle	0.78		
Sheep	0.32	0.46	
Pigs	1.00	0.41	0.87

Note: The unusual result of 1.00 occurs because there is no overlap
in ranks between two of the groups, which is the maximum
possible difference.

value equivalent to $p = 0.05$ (corrected to $p = 0.008$) jumps to 2.64, so only three of the comparisons would be considered statistically significant.

In this case, there are two other approaches that produce smaller 'families' of tests, resulting in a less stringent correction. The results for no grazing could be treated as a control, and all other 'treatments' compared only with these results, giving a 'family' of three tests. In this case, we would conclude that the increase in soil hardness caused by sheep grazing is not statistically significant, but cattle and pigs result in a strongly significant increase ($p < 0.001$). Alternatively, you could compare adjacent pairs rather than all possible pairs. In this case, the groups have to be re-arranged in order of mean ranks: none < sheep < cattle < pigs. In this case, we might conclude that no grazing results in the least compacted soil, but sheep grazing does not result in a statistically significant increase. Cattle grazing leads to a significant increase relative to sheep grazing and pigs are not significantly worse than cattle.

The effect sizes for the individual post hoc comparisons can also be examined (Table 9.24). They fall in the same order as the results of Dunn's test, as expected. It is unusual to see an effect size of 1.0 (between no grazing and pigs) but the reason for this result is clear from the table of ranks. The highest rank for no grazing is 32 and the lowest rank for pigs is 39, so in terms of the ranks on which these tests are based there is no overlap between these two groups. That is the maximum possible difference, hence the effect size value of one.

9.7.2.2 Example: Tourist Spending

You are interested in the impacts of tourism on the local economy and decide to interview tourists from different origins and ask them to estimate how much they spent the previous day. They fall into three groups (Table 9.25). Each group has a different average spend, the question is whether these differences are larger than we might expect to occur purely

TABLE 9.25

Estimated Spend for Three Groups of Tourists and the Same Data Translated into Ranks

	Estimated Spend (£)			Ranks		
	Europe	**America**	**Asia**	**Europe**	**America**	**Asia**
1	158	88	101	27	12	15
2	79	138	136	10	24	23
3	219	146	29	43	25.5	1
4	179	72	170	31	8	30
5	110	95	48	19	13	5
6	35	191	188	3	37	36
7	208	113	105	41	20	18
8	222	102	39	44	16	4
9	212	160	197	42	28	39
10	63	200	87	7	40	11
11	230	186	132	45	34.5	22
12	180	118	146	32	21	25.5
13	166		73	29		9
14	51		99	6		14
15	182		186	33		34.5
16	103			17		
17	34			2		
18	194			38		
Average	145.8	134.1	115.7	26.1	23.3	19.1

Note: Although the average spends are different, the results of the Kruskal–Wallis *H*-test suggest that there is about a 32% chance that differences as large as these could occur purely by luck. We would conclude that the differences are not statistically significant.

due to chance. Applying the Kruskal–Wallis *H*-test gives $H = 2.28$, $p = 0.32$. We can conclude that the differences between the groups are not statistically significant; there is about a 32% chance that differences as large as those observed could occur purely by chance. Given that the overall result is not statistically significant we do not proceed to compare individual pairs.

9.8 Jonckheere–Terpstra Trend Test

9.8.1 When It Is Useful

The Jonckheere–Terpstra trend test, sometimes called the Terpstra–Jonckheere test or Jonckheere's trend test, is similar to the Kruskal–Wallis *H*-test but it allows you to test for a trend across three or more categories (Terpstra 1952; Jonckheere 1954). The difference is that the *H*-test treats the categories as nominal, which means that they can be given names but they cannot be placed in any logical order. In geography, we often use categories that can be placed in a logical order, and which therefore are ordinal rather than nominal, and we can use the Jonckheere–Terpstra trend test to see if there is a trend rather than just an overall difference. Examples of classes that are ordinal rather than nominal include age, income, socio-economic groups, size, level of education, etc. Although rarely used, this a very useful test in both physical and human geography. It is a non-parametric test so can be used on small samples and the sample sizes do not have to be the same. There is no assumption of normality or equal variance, just that the data are continuous, which means that they are individual measurements rather than counts in categories.

A great advantage of this test is that it can be used in place of correlation and regression. Those methods require that the data be continuous, but it is quite common to see them applied to data that are clearly not continuous but fall into separate groups. Sets of data measured at several sampling stations down a river, for example. The Jonckheere–Terpstra trend test not only allows the statistical significance of a trend across categories to be calculated, it is also possible to calculate an effect size 'r' that is equivalent to Pearson's correlation coefficient and thus quantify the strength of the trend.

9.8.2 How It Works

The Jonckheere–Terpstra trend test can be seen as an extension of the one-tail Mann–Whitney *U*-test. In the *U*-test, the two samples are ranked, ignoring the groups, and then the sum of ranks in each group are compared, taking account of any differences in sample size by calculating the *U*-statistic. For the one-tail test, you predict in advance which sample will have the higher median and thus produce the lower of the two possible *U*-values. Although there are several different ways of making the calculations for the Jonckheere–Terpstra trend test, the simplest is to perform a one-tail *U*-test on all possible pairs and then take the sum of the *U*-values. The Mann–Whitney *U*-test is described in Section 8.5. If you do this be very careful, because it is critical that you take the correct set of *U*-values; they must be the ones that are predicted by your hypothesis of a trend across the categories. Some software and websites will give you a one-tail probability but it is based on the lower of the two *U* values without taking account of the direction of the prediction. Using the lower *U*-value, irrespective of which one was predicted to be lower, will produce nonsense for the Jonckheere–Terpstra trend test.

To perform the Jonckheere–Terpstra trend test, it is critically important that you put your categories in a sensible order, and that order has to be based on logic, not by looking at your data. For example, if you were interested in whether there is a general increase in exam results over the 3 years of a degree scheme, and have data from three different groups of students, then the logical order is 1–2–3 and your hypothesis is that the marks (actually it is the median of the marks) will fall in the order 1 < 2 < 3. If you were interested in the marks obtained in the same year by three different cohorts of students, for example Welsh, English and Scottish, and you observe that the marks fall in the order Welsh > Scottish > English, it is not valid to use the Jonckheere–Terpstra trend test, because this ordering is not based on some theory or logic (it is not *a priori*) and the categories are really nominal rather than ordinal (there is no logical order).

When your categories have been arranged in logical order, the test works by comparing the values in each category to the values in the categories to the left and the right. If there is no trend, then we might expect a rather random mix, with plenty of higher and lower values to either side. If there is a very strong rising trend across the categories, then we might expect the values in one of the middle categories to have mostly lower values to the left and higher values to the right.

With small samples, it is possible to calculate the Jonckheere–Terpstra trend test by hand, but it is tedious and easy to make a mistake, so I would not recommend it. I have not managed to find an online calculator that will perform this test and I cannot find a plug-in for *R* Commander either, though there is R code. The simplest option is to use SPSS, which works very well and also produces useful graphs. I have also produced a simple calculator, on the companion site, that can be used to compare up to six groups with sample sizes up to 100 and it will also perform post hoc tests and calculate the effect size 'r'.

9.8.3 One- and Two-Tail Testing, Post Hoc Tests and Effect Size

This test is only logical if there is some sensible order to the groups, so that they can be placed in that order without consulting the results. For example, age groups can be arranged from old to young or young to old but no other arrangement is sensible. For a one-tail test, you should be able to predict, in advance and on the basis of logic or theory (not by peeking at the data), whether the values should increase or decrease across the groups. If you cannot make such a prediction, then apply a two-tail test.

Since the simplest way to calculate the Jonckheere–Terpstra trend-test statistic (J) is by conducting one-tail Mann–Whitney U-tests on all possible pairs, those individual results would seem to be convenient post hoc tests. Each comparison can be given a probability and the Bonferroni correction (Section 9.2) used to take account of family-wise error resulting from multiple testing. It is also reasonable to produce an effect size for the overall analysis. If the samples are not very small (<8 perhaps), it is reasonable to convert the test statistic J into a z-score and from that calculate the effect size 'r' using the usual procedure of dividing the z-score by the square root of the total sample size. A worked example is given in Box 8.3. The effect size r is particularly appropriate in this case because it is equivalent to Pearson's correlation coefficient, so applying the Jonckheere–Terpstra trend test and calculating the effect size r is equivalent to using correlation, but on data that is arranged in groups, rather than being continuous.

9.8.4 How to Do It

9.8.4.1 In SPSS

To run the Jonckheere–Terpstra trend test in SPSS, you have to arrange your data in the 'long' format, which means that you place all of the data in one column and in the next column identify which group each data point belongs to. It is important to identify the groups in a way that SPSS will recognize as the logical order, such as numbers or sequential letters. Do not use names, otherwise SPSS will treat them in alphabetical order. I like to do this in a spreadsheet and then cut and paste (without the headers) into the 'data view' part of SPSS (the bit that looks like a spreadsheet). Now click the 'variable view' tab at the bottom left and you can give your data and groups a name. You also have to define the 'measure', which for your data is 'scale' and for your groups is 'ordinal', Now there are three clicks: 'Analyze', 'Non-parametric tests' and 'Independent samples'. As usual with SPSS do not let it do anything automatically. On the 'Objective' tab, choose 'Customize analysis' and on the 'Fields' tab, put your data in the 'test field' and your groups in the 'Groups'. On the 'settings' tab, choose 'Customize analysis' and tick the 'Test for ordered alternatives (Jonckheere–Terpstra for k-samples)'. Now you have to choose the 'Hypothesis order'. When you arranged your groups in the spreadsheet column, were you predicting that the values would increase from the top group to the bottom group? If so then choose 'Smallest to largest', otherwise 'Largest to smallest'. As with the Kruskal–Wallis H-test, you can also choose your post hoc tests, either comparing all possible pairs or, perhaps more logically, a 'stepwise step-down' procedure. It is easier to understand the output if you see a worked example.

9.8.4.2 Companion Site Calculator

The companion site calculator can cope with up to six groups and individual sample sizes up to 100. Simply enter or cut and paste your data into the columns, one column for each group, and the results will appear. The important numbers are the test statistic '*J*', which is the sum of the one-tail *U*-values, a *z*-score and one- and two-tail probabilities. The sign of the *z*-score indicates whether the trend is rising (*z* is positive) or falling (*z* is negative) so if you are using a one-tail test make sure that the trend is in the direction that you predicted. Also provided is the effect size '*r*' which can be treated as a correlation coefficient measuring the strength of the trend. This calculator ignores tied ranks, so if you have a lot of ties between groups the results may be slightly conservative.

A table is also produced giving the uncorrected one-tail probabilities for all possible pairwise comparisons conducted using the *U*-test. If you have predicted a falling trend, use the values in the upper right and if you predicted a rising trend, use the values in the lower left. When a comparison has median values that are in the opposite direction to the prediction, the probability is returned as one. If your original hypothesis was two-tail, in that you did not predict the direction of the trend in advance, and the result of the Jonckheere–Terpstra trend test is significant, you can still use these individual *U*-test values but you have to double all of the probabilities. Even though your overall test is two-tail, you still only look at one half of the *U*-values; the half that fit the overall trend indicated by the sign of the *z*-score.

If you decide to use the pairwise comparisons, it is a good practice to make a correction to take account of the family-wise error that results from making multiple comparisons (Section 9.2). The simple option of comparing all possible pairs results in a very harsh correction if there are more than three groups (critical threshold of probability divided by the number of tests). An alternative is to reduce the 'family' of tests by treating one of the samples as a control, if that is logical, or by comparing adjacent pairs in logical sequence. With six groups, for example, this leads to a family of five tests rather than 15 if all pairwise comparisons are made. The Bonferroni-corrected probabilities for these two options are provided as well. For each of the pairwise comparisons, the effect size '*r*' is also provided in a separate table.

9.8.5 Examples

9.8.5.1 Example: Age Group Opinions

You wish to investigate whether age has an influence on levels of anxiety about proposed changes to old-age pension provision. You are not sure whether concern will decline with age, because the older you are the lower the effect new changes will have, or whether they will increase with age because younger people have other things to worry about. Your hypothesis is thus two-tail. You take three age classes of people (<30, 30–60, >60) and ask them to mark their level of concern on a 10 cm straight line marked 'not at all concerned' at one end and 'extremely concerned' at the other. You measure the distance of their cross on the line in mm to give a score in the range 0–100. The data do not have to be normally distributed and sample sizes do not need to be the same, so the artificial data shown in Table 9.26 would be appropriate.

Ignoring the logical order of the three age classes, you could use the *H*-test (Table 9.27) to see if there is a significant difference between the three age classes overall. The result is not statistically significant ($p = 0.056$), which might lead you to conclude that age has no effect on level of anxiety. However, running the Jonckheere–Terpstra trend test produces a

TABLE 9.26

Where Data Are Arranged in Groups That Fall in a Logical Order, Such as Age Groups, the Jonckheere–Terpstra Trend Test Can Be Used to Test for a Rising or Falling Trend across the Groups

Age	<30	30–60	>60
	Young	Middle	Old
Count	21	30	26
Average	42.48	35.13	48.50
Median	36	32	47.5
Age	<30	30–60	>60
	24	12	47
	35	30	23
	58	46	49
	49	32	69
	62	36	82
	27	42	59
	83	17	76
	58	12	35
	10	18	42
	4	20	16
	28	47	73
	64	95	47
	50	34	23
	79	25	48
	34	2	96
	94	4	46
	59	35	49
	23	27	74
	12	43	69
	3	58	74
	36	42	72
		8	35
		16	4
		32	25
		68	16
		75	12
		23	
		54	
		75	
		26	

statistically significant result ($p = 0.029$). The results are different because they are testing different hypotheses. The *H*-test cannot make use of the order of the categories, so it cannot test for a trend. The Jonckheere–Terpstra trend test can use that extra information and so is the more powerful test in this situation.

Before getting too excited, you should look at the effect size, '*r*', which is equivalent to a correlation coefficient. In this case, $r = 0.25$ indicating a weak effect (<0.3). Also, although two of the pairwise comparisons (A:C and B:C) provided by the companion site calculator

TABLE 9.27

Summary Results for the Three Age Groups and Results of the Kruskal–Wallis *H*-Test and Jonckheere–Terpstra Trend Test

Age	<30	30–60	>60
	Young	Middle	Old
Count	21	30	26
Median	32	32	47.5
	H or *J*	Two-tail *p*	One-tail *p*
H-test	5.76	0.056	na
JT trend test	740.5	0.029	0.015

Note: The probabilities are different because they are testing different hypotheses. There is no one-tail probability for the *H*-test because it treats the categories as nominal, so a direction of trend cannot be predicted.

are statistically significant ($p = 0.017$ and $p = 0.022$), these are one-tail probabilities and have to be doubled to take account of the two-tail hypothesis. Ignoring family-wise error they are still significant ($p = 0.034$ and 0.044) but even taking just the two adjacent pairs as the family of tests requires that the critical probabilities be divided by two, so a threshold of 0.05 becomes 0.025, so that neither passes the threshold. A reasonable conclusion would be that there is a statistically significant but very weak increase in levels of anxiety with increasing age across the three groups.

9.9 Friedman's Test

Friedman's test, also known as Friedman's sum of ranks test, or to give it the full name 'Friedman's two-way analysis of variance by ranks' (Friedman 1937, 1940) is another non-parametric test that is rarely used in geography but is potentially very useful indeed, particularly in human geography.

9.9.1 When It Is Useful

It is used where you have three or more samples that are linked rather than independent. For example, if you have exam marks obtained by the same group of students in three different examinations, then the results are linked because each student has three results. It treats the groups as nominal, meaning they cannot be placed in any logical order. If you are interested in looking for a trend across groups that do fall in a logical order, use Page's trend test.

9.9.2 What It Is Based On

Like most non-parametric tests it works using ranks, so we do not need to have real measurements, just the relative order. The null hypothesis is that there is no difference between the groups, in which case we would expect the ranks to be randomly distributed between the groups, so that the sum of ranks values are all very similar. If, on the other hand, one or more groups contain unusually high or low values, then the sum of ranks values will be different.

A great advantage, especially in human geography, is that you can use this test to compare groups where ranking is easy but full measurement is difficult. For example, you may be interested in whether people view several landscapes as equally beautiful. Beauty is a difficult thing to measure, so asking someone to quantify the beauty or attractiveness of images of landscapes on some numerical scale is quite problematic. However, it is much easier to just place several different images in rank order, and that is all that is required for this test. This procedure also avoids the problem of different groups of people viewing the same numerical scale in different ways.

Unfortunately, there seems to be no standard shorthand for the number that is generated by this test, which is potentially very confusing. Real Stats uses 'H', SPSS uses 'test statistic', R Commander uses 'Friedman chi-squared', Neave and Worthington (1988) use 'M', Siegel uses χ^2 and Field (2013) uses F_r. Here, I will follow Field (2013) and use F_r on the basis that it hints at Friedman and ranks. Whatever it is called, this number has the advantage that if samples are not very small, it fits the chi-square distribution with k-1 degrees of freedom. Suitable tables are provided (Section 12.2).

9.9.3 One- and Two-Tail Testing, Post Hoc Tests and Effect Size

Friedman's test treats the categories as nominal, so the order in which they are arranged is irrelevant. This makes it impossible to predict any direction, such as a trend across the groups, in advance, so a one-tail test makes no sense and the probabilities are always two-tail. If there is a logical order to the categories and you wish to test for a trend, use Page's trend test.

As with the k-sample tests for independent samples, there are also some post hoc tests available so that you can check not just if there is an overall difference between your categories but also where the big differences lie. The most convenient way to perform the post hoc tests is using SPSS. You can either compare all possible pairs, or use a 'stepwise step-down' procedure to place the categories into homogeneous groups. In the companion site calculator, Wilcoxon's matched-pairs signed-ranks tests are used for the pairwise comparisons.

As with the H-test, it is not appropriate to calculate an effect size for Friedman's test because it is looking at an overall effect over several groups rather than a 'focussed comparison' (Field 2013), however, the effect size 'r' can be computed for the individual pairwise comparisons. SPSS does not do this automatically, but if Friedman's test is significant, double clicking on the output summary for the pairwise comparisons provides a table with further information including a 'standardized test statistic' which is a z-score. This can be translated into the effect size 'r' by dividing it by the square root of the sample size. The appropriate sample is double the number of differences between the two categories being compared, as for Wilcoxon's matched-pairs signed-ranks test (Section 8.2). If the overall Friedman's test is not significant, the pairwise comparisons are not shown. You could compute the pairwise comparisons individually using Wilcoxon's matched-pairs signed-ranks test (Field 2013). The companion site calculator provides pairwise comparisons and effect sizes.

9.9.4 How to Do It

With just a few groups, which is usually the case, the ranking procedure for Friedman's test is very easy, and the full sample size does not matter much because you just have to add up the ranks, so you can calculate the test by hand. The equation looks a bit daunting but step-by-step instructions are given in Box 9.2. The only simple online calculators I have been able to find are provided by Vassar stats (http://vassarstats.net/index.html), listed under 'ordinal data'. There are separate calculators for three or four groups ($k = 3$ or 4) but not for more than four. The Real Statistics Resource Pack for Excel is the most versatile solution and is very easy to use. Alternatively, you can use SPSS or *R* Commander. The only easy way to run post hoc tests to compare pairs of groups is SPSS or using the companion site calculator.

9.9.4.1 Real Statistics Resource Pack for Excel

If you have loaded the Resource Pack (Zaiontz 2015), use Ctrl-M to obtain a list of options and choose 'Non-parametric Tests'. Now identify the input range, which is three or more columns of data, note whether you have included headers, then identify where you want the results to appear. That is all there is to it. The output includes the *p*-value, which for this test is always two-tail. The *H*-stat number is the test statistic, which fits the chi-square distribution with $k–1$ degrees of freedom. Real Statistics does not currently provide post hoc tests for Friedman's ANOVA but new elements are constantly added, so check the most recent version. Run pairwise comparisons using Wilcoxon's matched-pairs signed-ranks test, which is supported.

9.9.4.2 Companion Site Calculator

The companion site calculator will accept up to six groups with sample sizes of up to 100. Simply enter the data in the columns marked A to F and leave any empty cells blank. The output includes Friedman's statistic 'T_r', the degrees of freedom (number of groups minus one) and the two-tail probability. The probability is calculated using the chi-square approximation. Note that this calculator does not deal with tied ranks and if there are a lot of ties, the results will be conservative (*p*-values slightly higher than they should be).

If the overall result of Friedman's test is significant, it is often useful to make some pairwise comparisons to see where the differences lie. The calculator uses Wilcoxon's matched-pairs signed-ranks test and two-tail probabilities are provided for all possible pairs. An effect size '*r*' is also provided for each pairwise comparison. When making several post hoc comparisons, it is a good practice to adjust the critical significance levels to take account of the 'family-wise error' that accompanies multiple testing (Section 9.2). For convenience, Bonferroni-corrected critical significance values are provided for the options of comparing adjacent pairs only (in order of median values) or for comparing all possible pairs. The procedure used here is not as good as that used by SPSS, which uses more sophisticated post hoc tests, so if you have access to SPSS this is a better approach. It is not difficult to prepare the data or run the test in SPSS and the output is not too complicated.

9.9.4.3 In SPSS

Enter or cut and paste the data as columns, without headers, into the 'Data View' spreadsheet page. Go to 'Variable view' and assign names and identify the 'Measure' as 'Scale'. Now click, 'Analyze', 'Non-parametric Tests' and 'Related samples'. Choose 'Customize analysis' and on the 'Fields' page move all of the data into the 'Test Field' box. Under 'Settings' choose the option 'Friedman's two-way ANOVA by ranks (*k*-samples)'. Also decide on the post hoc tests, 'all pairwise' or 'stepwise step-down'. The output summary gives the two-tail probability and only if this is significant can you access the post hoc test results. Double click on the 'Hypothesis Test Summary' box and you will see a table with some more results At the bottom of the page, there is a 'View' tab and one of the options there will be either 'Pairwise Comparisons' or 'Homogeneous Subsets'. If 'All Pairwise' was chosen, there will be a table showing all possible comparisons with a significance level and also an 'adjusted' significance level. The latter have been adjusted using a Bonferroni-type correction (Section 9.2), so the original values are multiplied by the number of tests. If 'Homogeneous Subsets' was chosen the samples are arranged in rank order and separate groups are placed in different columns.

9.9.4.4 R Commander

Place the data as columns in the top left corner of a clean spreadsheet. Use simple headers without spaces. Now save it as a text file (File, Save as, under the 'save as type' box choose 'text (tab delimited)'). In *R* Commander use 'Data', 'Import data' and the first option 'from text file'. Leave the default settings. Go to 'Statistics', 'Non-parametric tests' and choose 'Friedman rank-sum test'. Choose all of your columns and click OK. The output includes the median value, the test statistic which is called 'Friedman chi-squared', the degrees of freedom and the probability value.

9.9.5 Examples

9.9.5.1 Example: Exam Performance

You have taken a small but unbiased sample of 10 first year geography students and recorded their marks in three compulsory courses: physical geography, human geography and numerical methods. You wish to know whether there is a tendency for students to do equally well (or badly) in all three courses. Here, you can see the actual marks that they obtained (Table 9.28), but all you really need is the relative order of those marks. In this case, they are ranked from lowest mark (1) to highest mark (3), but you could rank them the other way, from high to low, and get the same result. Note that where two numbers are exactly the same they get the average of the two ranks. If three numbers were identical, they would also get the average rank (which would be $1 + 2 + 3 = 6$ divided by $3 = 2$).

Running Friedman's test using the data in Table 9.28 in *R* Commander or in SPSS gives $F_r = 8.895$ with two degrees of freedom and a two-tail probability of $p = 0.012$. It is clear from the sum or ranks values that the results for the physical and human geography courses are very similar but the methods course has a higher sum of ranks. Since the overall test is significant, it is appropriate to look at pairwise comparisons, which is easiest in SPSS. Choosing 'All pairwise' comparisons, the output includes the information in Table 9.29. The difference between physical and human courses is not significant, even ignoring the adjustment for multiple sampling, but the other two comparisons are significant even when the adjustment is made.

If Friedman's test is run using the same data but using the companion site calculator, the results are slightly different, with $F_r = 8.45$ and $p = 0.015$. The reason for the difference is that the simple calculator does not deal with the tied ranks, and so the probabilities are slightly conservative. If there are no ties, the results for Friedman's test will be identical. However, SPSS also performs more appropriate post hoc tests, so it is the best option. Although SPSS does not give effect sizes for the post hoc tests, they can be calculated by dividing the z-score (standardized test statistic) by the square root of the sample size,

TABLE 9.28

Example of Data That Are Suitable for Friedman's Test Because the Data Are Linked

	Marks Physical	Marks Human	Marks Methods	Ranks Physical	Ranks Human	Ranks Methods
Adrian	63	65	68	1	2	3
Cathy	32	23	33	2	1	3
Denise	68	77	74	1	3	2
Frank	64	60	66	2	1	3
Helen	52	54	60	1	2	3
Kath	57	57	62	1.5	1.5	3
Mike	78	58	60	3	1	2
Ruth	50	48	55	2	1	3
Tom	56	60	60	1	2.5	2.5
Wynn	78	67	84	2	1	3
F_r	8.45		Sum	16.5	16	27.5
$p =$	0.015		Average	1.65	1.6	2.75

Note: In this case, each individual student has produced one result for each examination. The sum (or average) of the ranks gives an immediate impression of the differences between the three exams.

which in this case is twice the number of samples. For the physical-methods comparison, the effect size 'r' is

$$r = \frac{z}{\sqrt{n}} = \frac{-2.46}{\sqrt{20}} = \frac{-2.46}{4.47} = -0.55$$

Since r is larger than 0.5 (the sign is ignored), this would generally be considered a large effect.

9.9.5.2 Example: Icons of Nationalism

You are interested in nationalism and the feeling of 'belonging' and are conducting a mainly qualitative study of what makes local residents feel Welsh. There are a few iconic images that people tend to identify with Wales and 'Welshness' and you are interested to see how they compare. You show an unbiased sample of Welsh people a set of five images and ask them to place them in order according to how strongly they associate them with

TABLE 9.29

Some of the Output from SPSS when 'All Pairwise' is Chosen for the Post Hoc Tests after Running Friedman's Test

Comparison	z-Score	Sig (p)	Adj. Sig.
Physical-human	0.0112	0.911	1.00
Human-methods	−2.57	0.01	0.03
Physical-methods	−2.46	0.014	0.042

Note: The z-scores are labelled 'standardized test statistic'. The adjusted significance values take into account the family-wise error resulting from multiple testing. In this case, there are three tests so the probability values are multiplied by 3 (up to the maximum of 1.0).

a feeling of being Welsh. They are allowed to say that two or more images are equal. The easiest way to do this would be to give them each five printed images and ask them to place them on the table in order, allowing two or more to be placed on top of each other. You can then record the ranks, including any tied ranks. You can see from some fabricated data in Table 9.30, which uses a sample of 18 because I ran out of nice Welsh names, that there are big differences in the way that people view these images. In this case, they were asked to rank from 'most' to 'least' and it is clear from the sum or average of ranks that the red dragon ranks very highly whereas Prince Charles is at the other end of the scale (on the basis that he has an English accent).

If you were to put 18 balls numbered one into a bag, together with 18 each of numbers 2–5, and then pick out the balls at random and just throw them into five equal piles, the most likely result is that all of the piles would add up to about 56 (the average rank is 3 so 3 times 18 is 56). It is extremely unlikely that you would end up with one pile that summed to 23 and another that summed to 78.5. Friedman's test tells us the probability of differences in sum of ranks of this magnitude occurring just by chance. In this case, the probability is millions to one ($F_r = 43.6$, $df = 4$, $p = 7.9E–09$), but quoting it as $p < 0.001$ is perfectly adequate.

Note that the sum or ranks now provides a simple way to put the five icons in rank order, which is dragon > feathers > leek > rugby > Charles. Running the analysis in SPSS and choosing 'Homogenous subsets' for the post hoc tests places the icons into overlapping groups based on similarity. Dragon falls in a group on its own followed by feathers and leek, which overlaps with leek and rugby, which overlaps with rugby and Charles.

TABLE 9.30

Results of a Group of Welsh People Placing Five Images in Rank Order According to How Strongly They Associate Them with Wales and 'Welshness'

	Dragon	Leek	Feathers	Charles	Rugby
Bethan	2	3	1	5	4
Ceri	1	3	2	4	5
Cerys	1	2.5	2.5	5	4
Darryl	1	4	3	5	2
Delyth	2	1	3	4	5
Eleri	1	3	2	5	4
Essllyt	1	2	3	5	4
Gavin	1	5	2	3	4
Gethin	1	4	2	5	3
Idris	2	3	1	5	4
Ieuan	1	3	3	5	3
Llinos	1	4	2	4	4
Meredith	1	2	3	4.5	4.5
Nerys	3	3	3	3	3
Osian	1	3	2	5	4
Pyrs	1	2	3	4	5
Rhiannon	1	2	3	5	4
Sian	1	3	4	2	5
Sum	23	52.5	44.5	78.5	71.5
Average	1.28	2.92	2.47	4.36	3.97

BOX 9.2 CALCULATING FRIEDMAN'S TEST BY HAND

The equation for calculating Friedman's test looks a bit daunting, but if you split it into three parts, calculate them separately, and then combine them at the end, it is not too bad. All you need to know is the sample size (N), number of groups (k) and the sum of ranks for each of the groups (R). In this example, we can use the results presented in Table 9.28, where the sample size was 10 and the sums of ranks for the three groups ($k = 3$) were: 16.5, 16 and 27.5. I will use 'F_r' as the test statistic.

The full equation is

$$F_r = \frac{12}{Nk(k+1)} \sum (R)^2 - 3N(k+1)$$

But we can split it into three parts

$$F_r = \frac{12}{Nk(k+1)} \times \sum (R)^2 - 3N(k+1)$$

$$F_r = (Part\ one) \times (Part\ two) - (Part\ three)$$

To solve part one

$$\frac{12}{Nk(k+1)} = \frac{12}{10 \times 3 \times (3+1)} = \frac{12}{10 \times 3 \times 4} = \frac{12}{120} = 0.10$$

Part two looks more complicated than it is, it just means take each sum or ranks value (R) and square it, then add them up, so

$$\sum (R)^2 = 16.5^2 + 16^2 + 27.5^2 = 272.25 + 256 + 756.25 = 1284.5$$

Part three is not too bad

$$3N(k+1) = 3 \times 10 \times (3+1) = 3 \times 10 \times 4 = 120$$

Now, we can put the three parts together, recalling the BODMAS rule that means you have to multiply before you subtract

$$F_r = (Part\ one) \times (Part\ two) - (Part\ three)$$

$$F_r = 0.10 \times 1284.5 - 120$$

$$F_r = 128.45 - 120 = 8.45$$

The test statistic here called 'F_r' fits the chi-square distribution with k-1 degrees of freedom, so in this case with $k = 3$ that means two degrees of freedom. You could look up the probability using tables of the chi-square distribution (Section 12.2) or just use an online calculator or spreadsheet to find the answer. In Excel, for example, you can use the CHISQ.DIST.RT function, enter 8.45 as 'X' and 2 as degrees of freedom and you get a probability value of 0.0146, the same as in Table 9.28. Note that in this case, there is no correction applied for the tied ranks, which means that (if there are ties) the results will be on the conservative side.

9.10 Page's Trend Test

This is a test that is rarely seen in statistical text books. I came across it in Neave and Worthington (1988), and I have not seen it used in geography yet. However, I think it has considerable potential in both physical and human geography as an alternative to the misuse of correlation and regression on small sets of data or where only the relative order of results is known.

9.10.1 When It Is Useful

It is similar to Friedman's test in that it is used where you have linked data for more than two groups, but in this case, the groups are treated as ordinal rather than nominal, which means that they can be placed in some logical order (Page 1963). It allows you to investigate whether there is a general trend across your groups by testing your prediction for a general increase or decline.

In physical geography, I often see student projects that involve looking at changes in several different parameters, measured at just a few points, but with a clear sense of direction. For example, you might measure levels of metal pollution at five points down a river, downstream of a source of pollution such as an old mine. Measurements of some microclimate parameters, or of vegetation with increasing altitude is another quite common example. The default method for investigating the data in such projects tends to be a set of x–y plots showing how each measured parameter changes along the profile, often with a regression line and R^2 value produced in a spreadsheet. With sample sizes as low as four or five points, however, these methods are not really suitable and it is also very difficult to summarise the results of a set of such scatter graphs, particularly where the unit or scale of measurement changes. Page's trend test provides a very useful alternative.

Human geography students also come to me with problems that are similar except that they tend to be dealing with change across time rather than space. For example, you might wish to compare how several indicators of poverty or wellbeing have changed over time, but data are only available for specific intervals, such as census or survey dates. Spatial trends can also be investigated of course, for example, you might be interested in changes with distance from the city centre, with the measurements taken as the average values in concentric circles.

9.10.2 What It Is Based On

The logic of this test is similar to that of Friedman's test in that the linked measurements are placed in rank order and the test uses the sum of the ranks for each group. If there was no difference between the groups, we would expect the sum of ranks values to be very similar. The difference in this case is that the sum of ranks value for each group is weighted according to the order in which you expect the trend to occur. The test statistic '*L*' is obtained by adding up the sum of rank values after they have been weighted. The first group has a weight of one, the second is multiplied by two, the third by three, etc.

$$L = (1 \times R_1) + (2 \times R_2) + (3 \times R_3) + (4 \times R_4) \text{ etc.}$$

When you use this test, it is very important that you rank your data in the correct way according to the direction of the trend that you expect. They must be arranged so that the smallest sum of ranks is expected to occur in the first group and the highest sum of ranks in the last group. You can make sure you get it the right way around by just deciding whether it is appropriate to rank from low to high (where you predict a rising trend) or from high to low (where you predict a falling trend). As with Friedman's test, you do not need actual measurements, it is enough to know the relative order of the results. For example, if you asked a sample of people to taste a range of wines, falling over a range of prices, you could just ask them to place the wines in rank order from worst to best. The logical order here is the price of the wines and the logical order is from cheapest to most expensive. If quality is linked to price, you would expect to see an increase in the sum of ranks.

9.10.3 One- and Two-Tail Testing, Post Hoc Tests and Effect Size

The fact that you have to predict the direction of the trend in advance, so that you can decide on the correct form of ranking, makes this test automatically one-tail. However, if you had no idea whether there should be a trend or not, I suppose you could run the test in both directions, by ranking up then ranking down, and then double the significance value that is provided by the normal approximation, so that it changes from one- to two-tail.

I am not aware of any specific post hoc tests for Page's test, but since the two-tail Wilcoxon matched-pairs signed-ranks tests are an acceptable alternative to more specific post hoc tests for Friedman's test (Field 2013) I see no reason why one-tail versions would not serve the same purpose in this case. The companion site calculator thus performs one-tail Wilcoxon matched-pairs signed-ranks tests, assuming that values should increase from left to right across the columns of data. Where the order is reversed, the use of one-tail tests requires that the null hypothesis be accepted, so in those cases, the probability is returned as one.

Since Page's trend test is testing for a trend across the groups, in a direction that is specified in advance, it seems reasonable to use an effect size and the most appropriate is the effect size '*r*' which is equivalent to a correlation coefficient. In this case, it can be considered equivalent to Spearman's rank correlation coefficient as an indicator of the strength (rather than statistical significance) of the trend. Since Page's *L* quickly becomes normally distributed, the effect size '*r*'; can be obtained from the *z*-score divided by the square root of the sample size.

9.10.4 How to Do It

I cannot find any online calculators for this test, it is not included in the Real Statistics Resource Pack for Excel, in any plug-ins to *R* Commander that I can find or even in SPSS. The only option seems to be to make the calculations by hand, or using a spreadsheet. I have also produced a simple calculator on the companion site that can cope with up to six groups and sample sizes up to 100. Fortunately, if the samples are small and you are happy to use tables of significance levels, then the calculation is quite easy. If you have larger samples or want an estimated probability value and effect size, then it is possible to use a normal approximation method which is described in Box 9.3. Although the equations look daunting, they are mostly functions of the sample size and number of categories, so a spreadsheet can be set up to make the calculations for you. If your sample sizes, or number of categories, are too large for the companion site calculator, there is another simple calculator that will apply the normal approximation and provide a probability and effect size. You just need to calculate L, which is not difficult at all, and enter the size of the samples and the number of groups.

9.10.5 Examples

9.10.5.1 Example: Metal Pollution in a River

You are interested in water quality and wish to test the hypothesis that metal pollution will decline with distance from an old mine and spoil heap. You decide to measure metal concentrations at five easily accessible points (A–E) down the river. The concentration values for different metals are very different, so this kind of data is difficult to summarise on a single graph (see Section 11.9 for a method). However, for this test, we only need to know the relative order of the concentrations along the river (Table 9.31).

In this example, our hypothesis is that there will be a general decline in concentrations downriver, so we need to rank in such a way that if that hypothesis were true, there would be an increase in the sum of ranks downriver. That means that the highest pollution values have to be given a rank of one and the lowest a rank of five. Where the concentrations for two or more sites are not distinguishable, using the measurement methods available, the ranks are shared (as for arsenic).

It is clear from the sum of ranks values that the data support our prediction that there will be a general decline in metal concentrations downstream. To test whether the trend that we see is statistically significant, we can calculate the test statistic L

$$L = (1 \times R_1) + (2 \times R_2) + (3 \times R_3) + (4 \times R_4) + (5 \times R_5)$$
$$L = (1 \times 7.5) + (2 \times 15.5) + (3 \times 23) + (4 \times 26.5) + (5 \times 32.5)$$
$$L = (7.5) + (31) + (69) + (106) + (162.5) = 376$$

TABLE 9.31

Metal Pollution Levels at Sites Arranged in Order of Distance Downriver of a Mine

	Site A	Site B	Site C	Site D	Site E
Lead	1	2	3	5	4
Zinc	1	2	4	3	5
Cadmium	1	3	3	3	5
Arsenic	1.5	1.5	4	4	4
Nickel	1	2	3	4	5
Copper	1	2	3	4.5	4.5
Chromium	1	3	3	3	5
Sum of ranks	7.5	15.5	23	26.5	32.5

Note: Only the relative order needs to be known not the actual values.

Using tables (Section 12.10) for $n = 7$ and $k = 5$, we obtain critical values for $p = 0.05$ and 0.01 of 338 and 346. Since our measured value of L is 'greater than or equal to' the critical value for $p = 0.01$, we can conclude that there is less than a 1% chance that the trend that we observe could have occurred just by luck. Using the normal approximation, as described in Box 9.3 and obtained using the companion site calculator the probability is given as $p = 0.0000002$ with a large effect size of 0.87.

Incidentally, this is not a great dissertation topic because it is really rather obvious that metal concentrations are going to decline as you move away from the source of pollution. It might be more interesting to look at how the metal concentrations behave in different sediment grain size fractions or under different river flow conditions.

9.10.5.2 Example: Rental Prices with Distance from City Centre

We wish to investigate whether there is a trend in accommodation rental prices with distance from the centre of a city. We obtain rental prices for several different types of property in four zones surrounding the centre. Our hypothesis is that prices will decline with distance, so the appropriate way to rank the data is from highest price (1) to lowest price (4). If our hypothesis is true, we would predict that the sum of ranks values will increase from zones 1 to 4.

In this case, the results are not quite so clear (Table 9.32). There is an increase across zones 1–3 but then the sum or ranks falls, but not by very much. Just by looking at these results, it is not easy to judge whether the trend could reasonably have occurred just by chance, because it is a small sample. This of course is precisely why inferential statistical tests are useful. In this case,

$$L = (1 \times R_1) + (2 \times R_2) + (3 \times R_3) + (4 \times R_4)$$
$$L = (1 \times 9) + (2 \times 13) + (3 \times 20) + (4 \times 18)$$
$$L = (9) + (26) + (60) + (72) = 167$$

Given a sample size of 6 ($n = 6$) and four groups ($k = 4$), the critical values from the tables in Section 12.10 for $p = 0.05$ and $p = 0.01$ are 163 and 167. Since our calculated value of L is 'greater than or equal to' the critical value for $p = 0.01$, we can conclude that there is less than a 1% chance that the trend that we observe could have occurred by chance. Using the

TABLE 9.32

Rank Order of Rental Prices in Four Zones Arranged by Increasing Distance from the City Centre

	Zone 1	Zone 2	Zone 3	Zone 4
Bedsit	2	1	4	3
One room apartment	1	3	2	4
Two room apartment	2	4	3	1
Two bedroom terrace house	2	1	3	4
Two bedroom semidetached	1	2	4	3
Three bedroom semidetached	1	2	4	3
	R_1	R_2	R_3	R_4
Sum of ranks	9	13	20	18

normal approximation, we obtain a probability of 0.008 and effect size of 0.57. Note that in this case, our measured value is the same as the tabulated critical value for $p = 0.01$ and if it had been just one point lower it would not have been significant at this level ($L = 166$ gives $p = 0.012$), confirming that the normal approximation is very good.

BOX 9.3   OBTAINING A PROBABILITY AND EFFECT SIZE FOR PAGE'S TREND TEST

The companion site includes a simple calculator that will calculate a one-tail probability value and effect size just by inputting the values for L, the sample size (n) and the number of groups (k). However, should you wish to have a go here is how it works. The normal approximation works better as the sample size increases, so for very small samples, it is safer to use the tables of critical values.

The probability is obtained from what is known as a 'normal approximation' and the equation has three parts to it: L, μ (mu) and σ (sigma). Apologies for the Greek letters, it seems to be a convention so I have kept them here. First, you have to calculate a 'z-score' using

$$z = \frac{L - \mu}{\sigma}$$

where
 z = a value from the standard normal distribution (z-score)
 L is the test statistic based on adding the weighted sum of ranks
 μ (mu) is the value of L that you would get if the sum of ranks values were all the same
 σ (sigma) is a function of the size of the sample (rows and columns)

Taking the second (human geography) example above, we have the following values to work with:
 $L = 167$, n (number of samples) is 6 and k (number of groups) is 4.

$$\mu = \frac{nk(k+1)^2}{4}$$

and

$$\sigma = \sqrt{nk^2(k+1)(k^2-1)/144}$$

Taking μ first we get:

$$\mu = \frac{nk(k+1)^2}{4} = \frac{6 \times 4 \times (4+1)^2}{4} = \frac{6 \times 4 \times 5^2}{4} = \frac{6 \times 4 \times 25}{4} = \frac{600}{4} = 150$$

And for σ we get

$$\sigma = \sqrt{nk^2(k+1)(k^2-1)/144} = \sqrt{6 \times 4^2 \times (4+1) \times (4^2-1)/144}$$

$$\sigma = \sqrt{6 \times 16 \times (4+1) \times (16-1)/144} = \sqrt{6 \times 16 \times 5 \times 15/144}$$

$$\sigma = \sqrt{\frac{7200}{144}} = \sqrt{50} = 7.071$$

And so z is equal to

$$z = \frac{L-\mu}{\sigma} = \frac{167-150}{7.071} = \frac{17}{7.071} = 2.404$$

To translate z into a probability value, you can use tables of the standard normal distribution (Section 12.14), an online calculator, or a spreadsheet function. In Excel, you can use the 'NORM.S.DIST' function where you add the z-value and in the second part say TRUE. This returns the cumulative distribution function of the standard normal distribution. What you need is one minus this number, so for this example

$$z = 2.404$$

$$\text{NORM.S.DIST } (2.404, \text{ TRUE}) = 0.992$$

$$p = 1 - 0.992 = 0.008$$

To obtain an effect size r (which apparently is equivalent to Spearman's rho in this case)

$$r = \frac{z}{\sqrt{n(k-1)}} = \frac{2.404}{\sqrt{6(4-1)}} = \frac{2.404}{\sqrt{6 \times 3}} = \frac{2.404}{\sqrt{18}} = \frac{2.404}{4.243} = 0.567$$

References

Dunn, O.J. 1959. Estimation of the medians for dependent variables. *Annals of Mathematical Statistics* 30: 192–197.

Dunn, O.J. 1961. Multiple comparisons among means. *Journal of the American Statistical Association* 56: 52–64.

Dunn, O.J. 1964. Multiple comparisons using rank sums. *Technometrics* 6: 241–252.

Friedman, M. 1937. The use of ranks to avoid the assumption of normality implicit in the analysis of variance. *Journal of the American Statistical Association* 32: 675–701.

Field, A. 2013. *Discovering statistics using IBM SPSS statistics* (4th ed.). London: Sage, 916pp.

Friedman, M. 1940. A comparison of alternative tests of significance for the problem of *m* rankings. *The Annals of Mathematical Statistics* 11: 86–92.

Jonckheere, A.R. 1954. A distribution-free *k*-sample test against ordered alternatives. *Biometrika* 41: 133–145.

Kruskal, W.H. and Wallis, W.A. 1952. Use of ranks in one-criterion variance analysis. *Journal of the American Statistical Association* 47: 583–621.

Neave, H.R. and Worthington, P.L. 1988. *Distribution-Free Tests*. London: Unwin-Hyman Ltd., 430pp.

Page, E.B. 1963. Ordered hypotheses for multiple treatments: A significance test for linear ranks. *Journal of the American Statistical Association* 58: 216–230.

Terpstra, T.J. 1952. The asymptotic normality and consistency of Kendall's test against trend, when ties are present in one ranking. *Indagationes Mathematicae* 14: 327–333.

Zaiontz, C. 2015. *Real Statistics Using Excel*. www.real-statistics.com.

10

Correlation

10.1 Introduction

One of the most common questions in geographical research and in many other areas of research is whether two or more variables are related in some way. Related in this sense means that, generally speaking, as one goes up the other one goes up as well, which we would call a positive relationship. A negative relationship occurs when one goes up and the other tends to go down. On several residential field courses I have noticed, for example, that Students' ability to chat to professors tends to be positively related to the amount of alcohol consumed, whereas the quality of the conversation shows a negative relationship.

Earlier in this book we have already met several tests that are used to look at relationships, but so far they have been restricted to data arranged either as counts in categories or where individual measurements have been taken from several groups that fall in some logical order. The chi-square test, for example, looks at the degree of 'contingency' between the parameters that form the rows and those that form the columns. Contingency is essentially a measure of the strength of a relationship. Where independent measurements are obtained from categories that have some logical order, the Jonckheere–Terpstra trend-test can be used to test whether there is a general increase or decline across the categories. For linked data the equivalent is Page's trend-test. This chapter deals with data that have been collected as individual pairs of measurements and where there are no discrete categories. When dealing with this kind of data the term correlation is used when describing the strength of a relationship.

Where the aim is to investigate the relationship between two variables the first step is to plot the data on an x–y plot or scatter-graph. This is simply a graph with a horizontal axis which, by convention, is called the x axis and a vertical axis known as the y axis. Strictly speaking, for correlation alone it does not matter which parameter is plotted on which axis, but if you want to put a line through your data, which is usually the case, then it actually matters a lot, for reasons that are explained in Chapter 11. The rule is that if you suspect that one parameter depends in some way on the other then the 'independent' measure goes on the horizontal axis and the 'dependent variable' goes on the vertical axis. For example, if you were plotting annual tree growth and annual mean summer temperature then you would put the temperature data on the horizontal (x) axis and the tree growth on the vertical (y) axis. They must be plotted this way because it is logical that summer temperature could influence tree growth but it is not logical that tree growth can change summer temperature. Similarly, if you were plotting salary against age you would place age on the horizontal because it is logical that salary might depend to some extent on age,

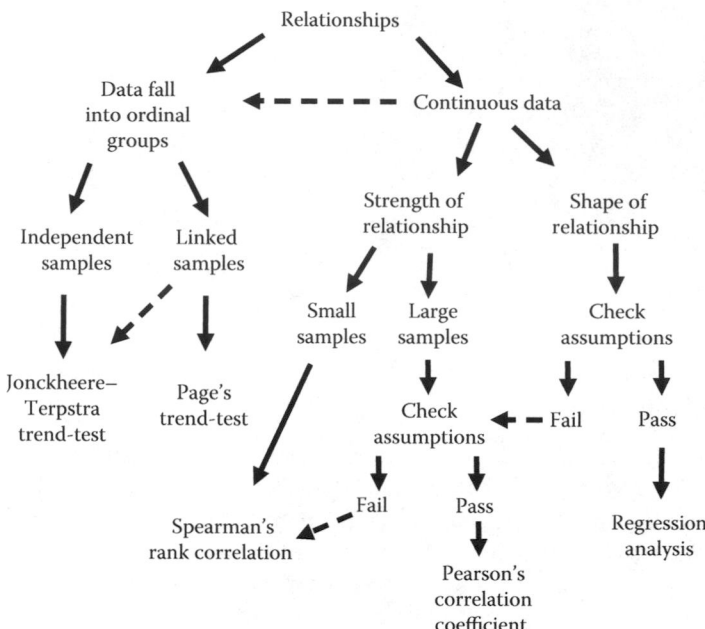

FIGURE 10.1
A logical series of steps in deciding which test to use to look at relationships.

but taking a pay cut, sadly, does not make you any younger. Sometimes there is no logical cause and effect inferred in the relationship. For example, you may be comparing a single property measured using two different methods, in which case it does not matter which axis they are plotted on (it still influences the line, but more on that later).

You can plot a scatter-graph by hand on graph paper of course, but it is much easier to do it in a spreadsheet. If you place the independent variable in the left column and the dependent variable to the right, highlight them and click 'insert' and 'x–y plot' it should all work automatically. Make sure that you label the axes properly and also adjust the axes to suit your data or you will surely lose marks. Sloppy graphs drive me mad, and that is true of many colleagues, so don't throw marks away through laziness.

Just by looking at a scatter-graph you get a good visual impression of the strength of a relationship. In Figure 10.2, for example, it is clear that there is some correlation between the two parameters and that the strength of relationship declines across the graphs from left to right. The top row shows positive correlations and the bottom row shows negative correlations. By far the easiest way to quantify the strength of the relationship is to click on the dots on the graph, right click and choose 'add trendline'. When you add a trend line one of the options is to also 'display R-squared value on chart'. The R^2 value is a measure of the strength of the relationship and it is related to the Pearson's correlation coefficient that is described in the next section. If all of your points lie very close to the line the R^2 value will be close to one. If you have a cloud of data with no apparent relationship between the two parameters the R^2 value will be close to zero.

In many cases that is as far as people go with quantifying the relationship. Unfortunately it is not actually that simple, and in many cases the R^2 value is virtually meaningless. If you want to use this method, as with any method, you need to understand what it actually means and what assumptions are involved in the calculation.

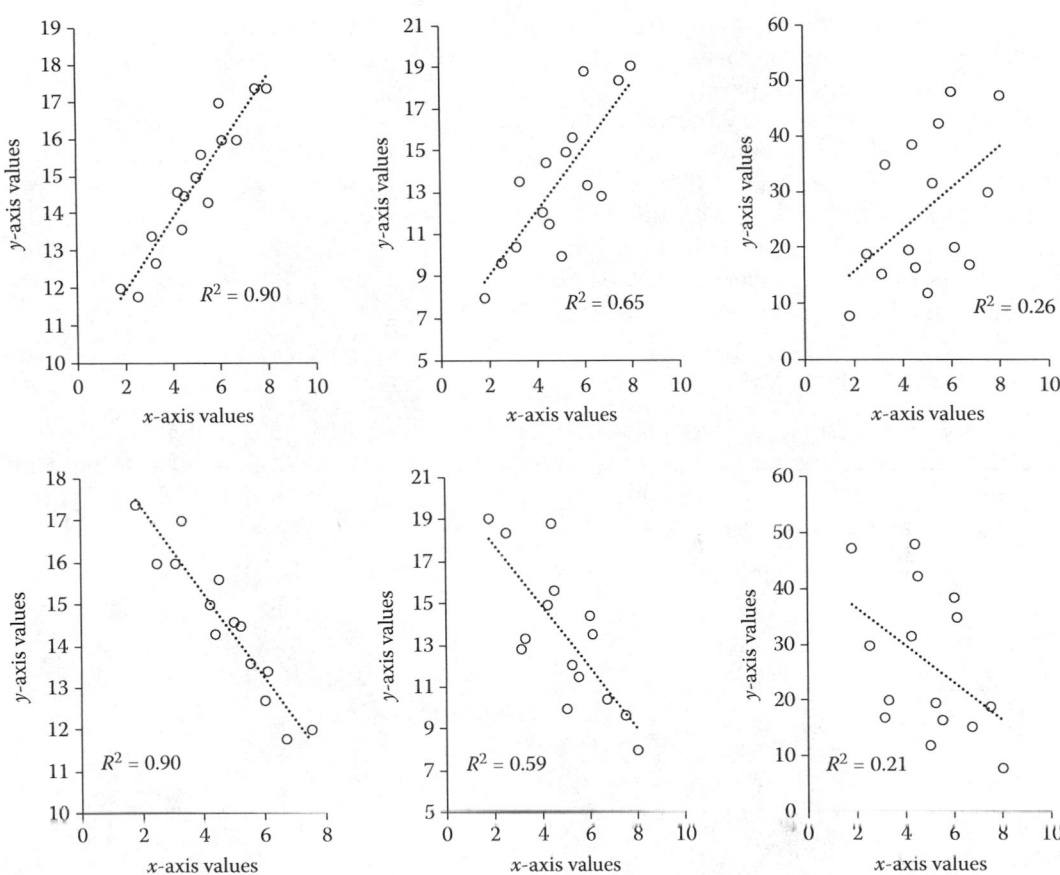

FIGURE 10.2
Examples of scatter-graphs, or x–y plots with trend lines and R^2 values. Note that the greater the spread the lower the R^2 value.

10.2 Assumptions of Correlation Analysis

The R^2 value that Excel provides to accompany the 'trend line' on a scatter-plot is given a variety of names, one of the most common being the coefficient of determination. It is usually just called R^2 (or R-squared). That is convenient because it is literally the square of another statistic, the full name of which is Pearson's product moment correlation coefficient but which is usually just called 'the r-value'. It is the r-value that provides the most useful measure of the strength of a relationship, but only when certain assumptions are met. If the assumptions are not met it does not mean anything at all. To obtain the r-value you take the square root of R-squared. In Excel you can use the function 'SQRT' or you can use 'to the power of a half' (^0.5), which means the same thing. Having obtained the r-value as a square root you then need to decide if it is positive (for a rising trend) or negative.

I will not give the equation for calculating the r-value because I doubt anyone actually uses it any more. Most modern calculators will give the r-value and it is very easy in a spreadsheet or using an online calculator. Nowadays obtaining an r-value is the easy bit;

it is ensuring that it means something that is more complicated. The problem is that the *r*-value is only a sensible measure of the strength of a relationship, or correlation, when certain assumptions are met.

The assumptions are very important, but checking that they are satisfied is not always straightforward. This is a case where plotting the data and applying a bit of common sense is actually a lot more useful than trying to do any fancy statistical checks. The graphs below (Figures 10.3 to 10.11) outline the main assumptions and, where they are clearly violated, give some suggestions for alternative approaches. Note that there is no assumption that your two variables have to be measured in the same units (Figure 10.3), on the contrary the units of measurement make no difference at all.

10.2.1 Assumption 1: Data Are Continuous

This is by far the most important assumption and if seriously violated your results will be nonsense. You should not use correlation analysis on data that fall into discreet groups (Figures 10.5 and 10.6), or where they form tight clusters (Figures 10.4 and 10.7); they have to be spread out continuously.

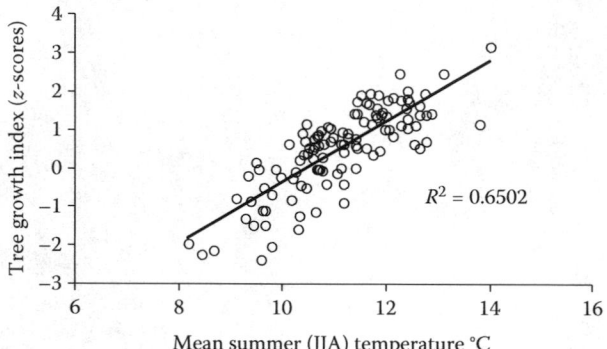

FIGURE 10.3
These data are well-suited to parametric (Pearson's) correlation analysis. A large sample, more values towards the middle than at the ends, no big outliers and a straight line fits well.

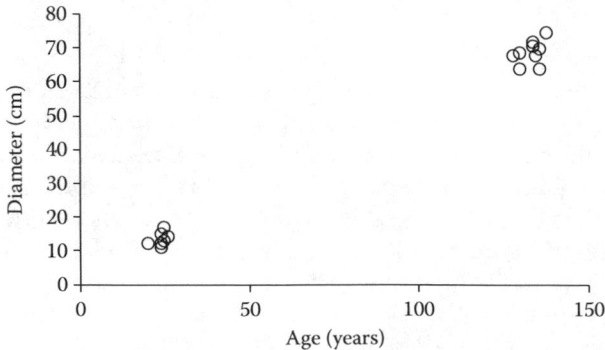

FIGURE 10.4
Tree diameter plotted against age. These data are not suitable for any kind of correlation analysis because they are not continuous; there is a group of young trees and a group of old trees. You could demonstrate that the older trees are significantly larger than the young trees using the Mann–Whitney *U*-test or Student's *t*-test, but it is obvious.

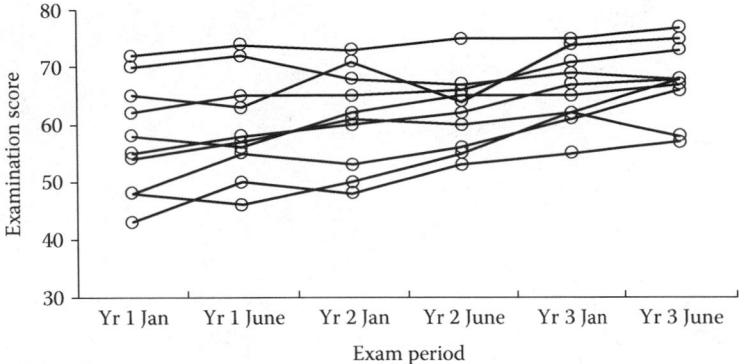

FIGURE 10.5

Examination scores of 10 individual students obtained in six examination periods. These data are not suitable for correlation analysis because they are not continuous; the data fall in six discreet categories. The results are linked, rather than independent, because each student obtained a mark in each period. To test for a general increase or decline over the six examination periods you could use Page's trend-test (Section 9.10).

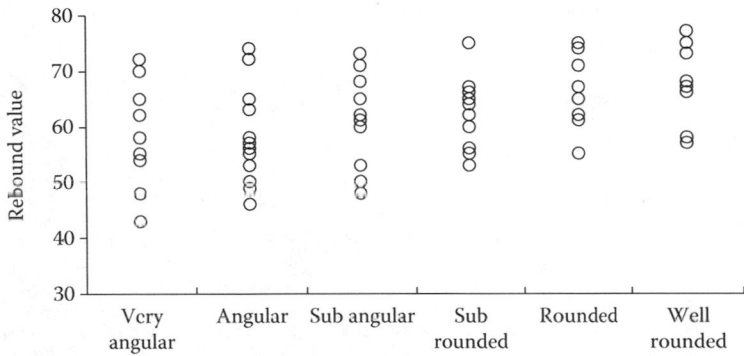

FIGURE 10.6

Schmidt hammer rebound values obtained from boulders of the same age but different roundness. We wish to know whether there is a relationship between rebound value and boulder roundness. These data are not suitable for correlation analysis because they are not continuous, they fall in six groups, and the horizontal axis is not a numerical scale, they are just categories that fall in a logical order (ordinal scale). The data are independent, rather than linked, because there is no reason to link one particular boulder in the very angular category with any particular boulder in any other category. An appropriate test here would be the Jonckheere–Terpstra trend-test (Section 9.8).

10.2.2 Assumption 2: Most of the Data Are Near the Middle, or They Are Evenly Distributed

The underlying assumption here is that the data are near normally distributed, but that is difficult to test (see Section 3.3). If the data are near normally distributed there will be more points in the middle and fewer at the two ends (e.g. Figure 10.3). If the data are evenly distributed correlation can still be used, and often is, but a non-parametric approach will be more robust. If the data are concentrated at the two ends (e.g. Figure 10.4) it is not valid to use any kind of correlation.

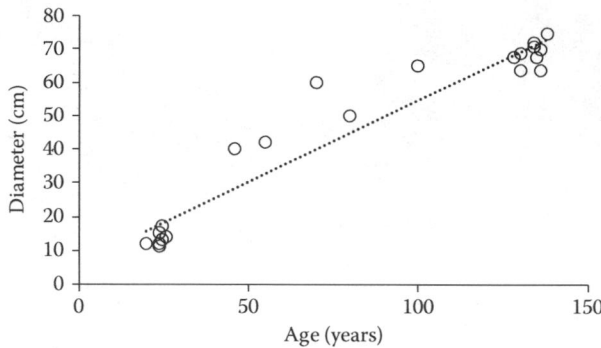

FIGURE 10.7
These data are more continuous than those in Figure 10.6 but there are too many measurements at the extremes and not enough in the middle. The trend-line is strongly controlled by the data clusters at the end. These data are not suitable for any kind of correlation or regression analysis, but see Figure 10.8 for a solution.

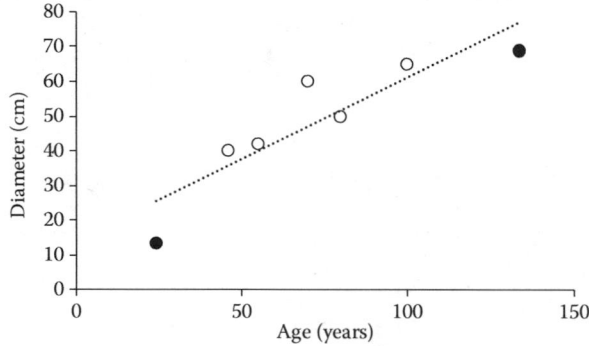

FIGURE 10.8
These are the same data as Figure 10.7 but now the two clusters at the extremes have each been replaced with a single average value so that the points are more evenly distributed. Given the small sample size, and possibly curvilinear relationship, a non-parametric correlation by ranks, such as Spearman's rho, would be safer than using Pearson's parametric correlation. Despite the small sample size the correlation is strongly significant (rho = 0.96, one-tail $p < 0.001$).

10.2.3 Assumption 3: The Relationship Forms a Straight Line Rather Than a Curve

This assumption applies only to the parametric form of correlation analysis (Pearson's product moment correlation coefficient). If the relationship describes a simple curve, rather than a straight line (Figure 10.9), you can still use a non-parametric approach such as Spearman's rank correlation. However, even the non-parametric methods require the relationship to be 'monotonic', which means that points should not go up and then back down or down and then back up. If it is not clear whether the relationship is linear or not, the number of runs test, based on the number of times that the data cross the 'trend line', can be used (Box 5.2).

10.2.4 Assumption 4: No Outliers or Extreme Values

This assumption is critical for parametric correlation (Pearson's) because outlying values have a disproportionate effect (Figure 10.10). It is not such a problem for correlation methods based on the ranks, including Spearman's rho.

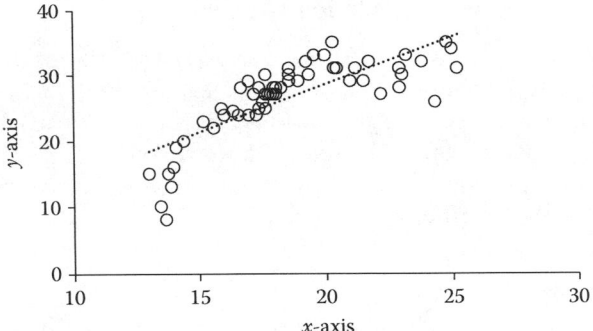

FIGURE 10.9

These data form a curve rather than a straight line. Note that the fitted trend-line is not repeatedly crossed by the data all along its length; the data are mostly below, then mostly above, then mostly below the line. These data are not suitable for Pearson's correlation. However, the data are monotonic in that they rise and then level off, rather than falling back down again, so they are suitable for a rank correlation such as Spearman's rho.

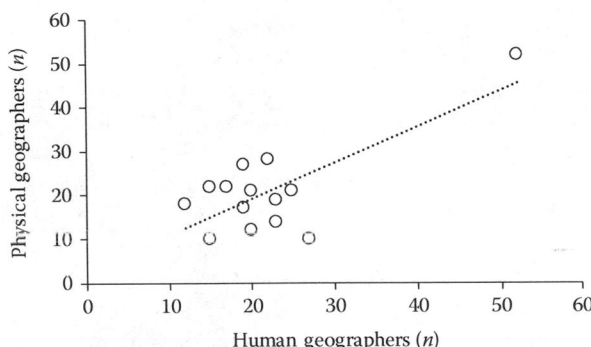

FIGURE 10.10

There is one clear outlier here and it will have a hugely disproportionate effect on Pearson's correlation coefficient. Rank correlation is much less sensitive to the effect of outliers because it does not matter how far out it is, it still just counts as the highest rank, so Spearman's rho is appropriate (see example below).

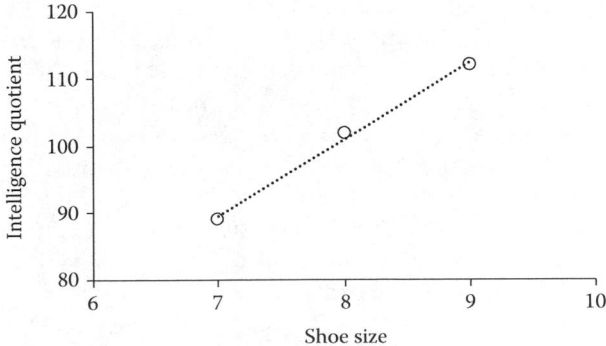

FIGURE 10.11

Three points is too few for any kind of correlation analysis.

10.2.5 Assumption 5: Sample Size Is Large Enough

It is surprisingly difficult to get a straight answer on how many pairs you need for a reasonable correlation analysis. Some tables of critical values go as low as a sample size of 3, but then it is impossible to check the assumptions (Figure 10.11). I advise that when sample size is less than 10 it is safer to use non-parametric correlation, because there are fewer assumptions, or at least to report both parametric and non-parametric results. For non-parametric correlation the minimum size is four pairs for a one-tail test and five pairs for a two-tail test.

10.3 Pearson's Correlation Coefficient

If the scatter-graph looks sensible, according to the criteria above, it is acceptable to use the r-value (Pearson's product moment correlation coefficient) as a measure of the strength of the correlation and you can also then obtain a probability value, which is effectively the chances of obtaining an r-value this far from zero, given a sample of the size you have, just by luck. The logic is similar to taking your values from the x axis and putting them in one bag, and the values from the y axis and putting them in another bag, and then pulling out random pairs and plotting them on a scatter-graph. The most likely outcome, of course, is no correlation at all and an r-value of close to zero. With just a few pairs, however, there is quite a good chance that the points will, just by luck, fall close to a straight line, giving a high r-value. As the number of points increases the chances of a lucky high r-value declines. With just 10 points, for example, the chances of obtaining an r-value of plus or minus (i.e. two-tail) 0.4 is $p = 0.25$, with 20 points it is $p = 0.08$ and you need 25 points before the probability drops below 0.05 ($p = 0.048$) and 41 points before it reaches $p = 0.01$. With 100 points the probability of obtaining an r-value of plus or minus 0.4 just by luck is less than one in ten thousand ($p = 0.00004$). This is a simplification of the way the probabilities are really calculated because rather than just using the values you have collected the procedure assumes that they have been drawn from a population that is 'normally distributed', which is one reason why the assumptions are so important.

To determine whether the r-value is 'statistically significant', given the size of your sample, the easiest option is to use tables of critical values such as those included here (Section 12.11). You need to decide whether you want a one-tail or two-tail probability and that depends on the hypothesis that you are testing. If your hypothesis predicted the direction of the trend line (rising or falling) in advance, then you can use the one-tail probabilities, otherwise, or if you are not sure, use the two-tail probabilities. An r-value is significant at the tabulated level of probability if it is greater than or equal to the value listed for the given sample size. Some tables use degrees of freedom rather than sample size, which for a correlation is the number of pairs minus two.

If you want an actual probability value then things are a little bit more complicated because Excel, for example, does not automatically give the probability associated with an r-value. You can of course calculate the probability by hand, or in a spreadsheet, and there are two different methods for doing this. The most commonly used is based on Student's t-distribution (Box 10.1), the other on the normal distribution after a 'Fisher transformation' of the r-values. If you are performing correlation using the 'Real Statistics Resource Pack for Excel, R Commander, SPSS or online calculators then the probability is given alongside

---

BOX 10.1   CALCULATING THE SIGNIFICANCE OF PEARSON'S *r*-VALUE

The most common way to calculate the statistical significance of a correlation coefficient is to use a *t*-test. The equation is

$$t_r = \frac{r\sqrt{N-2}}{\sqrt{1-r^2}}$$

The statistical significance is found using the *t*-distribution with $N-2$ degrees of freedom. You can use tables of critical values (Section 12.7) but if you are working in a spreadsheet you can use the TDIST function which requires the *t*-value, degrees of freedom and the number of tails that your hypothesis requires. Note however that this function only works if it is a positive number, and the equation above will give a negative value of *t* if the correlation is negative (declining) rather than positive. You can change the *t*-value to a positive number by using the ABS function, which means take the absolute value (ignore the sign).

The probability of a Pearson's correlation coefficient can also be calculated using a normal approximation which requires a 'Fisher correction' of the *r*-values to make them normally distributed. It is fiddlier than the *t*-test option, and gives the same result, so just be aware that it is an option that is used by some software. SPPS and *R* Commander both use the *t*-test, the Real Statistics Resource Pack for Excel gives the results of both methods.

---

the *r*-value. The companion site calculator also returns the one and two-tail probabilities for both parametric (Pearson's *r*-value) and non-parametric (Spearman's rho) correlation.

10.3.1 *R*-Squared, *r*-Values and the Effect Size

When you use correlation there is no need to calculate an effect size because that is precisely what a correlation coefficient is. Pearson's *r*-value falls very conveniently on the scale of minus one (perfect negative correlation), via zero (no correlation) to plus one (perfect positive correlation). In fact in the other tests in this book I have generally tried to produce the effect size *r*, rather than alternative metrics, so that they are always equivalent to Pearson's *r*. By convention (i.e. no logical reason, just someone's opinion) we assume that an absolute (that means plus or minus makes no difference) *r*-value of less than 0.3 is a 'small effect' or 'weak' correlation, values of 0.5 and above represent a 'large effect' or 'strong' correlation and intermediate values are 'medium'. The square of Pearson's correlation coefficient is the 'coefficient of determination' or the *R*-squared (R^2) value. You should not use this for correlation analysis, it is used in regression which is described later. Confusing *r*-values and *R*-squared values is a common source of error in student work, so be careful.

10.3.2 How to Do It

10.3.2.1 Companion Site Calculator

There are two companion site calculators for correlation analysis. The simple one ('correlation $n = 100$') will allow you to enter two sets of data with sample sizes up to 100 in two columns. Just cut and paste the data in and leave any empty cells blank. A scatter-graph is provided and you should use this to check that the assumptions are met. The results include a list of descriptive statistics, Pearson's r-value and the one and two-tail probabilities calculated using the t-distribution. If you suspect that the assumptions of parametric correlation are not met, due to non-linearity or outliers, for example, then you can use the results for Spearman's rank correlation instead, which are also provided. If the results of the Pearson's and Spearman's correlations are not very similar it is usually a sign that there is a problem with the assumptions. A second scatter-plot shows how the ranks, rather than the original measurements, are correlated.

In some cases it can be difficult to decide whether the relationship is really linear or not, in which case there is another calculator that may help ('runs test for linearity'). It is important that you enter the data so that the x-axis values are in rank order from lowest to highest. There is a warning to check this is true. Now it will apply the non-parametric 'one-sample number of runs test for randomness' (Section 5.5) to check how many times the data cross the linear 'trend line'. If the probability drops below $p = 0.5$ you should not use Pearson's correlation. As long as the trend is monotonic (does not go up then back down or down then back up) you can still use Spearman's rank correlation.

10.3.2.2 In a Spreadsheet

There are two Excel functions that return Pearson's correlation coefficient (CORREL and PEARSON). The problem is obtaining a probability value to go with it. If you have installed the 'data analysis tools' (which is often hidden away, see Section 4.4) you can click on 'data' and 'data analysis' to access a list of functions that includes 'correlation'. For a single correlation this does not help at all, it just returns the r-value. To find the two-tail probability for the r-value go instead to 'regression' and use the mouse to put one series in the y-axis and the other in the x-axis boxes. The output is a bit bewildering, but the number you need is in the bottom table and it is the p-value for the X-variable (not the intercept). If your hypothesis is one-tail (you predicted a positive or negative correlation) divide the probability value by two. If you cannot access the data analysis tools you can calculate the probability from the t-distribution following the instructions in Box 10.1. Alternatively use an online calculator (Table 10.1).

The 'correlation' option in the 'data analysis toolpack' is useful if you have a larger set of data, not just two samples, and want to compare all possible pairs. This kind of 'scatter-gun' approach to data analysis is not really to be recommended, but it can be useful as a way of scanning your data for patterns or for errors. For example if you have taken many

TABLE 10.1

Selection of Websites Allowing Pearson's r-Values to be Converted into Probabilities

Provider	Website	Comments
Vassar stats	http://vassarstats.net/tabs_r.html	Enter N and r to obtain one and two-tail p-values. Also gives t-value. Good site.
Social Science Statistics	http://www.socscistatistics.com/pvalues/pearsondistribution.aspx	Enter r and N and it returns the two-tail p-value
Graphpad	http://graphpad.com/quickcalcs/PValue1.cfm	Enter r and degrees of freedom $(n - 2)$. Returns two-tail p-value.
EasyCalculation.com	https://www.easycalculation.com/statistics/r-to-p.php	Enter N and r to obtain one-tail and two-tail p-values. Also gives t-value.

samples of the same thing and you expect them all to be correlated this method will identify samples that are in some way unusual, due to values that have been mistyped for example. You might use this method to compare temperature records from several different stations, for example, before combining them to produce a regional average. In this case the *r*-values are being used as an effect size. If you want to convert the *r*-values into probabilities you need to be very careful because of the 'family-wise error' that accompanies multiple testing (Section 9.2).

10.3.2.3 Real Statistics Resource Pack for Excel

Use Control-M to obtain a list of functions and choose 'correlation (one sample)'. Use the mouse to identify the range of data for each input range. It does not matter which way around they go. Alpha is the critical threshold of probability you are using, the default is $p = 0.05$ (5%). Note that you have to decide whether your hypothesis is one or two-tail. The output includes three correlation coefficients: Pearson's is the usual parametric *r*-value, the other two are non-parametric methods based on ranks (Spearman's and Kendall's rank correlation). The probability value is then calculated using both the *t*-distribution and Fisher's transformation. The important number is the *p*-value. It does not tell you if it is one or two-tail, it depends on what you asked for. It is best to write one or two-tail next to it in case you forget. The last two numbers are the 95% confidence limits for the *r*-value.

10.3.2.4 Online Calculators

There are plenty of free online calculators for Pearson's correlation coefficient (Table 10.2) but most of them are not very good. Some only give the *r*-value without a probability, which is a bit pointless, and others give so much output that it is all very confusing.

TABLE 10.2

Selection of Free Online Calculators That Will Compute Pearson's Correlation Coefficient

Provider	Website	Comments
Vassar stats	http://vassarstats.net/index.html Listed under correlation and regression	Choose 'basic linear correlation and regression', and 'data import version'. Paste both data columns in left box. Gives *r*-value and one-tail and two-tail *p*-values.
Social Science Statistics	http://www.socscistatistics.com/ tests/pearson/Default2.aspx	Plots a scatter-graph and returns two-tail *p*-values hidden amongst too much output.
Wessa.net	http://www.wessa.net/rwasp_ correlation.wasp	Produces a scatter-graph with histograms. The QQ plots are not scatter-graphs, they are for checking normality! Look for correlation (= *r*-value) and the one-tail and two-tail probabilities. Can be confusing.
Endmemo	http://www.endmemo.com/ statistics/cc.php	Plots a scatter-graph and gives *r*-value but no probability.
EasyCalculations.com	https://www.easycalculation.com/ statistics/correlation.php	Enter data one at a time to get *r*-value and no *p*-value. Not easy.
Alcula.com	http://www.alcula.com/calculators/ statistics/correlation-coefficient/	Cut and paste both columns into the box. Returns *r*-value but no probability.

10.3.2.5 In R Commander

To perform correlation analysis in *R* Commander enter the data in the usual way, as two columns with headers. Before performing any kind of correlation, first check that the assumptions are met by plotting a scatter-graph. Go to 'graphs' and 'scatterplots' and choose the independent variable for the *x* axis and dependent variable for the *y* axis. On the Options tab choose 'least-squares line' and leave the others unchecked. If you are happy that the assumptions are not violated then go to 'statistics', click on 'summaries' and 'correlation test'. You also have to either choose a two-tail test or predict the direction of the correlation (positive is > 0 and negative is < 0). The output includes the *r*-value (final number) and the probability as well as the *t*-value and degrees of freedom used to calculate the probability and the 95% confidence intervals for the *r*-value.

10.3.2.6 In SPSS

Enter the data in two columns and identify the 'measure' as 'scale'. Before performing correlation analysis plot a scatter-graph to check the assumptions. The easiest option is to use 'graphs', 'legacy dialogues' and choose 'scatter/dot' then 'simple scatter'. Now click 'analyze', 'correlate' then 'bivariate'. Put both data sets in the variables box and choose 'Pearson's' and either one-tail or two-tail. The output is the Pearson's correlation coefficient, probability value and sample size (all twice for some reason).

10.3.3 Examples

10.3.3.1 Example: Tree Growth and Summer Temperature

Several different indicators of annual tree growth (ring width, density and height increment) were combined for many trees at several sites close to the northern timberline of Europe. Meteorological data from that region are available between AD1890 and AD2005, giving a sample size of 116 (Figure 10.12). We know from previous research that tree growth close to the timberline is sensitive to summer temperature so we predict that there will be a positive correlation with summer (June to August) mean temperature.

The data are plotted as an *x*–*y* scatter-graph with tree growth on the vertical (*y*) axis because that is logically the dependent variable. A 'trend line' has been fitted and we can see that the assumptions of Pearson's correlation are met. It is a large sample, the data are

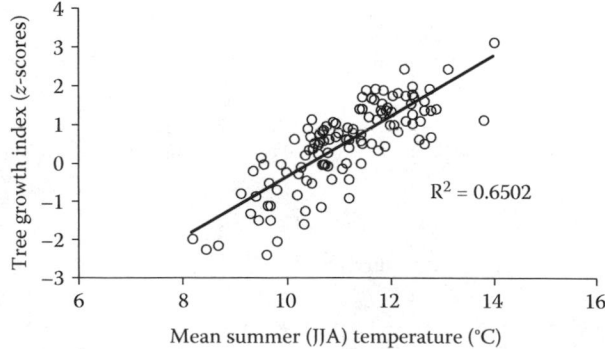

FIGURE 10.12
A tree growth index plotted as a function of mean summer temperature.

continuous, fall on a straight line rather than a curve, and there are no big outliers. The R^2 value tells us that about 65% of the variation in tree growth index is explained by variability in mean summer temperature. The Pearson's product moment correlation coefficient, or r-value, is obtained by taking the square root of R^2, giving $r = 0.81$. It is positive rather than negative because it is a rising trend. Looking at the tables of critical values (Section 12.11) for a one-tail test, because we predicted in advance that the relationship would be positive, we can see that the r-value is much higher than the tabulated value for $p < 0.001$. Checking the significance value using an online calculator gives $p = 1.4\text{E-15}$, which is a very small number indeed. We can confidently conclude that the strong correlation that we see is very unlikely to have arisen just by chance.

Even with this very high and strongly significant correlation some care in needed in interpreting the results. It is possible that summer temperature is strongly controlling tree growth, but another possibility is that tree growth is actually controlled by a different factor that is also strongly correlated with summer temperature, such as summer sunshine.

Don't immediately jump to the conclusion that correlation means causation; try to think 'outside the box' as well. Actually in this case I suspect that it really is temperature that is the controlling factor because although sunshine controls the amount of photosynthesis, tree growth requires cell division, and that is strongly controlled by temperature. Carbon isotope values from the same trees are also very strongly correlated with summer temperature, but they are probably controlled by sunshine (Gagen et al. 2011; Young et al. 2012).

10.3.3.2 Example: Opinions on Global Issues

You are interested in students' views on global warming and energy security and as part of a mixed-methods approach you ask a small unbiased sample to mark their level of agreement to a series of questions, including

1. To what extent do you agree that nuclear power is too dangerous to be used?
2. To what extent do you agree that anthropogenic global warming is real?

You are interested to see whether people who have strong views on one of these questions also have strong views on the other, but you have no idea whether there should be a positive or negative (or any) correlation. This calls for a two-tail test. By marking their level of agreement on a line, rather than ticking a box, you have continuous data rather than counts in categories, so you can use correlation. You could just measure the position of the mark along the line and use mm, or you could express the distance as a percentage of the total length of the line (as here), it makes no difference for correlation analysis. When you plot the data as an x–y scatter-graph (Figure 10.13) it does not matter which answers go on which axis because you are not inferring that the answer to one question is in any way dependent on the answer to the other.

Despite the small sample size, of only 15, the scatter-graph suggests that the assumptions of Pearson's correlation are not strongly violated. The data are continuous, the relationship looks linear and there are no strong outliers. The square root of R^2 gives the r-value of 0.88 suggesting a strong correlation. Even with only 15 points this is strongly significant, with a two-tail probability of 1.7E-05 (move the decimal place five points to the left, so = 0.000017). You can confidently conclude that there is a strong positive correlation between the answers to these two questions.

Note that this result does not mean that people tended to have the same level of agreement to the two questions; the answers to question one span the range 40% to about 85%

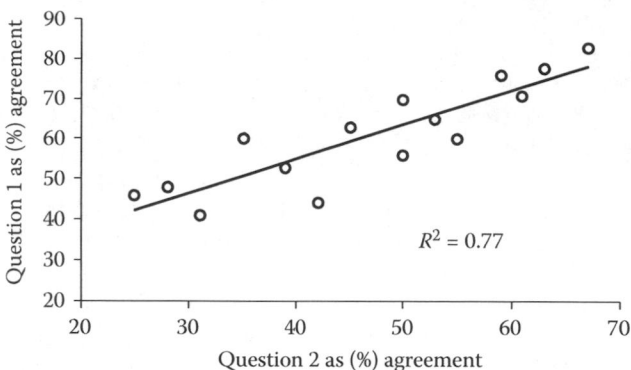

FIGURE 10.13
Level of agreement with two questions recorded as percentages.

whereas the range for question two is about 25% to less than 70%. The average scores for the two questions are about 61% and 47%. The units do not matter in correlation so you could multiply one of the scales by 100 and still not change the R^2 or r-value. If you wanted to know if there was a significant difference in level of agreement to the two questions you would have to use a different kind of test. Your data are paired, because each individual has supplied one answer to each question, so if you wanted to know whether individuals tend to provide a higher or lower level of agreement to one of these questions you could use the non-parametric Wilcoxon's matched-pairs signed-ranks test (sample size is a bit small for the parametric paired t-test). However, a simpler alternative here is the sign test (Section 6.2), because it is clear from Figure 10.13 that all 15 gave a higher agreement score to the first question, and the probability of that occurring just by luck is about $p = 0.0001$. If you do not care about comparing the views of individuals, and just want to know if there is a general difference in the level of agreement, then you could treat the data as independent and use the non-parametric Mann–Whitney U-test ($U = 53$, two-tail $p < 0.05$, effect size $r = 0.45$).

10.4 Point-Biserial Correlation

Pearson's correlation analysis is sometimes applied when one of the variables is continuous but the other is dichotomous. For example if you wanted to know to what extent exam performance was related to gender you could use male/female as one of the variables by coding the two options as 0 and 1. This procedure is known as point-biserial correlation. However, it is actually no different to applying an independent samples t-test, which is probably easier and, to me at least, seems like the more logical procedure. The effect size produced by the 'two independent samples' companion site calculator includes an effect size r for the independent samples t-test that is identical to the point-biserial correlation coefficient. If there is a clear causal relationship you can square the r-value to give the proportion of variance explained. For example, if the effect size r (point-biserial correlation) is 0.71 then you could conclude that gender explains about 50% (0.71 squared = 0.50) of the variance in examination performance.

10.5 Spearman's Rank Correlation or Spearman's Rho

10.5.1 When It Is Useful

There are many circumstances when it is not really appropriate to use the Pearson's correlation coefficient because the assumptions are not met. Spearman's, in contrast, is a non-parametric method, so there are fewer assumptions and it can often be used when the data are not suitable for the parametric method. As with other non-parametric methods it copes well with small samples and there is no assumption of normality, so the data can be skewed and outliers are not such a problem. An important advantage of Spearman's approach is that it does not matter if the relationship between the two parameters is linear or curved. The relationship still has to be monotonic, however, which means that it cannot go up and then back down, or vice versa.

10.5.2 What It Is based On

The difference between Pearson's r-value and the equivalent Spearman's rho (Greek letter that looks so confusingly like the letter p in most fonts (ρ) that I will just spell it), is that Spearman's approach uses the ranks of the two data sets rather than the original values (Spearman 1904). In every other respect the two methods are identical, which allows us to use a sneaky shortcut when calculating Spearman's rho in a spreadsheet.

10.5.3 How to Do It

If you have to calculate Spearman's rank by hand you use a rather long-winded method involving ranking, taking the squared differences, summing them and then putting that value into a formula. I have given the method in Box 10.2 in case your evil professor forces you to show the workings. In a spreadsheet all you really have to do is rank them, ensuring that any shared ranks are given the average of the shared ranks (Box 8.1), and then perform a Pearson's correlation on the ranks, using the 'CORREL' function. The result is Spearman's rho. You can check for significance using tables of critical values (Section 12.12) using the same procedure as for the parametric method.

The Real Statistics Resource Pack includes Spearman's rho as one of the options listed under correlation and it will give you a probability value too, or you can use the SCORREL function, listed under 'user defined' functions, just to obtain Spearman's rho. There are a few online calculators that will allow you to cut and paste your data in to perform the test (Table 10.3), but it is no easier that running it in a spreadsheet. In *R* Commander the procedure is the same as for Pearson's and the correlation option is hidden away under 'summaries'. You can run Spearman's rank correlation in SPSS in exactly the same way as running Pearson's; put your data in two columns, identify the data as 'scale' then click 'analyze', 'correlate' and 'bivariate' then choose the Spearman option and a one-tail or two-tail test. The companion site calculators for 'correlation' give the results for Spearman's rank analysis alongside those for Pearson's correlation.

10.5.4 Examples

10.5.4.1 Example: As Used for Pearson's Correlation

Spearman's rho can be calculated for the two examples used above for Pearson's correlation. The comparison of tree growth index and summer temperature (Figure 10.12),

TABLE 10.3

Selection of Free Online Calculators That Will Perform Spearman's Rank Correlation

Provider	Website	Comments
Social Science Statistics	http://www.socscistatistics.com/tests/ spearman/default2.aspx	Easy, just cut and paste. Only gives two-tail *p*-value so divide by 2 for one-tail. Option to show full workings.
Vassar stats	http://vassarstats.net/ Under 'correlation and regression' look for 'rank order correlation'.	Cut and paste both columns, without headers into the 'data import' window on the right then click 'import data'. Easy and gives one-tail and two-tail *p*-values.
Maccery.com	http://www.maccery.com/maths/	Gives rho, with all of the workings, but no probability value. Lots of other output. Too complicated.
EasyCalculations.com	https://www.easycalculation.com/ statistics/spearman-rank-correlation- calculator.php	No cut and paste, type in values. No *p*-value given. Not easy.
SciStatCalc	http://scistatcalc.blogspot.co.uk/2013/10/ spearman-rank-correlation-calculator.html	Data need to be comma separated. No *p*-value.
Wessa.net	http://www.wessa.net/rankcorr.wasp	Draws an *x*–*y* plot and gives lots of output. Too complicated.

with a sample size of 116 gave an *r*-value of 0.81 and gives a Spearman's rho value of 0.80. For the comparison of answers to two questions (Figure 10.13), the *r*-value was 0.88 and Spearman's rho is 0.87. You can see that when the data are suitable for the parametric test the non-parametric test gives almost identical results. When samples are small and/or not normally distributed Spearman's rho is more powerful than Pearson's correlation. If in any doubt about the suitability of your data for parametric correlation just use Spearman's rho instead.

10.5.4.2 Example: Coastal Zone Vegetation

You are interested in biogeography and wish to examine how species composition changes on a transect away from the coast. You expect fewer species to occur close to the shore because of the harsh environmental conditions, calling for a one-tail test. You have recorded the number of species in quadrats at increasing distance from the shore and plotted them as a scatter-graph (Figure 10.14). It should be clear that these data are not at all suited to parametric correlation analysis. The data are skewed to the left but more seriously the relationship is clearly not linear. The linear 'trend line' does not follow the data; in fact they only cross the line twice. It is clear that the relationship is curvilinear. If these data are entered into the companion site correlation calculator that includes the 'number of runs' test (Box 5.2) the result is a one-tail probability of $p = 0.002$ and an effect size of $r = 0.86$, confirming that a straight line is not appropriate.

Although the relationship is not a straight line it remains monotonic, in that the number of species rises rapidly and then stabilises, it does not fall back down. Excel will allow you to calculate an R^2 for the linear trend-line, or to perform parametric correlation but in this case it would be meaningless. The data are well-suited to analysis using Spearman's rank correlation, however, and the rho value is 0.92. Given a sample of 11 pairs this far exceeds

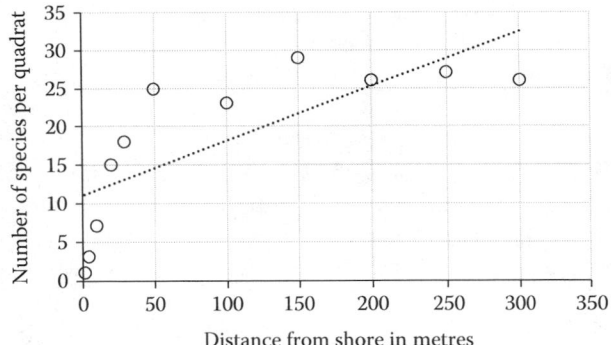

FIGURE 10.14
Number of plant species per quadrat with distance from the shore. The relationship is strongly non-linear, so Pearson's correlation would not be appropriate, but since the relationship is monotonic (does not rise then fall) Spearman's rho is a good choice.

the one-tail (we predicted the rise in advance) tabulated critical value at $p < 0.001$ so we can conclude that it is strongly statistically significant. No further information is really needed, but using the t-distribution the probability can be calculated as $p = 3.0E\text{-}05$, or $p = 0.00003$.

10.5.4.3 Example: Use of Space

You are interested in the use of space and decide to study the different ways that physical and human geography students use the library. You visit the library on 14 different occasions and count the number of physical geographers and human geographers (Figure 10.15). You have no idea what to expect and in reality you are only doing this because the author has run out of ideas for examples. Anyway you plot the pairs of data on an x–y plot and add a trend line and R^2. The R^2 value of 0.55 translates into a Pearson's r-value of 0.74 which would suggest a very strong positive correlation that is statistically highly significant ($p = 0.002$). However these data are not at all suitable for Pearson's correlation because there is one clear outlier that is having a hugely disproportionate effect on the correlation coefficient. If that one point is removed the Pearson's r value drops to almost zero

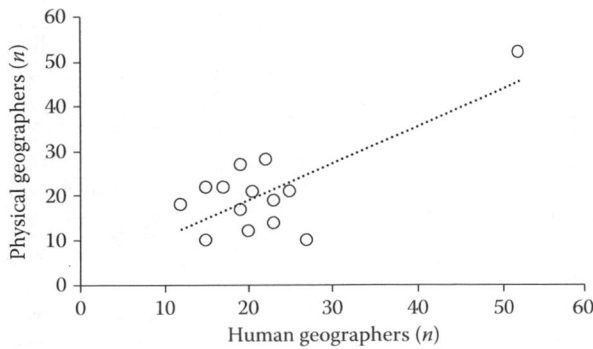

FIGURE 10.15
In this case one of the values is a clear outlier. This will have a hugely disproportionate effect on Pearson's correlation coefficient (r-value) but very little effect on Spearman's rho.

(−0.09). The outlier does not have a big effect on the Spearman's rho value because it uses rank order, not the actual values. The Spearman's rho value without the outlier is −0.14 and with the outlier it changes only slightly to 0.09, which is still nowhere near statistically significant (two-tail critical value of $p = 0.05$ is rho = 0.54). There is no correlation between the numbers of human and physical geography students visiting the library.

When I see a big outlier like this the first thing I do is go back to the original notebooks and check the numbers. When you type numbers into a spreadsheet it is easy to make a mistake and transposing two digits is a common error. The big outlier has two values of 52, and if they were transposed it would be 25 and 25, which would place that dot close to the other data.

10.5.4.4 Example: Using Rank Order Rather Than Numbers

One of the advantages of rank correlation methods, including Spearman's, is that they do not require measurements on any numerical scale, they will work perfectly well with 'variables' that are just placed in rank order. This can apply to one of the axes or to both. Imagine, for example, that you conjured the bizarre hypothesis that there might be a positive correlation between how well known your lecturers are for their research and how good they are at teaching. You could use something like the citation index to quantify their research 'impact' and then place them in rank order using that metric. To deal with their teaching you could simply place them in rank order based on your own opinion.

In this case (Table 10.4), although there are some good researchers who are also good teachers and some useless researchers who are also useless at teaching, overall there is no significant correlation (rho = 0.23, $p < 0.05$).

TABLE 10.4

Spearman's Rank Correlation Does Not Require Individual
Measurements, Only the Rank Order of One or Both
Parameters

	Teaching Rank	Research Rank
Dr. N Spire	1	8
Dr. B Spark	2	3
Prof SO Humer	3	12
Dr. T Rieshard	4	1
Dr. M Umbler	5	13
Dr. Dull	6	11
Dr. B Ore	7	7
Dr. T Diem	8	6
Dr. Y Awner	9	8
Dr. HE Dunnit	10	5
Dr. Duller	11	4
Prof S Nooze	12	10
Prof Caresnot	13	2
Prof D'Umas	14	14
Prof SF Brains	15	15
Spearman's rho	0.23	

Note: This is very helpful when the thing being measured is difficult to
quantify precisely.

10.5.4.5 Example: Grumpy Old Men

As a visitor to Britain you are interested in the interactions between environmental conditions and people's feeling of wellbeing, or more specifically in the way that grumpy old men constantly moan about the weather. You decide to interview such people throughout the year and ask them how happy they are with the temperature. They can mark their level of happiness on a line and you can read it off as a percentage. If you were to simply run a correlation on the data, without first plotting it, you would obtain a Pearson's *r*-value of 0.22 and Spearman's rho of 0.19, suggesting that there is no relationship between temperature and the happiness of grumpy old men. Plotting the data, however (Figure 10.16), it is clear that there is actually a very strong relationship, but it is neither linear not monotonic; happiness rises to a peak at about 21°C and then falls again. Neither method of correlation can cope with a relationship that rises and falls like this. You could fit some sort of fancy curve to the data, the one on the graph is a third-order polynomial, but the simple solution is to just split it at the optimum temperature and demonstrate that that there is in fact a very strong correlation, but it is positive below 21°C and negative above 21°C (Figure 10.17). To produce two trendlines like this in Excel you have to treat the two halves as separate samples.

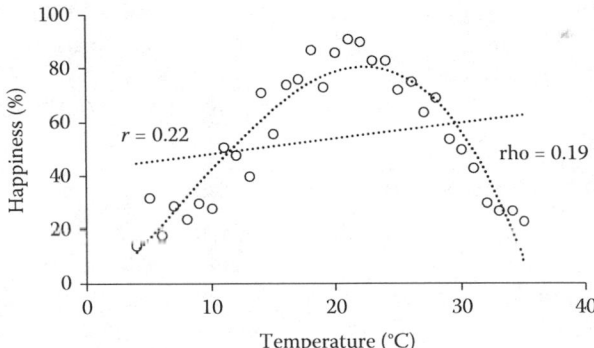

FIGURE 10.16
The relationship shown here is strong but it is not monotonic; it rises then falls. Correlation analysis (parametric or non-parametric) cannot cope with this so both Pearson's *r* and Spearman's rho yield very low values suggesting little or no correlation.

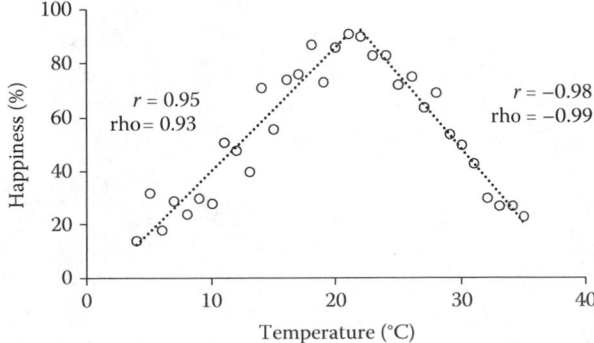

FIGURE 10.17
A simple solution is to split the data into two sections, either side of the optimum. Now there are two strong linear relationships, one positive and one negative, and Pearson's and Spearman's approaches yield very similar results.

This example is fabricated of course, but there are many real relationships that are not monotonic. The rate of plant growth, for example, tends to increase with temperature until it reaches an optimum and then it declines. In such circumstances you can only assume a linear or even monotonic relationship when all of your samples lie to one side of the optimum. The strong linear relationship between pine tree growth and summer temperatures in northern Fennoscandia, for example (Figure 10.12), occurs because in that area mean summer temperatures very rarely if ever reach levels that restrict rather than encourage tree growth.

10.6 Tests for Comparing Two Correlation Coefficients

When conducting research that involves correlation analysis it is often useful to compare two correlation coefficients. It is a common mistake to assume that two correlation coefficients are very different, and to draw far-reaching conclusions, when in reality the difference between them could be due purely to chance. The probability of two correlation coefficients being significantly different depends on the magnitude of the difference between them and on the size of the two samples. With small samples quite large differences might be expected purely due to chance, but with large samples large differences become increasingly unlikely to be just luck. The method used for the comparison depends on whether the two correlations are independent or are related because they involve some of the same numbers.

10.6.1 Independent Correlation Coefficients

There is a slight complication when attempting to compare two independent correlation coefficients because neither the Pearson's r-value nor Spearman's rho are normally distributed. However, Ronald Fisher devised a transformation, now called Fisher's z-transformation, which changes correlation coefficients to make them normally distributed. It has a small effect on low values and a much larger effect on high values. The method is normally used to compare two r-values, but since Pearson's r-value and Spearman's rho fall on the same scale I can see no reason why the same method could not be used for both if the samples are large (>25). Cohen (1988) suggests the difference between the two transformed r-values as the effect size.

10.6.1.1 How to Do It

The equation aims to produce a z-score, which can be translated into a probability using the normal distribution. If we call the two correlation coefficients A and B, based on sample sizes of N_A and N_B, then applying Fisher's transformation gives F_A and F_B and the z-score for the difference (z_{diff}) between them becomes

$$z_{diff} = \frac{F_A - F_B}{\sqrt{(1/N_A - 3) + (1/N_B - 3)}}$$

Once you have values for the Fisher transformation, which can be obtained from tables, it is not difficult to perform the test by hand. Using a spreadsheet is easier because there is a function (FISHER) that will make the transformation of the correlation coefficients. The resulting z-score for the differences can also be translated into a one-tail probability using the NORM.S.INV function. If z is a negative number it gives the p-value but if z is positive use one minus the result. The example below shows, step by step, how to make the calculations.

I can only find one free online calculator that will compare two independent correlation coefficients, provided by Vassar Stats (http://vassarstats.net/rdiff.html). However, it is easy to use so one is enough. I am afraid I have no idea how to run this test in SPSS or in R Commander. I do not think it is yet included in the Real Statistics Resource Pack for Excel, but that resource is constantly improving so check the website. A spreadsheet-based calculator for testing the difference between two correlation coefficients is provided on the companion site. Simply enter the two correlation coefficients and their sample sizes to obtain the z-score for the difference and one-tail and two-tail probabilities.

10.6.1.2 Worked Example: Incompetent Marking

Imagine that you are the student representative for your department and some of your colleagues have complained that a course taught only to human geography students is not being marked properly, whereas the equivalent course sat only by physical geographers has raised no concerns. The problem is not that the average marks for the two courses are different, only that students feel that they are not getting the marks they deserve because the papers are marked by a professor affectionately known as 'rambling Ron the random number generator'. A possible solution is to use correlation analysis to compare the mark that each student received for the module in question with the average mark they received for all other modules. You can do this for the physical and human geographers separately and compare the two correlation coefficients. Note that these two correlation coefficients are independent because the two groups of students are separate. The results for the two groups are

Human geography: Pearson's r-value $= 0.38$, Fisher's transformation $= 0.400$, $N = 64$

Physical geography: Pearson's r-value $= 0.72$, Fisher's transformation $= 0.908$, $N = 68$

The equation becomes

$$z_{diff} = \frac{F_A - F_B}{\sqrt{(1/N_A - 3) + (1/N_B - 3)}} = \frac{0.40 - 0.908}{\sqrt{(1/64 - 3) + (1/68 - 3)}} = \frac{-0.508}{\sqrt{(1/61) + (1/65)}}$$

$$z_{diff} = \frac{-0.508}{\sqrt{(1/61) + (1/65)}} = \frac{-0.508}{\sqrt{0.016 + 0.015}} = \frac{-0.508}{\sqrt{0.032}} = \frac{-0.508}{0.178} = -2.85$$

The z-score for the difference can be translated into a probability value using tables of the standard normal distribution, an online calculator or in a spreadsheet using the NORM.S.INV function to produce a one-tail probability. If we give the good professor the benefit of the doubt and conduct a two-tail test the outcome is strongly significant (two-tail $p = 0.005$). There is about a half a percent chance that the two correlation coefficients could be so different simply due to chance, so you have good grounds for complaint. Doubtless the useless creature will claim that he is testing something that the other courses do not test and that his marking is not just as good as everyone else's, it is much better. I used to have a colleague just like this; we called him 'Captain Chaos'.

10.6.2 Linked Correlation Coefficients

This method is a bit complicated for this book, but I am including it because it is something I have been doing wrong for years. It is nicely explained by Andy Field (2013) who attributes the solution he proposes to Chen and Popovich (2002).

Often in geography the two correlation coefficients that we want to compare are not independent because one of the series is included in both calculations. My close colleagues and I are particularly interested in using tree rings to reconstruct the climate of the past. If there is a good correlation between something we can measure in the rings and some climatic 'target' measured at a nearby meteorological station, then we go on to use regression methods to reconstruct the climate back in time, to long before any meteorological instruments existed. A recurring problem for us is to identify the most suitable climate target. For example, we often find good correlations with the mean temperature of one particular month, often July, or with some combination of months such as the whole summer (JJA). The question is should we take the highest correlation, or could the difference be just due to luck, in which case we could use either? In the past I have used the method described above, for independent correlation coefficients, to determine whether the two *r*-values are significantly different. However, this is not a fair test, because the two potential 'targets' are themselves correlated. The solution, apparently, is to compute a *t*-statistic that incorporates the correlation between all three sets of data.

In a case like this we have one dependent variable (y = tree rings) and two potential independent variables (x_1 = summer temperature and x_2 = July temperature), so there are three correlation coefficients to consider. I will try to make the equation easier by just giving them letters

Correlation of y with $x_1 = A$

Correlation of y with $x_2 = B$

Correlation of x_1 with $x_2 = C$

We are interested in the difference between A and B but we have to take C into account. Since all three data sets are related they must have the same sample size $= n$.

Given this notation the equation is

$$t_{diff} = A - B \sqrt{\frac{(n-3)(1+C)}{2(1-A^2-B^2-C^2+2ABC)}}$$

Using some real data from northern Fennoscandia the relevant values are

$$A = 0.70, B = 0.81, C = 0.77 \text{ and } n = 116$$

So the equation becomes

$$t_{diff} = 0.70 - 0.81 \sqrt{\frac{(116-3)(1+0.77)}{2(1-0.70^2-0.81^2-0.77^2+2\times0.70\times0.81\times0.77)}}$$

$$t_{diff} = -0.11 \sqrt{\frac{(113)\times(1.77)}{2(1-0.49-0.656-0.593+2\times0.70\times0.81\times0.77)}}$$

$$t_{diff} = -0.11 \sqrt{\frac{200}{2(-0.739+0.873)}} = -0.11 \sqrt{\frac{200}{2(0.134)}} = -0.11 \sqrt{\frac{200}{0.268}}$$

$$t_{diff} = -0.11 \sqrt{746.27} = -0.11 \times 27.32 = -3.00$$

Using the T.DIST.2T function in Excel (remove the minus sign to give $t = 3$) with 113 degrees of freedom gives a two-tail probability of $p = 0.003$. That is very different to the result that is obtained if the two correlations are treated as if they are independent ($p = 0.051$). The reason for the big difference in this case is the strong correlation between the two potential dependent variables. As that correlation declines the results of the two methods converge.

BOX 10.2 CALCULATING SPEARMAN'S RANK CORRELATION COEFFICIENT BY HAND

If you are using a spreadsheet you do not need to use this method, just rank the two series and then use the 'CORREL' function on the ranks rather than the raw values. If you really have to do it by hand follow these steps (Table 10.5).

Step 1: Rank the two series individually. It does not matter whether you go from lowest to highest or highest to lowest as long as you are consistent and treat both series the same way. If there are shared ranks give them the average. For example if ranks 5 and 6 are shared they each get 5.5 and the next one gets a rank of 7. If 8, 9 and 10 are shared they each get a rank of 9 (the average of 8, 9 and 10) and the next one gets 11. If you get it right your last rank should be the same as the sample size.

Step 2: Calculate the difference (d) between the two ranks. It does not matter which series goes first in the calculation, that just changes the sign of the difference and the next step gets rid of the sign.

Step 3: Square the differences (d^2). Squaring the difference gets rid of the minus signs, otherwise when you add up the differences the plusses and minuses would cancel each other out.

TABLE 10.5

The Steps Required to Calculate Spearman's Rho by Hand

A raw	B raw	A rank	B rank	d	d²
25	44	1	2	−1	1
28	48	2	4	−2	4
31	44	3	2	1	1
39	53	5.5	5	0.5	0.25
39	44	5.5	2	3.5	12.25
45	63	7	9	−2	4
53	65	10	10	0	0
55	60	11	7.5	3.5	12.25
50	56	8.5	6	2.5	6.25
61	71	13	12	1	1
63	78	14	14	0	0
67	83	15	15	0	0
35	60	4	7.5	−3.5	12.25
50	70	8.5	11	−2.5	6.25
59	76	12	13	−1	1
			$n = 15$	Σd^2	61.5

Step 4: Add up the squared differences to give 'the sum of the squared differences' (Σd^2).

Step 5: Enter your value for Σd^2 into the formula below to give Spearman's rho

$$\text{rho} = 1 - \frac{6\Sigma d^2}{n^3 - n}$$

For the data shown in Table 10.4 the calculations are

$$\text{rho} = 1 - \frac{6\Sigma d^2}{n^3 - n} = 1 - \frac{6 \times 61.5}{15^3 - 15} = 1 - \frac{369}{3375 - 15} = 1 - \frac{369}{3360} = 1 - 0.11 = 0.89$$

To check for statistical significance use the tables of critical values (Section 12.12). For a sample size of 15 the value of rho is well above the tabulated value at $p < 0.01$ for both one-tail and two-tail tests. We can reject the null hypothesis and conclude that there is a strong positive correlation between the two sets of data.

Strictly speaking, you should apply a correction to deal with any tied ranks, but it actually has very little effect. If you have a lot of tied ranks make the calculation using a Pearson's correlation applied to the ranks, which deals with the problem properly. In the example above, even though there are a few tied ranks, the results of the two methods are almost identical.

Since Spearman's rho falls on the same scale as Pearson's correlation coefficient the probability can be calculated in the same way, as long as the sample is not too small (>10) using the t-distribution. The equation appears in a variety of formats but one of the easiest to follow is

$$t = \frac{\text{rho} \times \sqrt{n-2}}{\sqrt{1 - \text{rho}^2}}$$

which using the example above gives

$$t = \frac{0.89 \times \sqrt{15-2}}{\sqrt{1 - 0.89^2}} = \frac{0.89 \times \sqrt{13}}{\sqrt{1 - 0.79}} = \frac{0.89 \times 3.61}{\sqrt{0.21}} = \frac{3.213}{0.458} = 7.01$$

A t-value of 7.01 with $n - 2 = 13$ degrees of freedom gives a one-tail significance of about 4.5E-06, so we can conclude that $p < 0.001$.

10.7 Kendall's Tau and Other Approaches to Correlation

In geography, Spearman's rank correlation seems to be the only non-parametric form of correlation that is used regularly. I like it for two reasons: it is easy to compute in a spreadsheet, by simply running a normal (Pearson's) correlation on the ranks rather than the original values and also because the scale of Spearman's rho is virtually identical to the scale of Pearson's r, so they are very easy to compare.

The only other type of correlation that you are likely to come across is Kendall's rank correlation coefficient, or Kendall's tau (Kendall 1938). In the past, when calculations were made by hand, there was some advantage to this method because with small samples it could be computed using a simple graphical method. However, in a spreadsheet it is more difficult than Spearman's so that advantage has gone. Both methods are based on the ranks, and since they use equal amounts of information they must have similar power in terms of recognising a correlation when it exists. The big difference between Spearman's and Kendall's approach is that the latter uses a scale that is very different to Pearson's *r*-value, which means that, for example, whereas a Pearson's *r*-value of 0.6 and a Spearman's rho of 0.6 can be considered as equivalent, a Kendall's tau value of 0.6 means something completely different. I think for most purposes this is a distinct disadvantage, so I never use Kendall's tau. It is sometimes argued, however, that it has some advantage over Spearman's rho because it is normally distributed. I have never seen this as a particular advantage, but meteorologists, for example, seem to regard it as important and prefer Kendall's approach over Spearman's. According to Andy Field (2013) it is more appropriate than Spearman's when you have a small sample with a lot of tied ranks.

If you need to calculate Kendall's tau the easiest way is to download the Real Statistics Resource Pack for Excel. It is one of the options listed under correlation and also has its own function (KCORREL). There are several boxes to complete for this function but you can ignore them apart from the first two, where you put the data. Kendall's tau is also one of the options in *R* Commander. Correlation is hidden away under summaries, but in there is an option to choose Kendall's tau. The same is true for SPSS, just follow the instructions as for Spearman's.

References

Chen, P.Y. and Popovich, P.M. 2002. *Correlation: Parametric and non-parametric measures*. Thousand Oaks, CA: Sage.

Cohen, J. 1988. *Statistical Power Analysis for the Behavioural Sciences* (2nd ed.) New York: Academic Press.

Field, A. 2013. *Discovering statistics using IBM SPSS statistics* (4th ed.). London: Sage, 916pp.

Gagen, M., Zorita, E., McCarroll, D. et al. 2011. Cloud response to summer temperatures in Fennoscandia over the last thousand years. *Geophysical Research Letters* 38. doi:10.1029/2010GL046216.

Kendall, M.G. 1938. A new measure of rank correlation. *Biometrika* 30: 81–83.

Spearman, C. 1904. The proof and measurement of association between two things. *American Journal of Psychology* 15: 72–101.

Young, G.H.F., McCarroll, D., Loader, N.J. et al. 2012. Changes in atmospheric circulation and the Arctic Oscillation preserved within a millennial length reconstruction of summer cloud cover from northern Fennoscandia. *Climate Dynamics* 39: 495–507.

11

Regression Analysis

11.1 Simple Linear Regression

Simple linear regression is a method that allows a 'best-fit' line to be added to a set of points on an x–y plot or scatter-graph. There are many uses for regression in geography and related disciplines. For example, when you know the value on the horizontal axis it allows you to define the most likely value on the vertical axis. Where one of the axes represents space or time the best-fit or regression line can be used, with care, to make predictions that go beyond the range of the measurements, providing a method of prediction. Relationships defined using regression can also be extended into the past allowing, for example, the reconstruction of past climate and environmental change.

When I was a student, simple linear regression was a method that was just touched on at the end of a typical geography course on statistical methods and the complexity of the mathematics made it very difficult to use. With modern computers, however, all of that has changed and performing regression analysis is remarkably simple and requires no mathematics at all. In fact you have already seen it performed in the last chapter, because the 'trendline' that is fitted to an x–y plot or scatter-graph in a spreadsheet is actually a 'regression line'. It appears at the click of a mouse. The ease with which regression can now be conducted is both a blessing and a curse for geography students. It is very easy to do it but it is also very easy to do it wrong and produce absolute nonsense. If you want to use regression it is really important that you understand how it works (Figure 11.1). Only then will you be able to check the assumptions have been met and sensibly interpret your results.

11.2 The Straight Line Equation

In the chapter on correlation analysis x–y plots were used to illustrate the shape of the relationship between two parameters or variables. Each point on such a graph represents a pair of numbers, one representing the variable on the horizontal (x) axis and the other the variable on the vertical (y) axis. The shape of the relationship was used to decide whether it was reasonable to use parametric or non-parametric approaches to correlation. Where the data proved suitable, correlation analysis was used to quantify the strength of the relationship between the two variables. The aim of regression analysis, in essence, is not to quantify the strength but to define the nature or 'shape' of the relationship. The simplest shape is a straight line and most applications of regression are essentially trying to define the straight line that best describes the relationship between the two variables. To understand

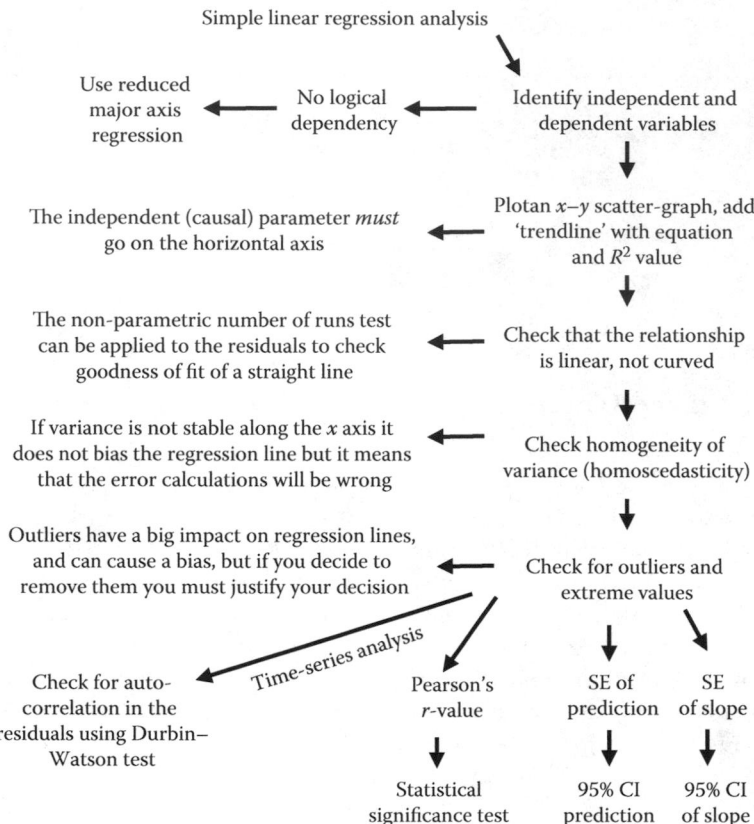

FIGURE 11.1
A logical procedure for performing simple linear regression using a spreadsheet.

regression it is essential that you understand how a straight line on an x–y plot is defined, and I am afraid that in this case there is no escaping a little bit of mathematics.

When a straight line is plotted somewhere in an x–y plot its exact position can be defined using a formula that effectively translates the units on the x axis into those on the y axis

$$y = a + bx$$

where y is the value of the parameter plotted on the vertical axis, x the value of the parameter plotted on the horizontal axis, a the 'intercept' of the line and b the 'slope' of the line.

The simplest case of a linear relationship applies when the values on the two axes are the same. For example, take the English and Scottish currencies. Both nations mint and print pounds, using their own banks, but an English pound is worth exactly the same as a Scottish pound. If we place the English pound on the horizontal axis and call it x and the Scottish pound on the vertical axis and call it y then we can describe the relationship using the simple formula $y = x$, because a Scottish pound is equal to an English pound (Figure 11.2a).

The full formula for this straight line is

$$y = 0 + 1x$$

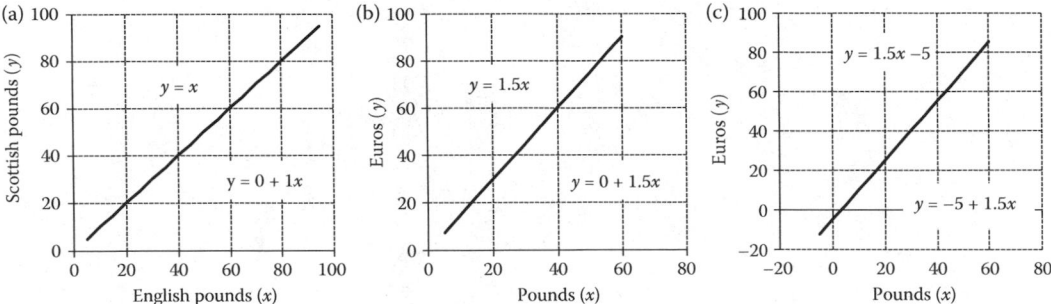

FIGURE 11.2
(a–c) Three examples of straight lines with different equations. The equation translates the values on the horizontal (x) axis into the values on the vertical (y) axis.

So looking at the equation above the intercept 'a' is equal to zero and the slope 'b' is equal to one.

The intercept 'a' is the value on the y axis when the value on the x axis is zero. If both axes start at zero then it is the point where the line crosses the vertical axis. The slope refers to the angle of the line and it tells you how much of an increase in the values of y are associated with a given increase in x. In this case an increase of one unit on the x axis, or one English pound, results in one unit increase on the vertical axis, or one Scottish pound.

Now consider two currencies that are not exactly equivalent. If I decide to visit Dublin I have to buy some euros, and at the moment for every British pound I get about 1.5 euros, and if I use my own bank I am not charged a fee for exchanging currency. The relationship between pounds and euros can be described by a straight line on an x–y plot (Figure 11.2b). I have placed euros on the vertical axis because I am interested to know how many euros I will get for my pounds, and the equation translates the values on the horizontal axis (x = pounds) into those on the vertical axis (y = euros). Since one pound is equal to 1.5 euros the relationship can be described as

$$\text{Euros} = 1.5 \times \text{pounds}$$

or

$$y = 0 + 1.5x$$

The slope of the line (b) is the value 1.5 and the intercept value must be zero because zero pounds is equivalent to zero euros.

Now consider a slightly more complicated situation where I run out of cash and need to buy some more euros in Dublin but have to pay a fee to the bank. They charge a fee of five euros irrespective of how much I exchange. Now the relationship between my pounds and my euros becomes

$$\text{Euros} = -5 + 1.5\,\text{Pounds}$$

or

$$y = -5 + 1.5x$$

Adding a negative number is the same as just subtracting it, so the equation can be re-written as:

$$\text{Euros} = 1.5 \times \text{Pounds} - 5\,\text{Euros}$$

or

$$y = 1.5x - 5$$

I can use this equation to translate the number of pounds I have in my pocket now into the number of euros I will have in my pocket after the transaction. For 10 pounds I will get 10 euros and for 100 pounds I will get 145 euros. The intercept value (*a*) tells me the value on the vertical axis (euros) when the horizontal axis value (pounds) is zero and the 'slope' (*b*) of the line is the exchange rate (Figure 11.2c). It tells me that for every unit increase on the horizontal axis (for every pound) the increase on the *y*-axis is 1.5 units.

Figure 11.3 shows a variety of straight lines drawn on *x–y* plots together with the equations that exactly describe them. Note that although in some cases the equation can be shortened for simplicity, for example when the intercept (*a*) is zero or the 'slope' (*b*) is one, or re-arranged, for example when the intercept is a negative number, you can always re-write the equation in the standard format where $y = a + bx$.

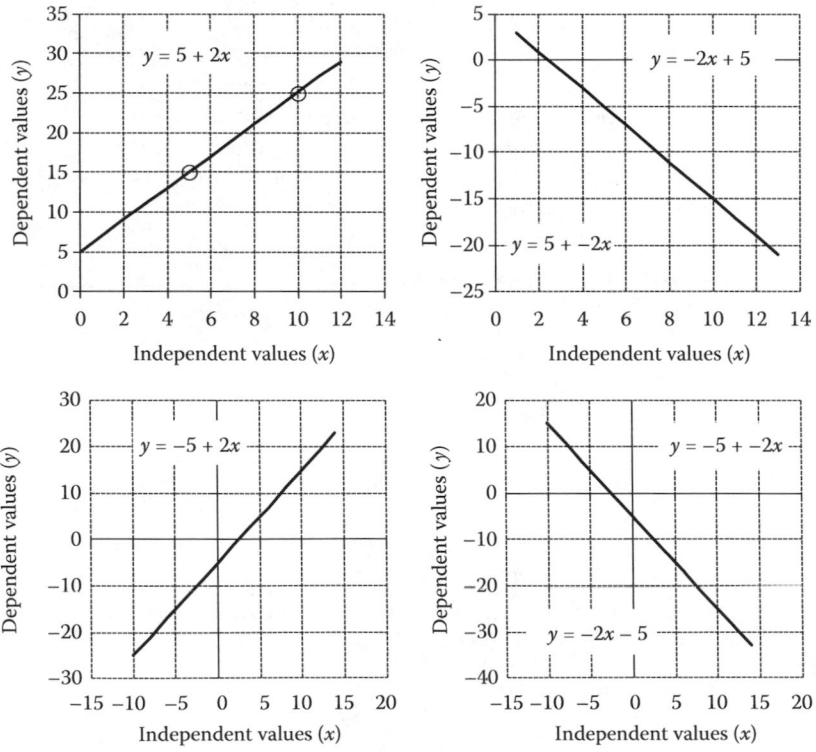

FIGURE 11.3
Four different straight line equations. Note that when the slope value is positive the line rises and when it is negative it falls. The intercept is the point on the *y* axis when the *x*-axis value is zero.

The equations given for the lines shown in Figure 11.3 are just mathematical definitions that allow you to place the line in the right place on the x–y plot. There are two ways to use the equation to position the line. The first is to plot the intercept value on the vertical axis and then use the slope of the line to propagate it, so that if $b = 2.0$ you would add 2.0 units on the vertical axis for every one unit on the horizontal axis. An easier way to do it is to use the equation to find two points and join them together by a straight line. The equation effectively translates whatever is on the horizontal x axis into whatever is on the vertical y axis, so if you enter a high and a low value of x into the equation you get two values for y. For example, if the equation is

$$y = 5 + 2x$$

Then for x values of 5 and 10 you would get

$$y = 5 + 2 \times 5 \quad \text{and} \quad y = 5 + 2 \times 10$$

Now recall the BODMAS rule (Section 4.2) that insists that you multiply before you subtract and you get

$$y = 5 + 10 = 15 \quad \text{and} \quad y = 5 + 20 = 25$$

Those two pairs of values (5:15 and 10:24) are included in the top left graph of Figure 11.3.

11.3 Best-Fit Regression Lines

When using simple linear regression our aim is to 'fit' a straight line through some data points on an x–y plot or scatter-graph. If the points fall perfectly on a straight line then of course it is easy, but that is rarely the case. A pragmatic solution is to just draw a line by eye, but that is not very satisfactory because we might all choose slightly different lines. What we need is some clear 'rule' or 'criterion' that allows everyone to fit exactly the same line. Put simply, the 'rule' is that the line should lie as close as possible to all of the points, or should minimise the distance between the points and the line. Since the line has to pass through the middle of the set of points it is obvious that some points will lie above it and others below it, so simply adding together all of the distances is not an option because the positive and negative values will cancel each other out. A simple solution to get rid of the signs of the differences is to square them. Of all the possible lines that might be drawn, the 'best-fit' one is that which minimises the squared distance between the points and the line. That is why this kind of line fitting is called 'least squares regression'.

There is a further complication, however, because 'distance' from the points to the line can be defined in different ways. You can choose to minimise the distance using the values on the vertical axis or the values on the horizontal axis (Figure 11.4). These options are the basis for two different approaches to regression. It is very important that you are aware of which one you are using because they do not give the same results (Figure 11.5).

In geography we nearly always use the first option, where the 'least squares' is defined using only the distance on the vertical axis. It is given a variety of names in different fields

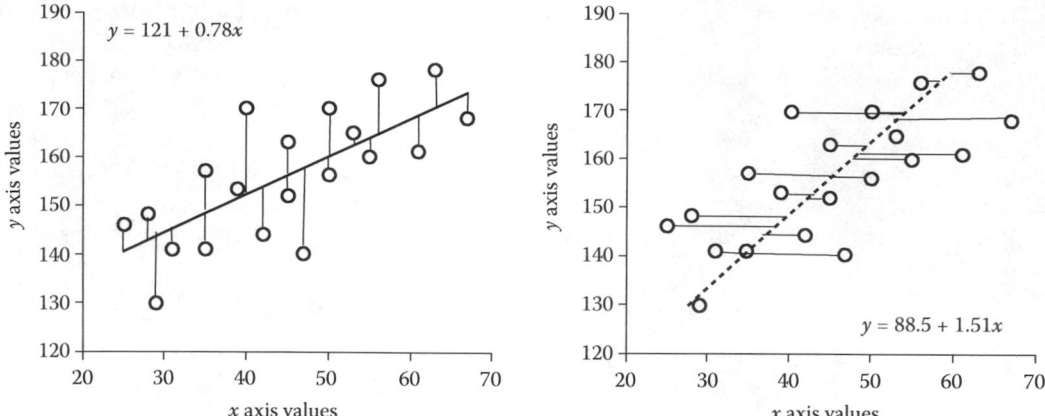

FIGURE 11.4

Two different approaches to regression. The most common approach is to minimise the distance between the points and the line using the vertical axis. If the horizontal axis is used the line is different. This is why it is important to put the data on the correct axes. The independent (causal) parameter goes on the horizontal axis and the dependent (effect) parameter goes on the vertical axis.

but the most common is simple linear regression. Fortunately this is also the method used by Excel and other spreadsheet packages and is also the default in most other software that fits a regression line. Because we generally use this specific approach to regression it is extremely important that you plot your data on the correct axes. The independent variable (the cause) goes on the horizontal axis and the dependent variable (the effect) goes on the vertical axis. When we use simple linear regression we are effectively assuming that the values on the horizontal axis are known with certainty, and it is the values on the vertical axis that are uncertain.

As an example of the difference between the two approaches to regression consider Figure 11.6 which shows the (fabricated) relationship between age and salary for a sample

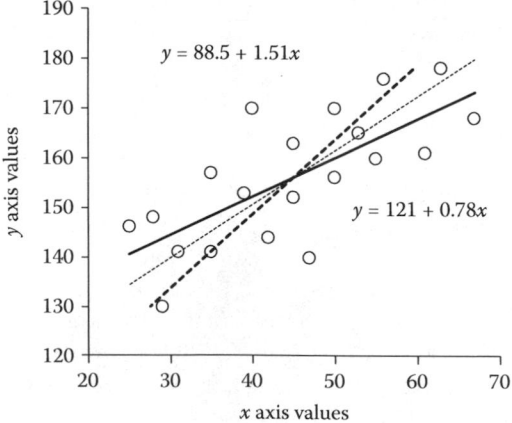

FIGURE 11.5

The two regression lines shown in Figure 11.4 plotted on the same graph. The most common approach, minimising the squared distance between the points and the line on the vertical axis, is the solid line. Minimising the squared distance on the horizontal axis gives the more steeply inclined heavy dashed line. The central fine dashed line is produced using 'reduced major axis' regression, which is described later.

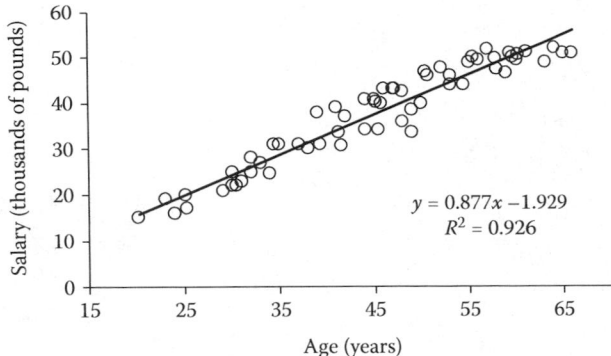

FIGURE 11.6
The relationship between age and salary. Salary is placed on the vertical axis because it is logical that salary might depend, to some extent, on age.

of employees at a large accountancy firm. The axes are plotted in logical order, with age on the horizontal and salary on the vertical. They should be arranged in this way because it is logical that, in a skilled profession, salary might depend to some extent on age and therefore experience. Logically, age is the independent variable and salary is the dependent variable. Note that in this case the correlation between age and salary is very high indeed and the R^2 value of 0.93 means that 93% of the variability in salary is accounted for by age. If we use simple linear regression to fit a straight line through the data points we obtain a line with the formula

$$y = 0.877x - 1.929$$

which means

$$\text{Salary} = 0.877 \times \text{Age} - 1.929$$

We can now use this line to estimate the most likely salary for employees of different age. Of course there is some variability in salaries, but the correlation between age and salary is very strong and regression provides the 'best estimates'. For example, the most likely salary for employees aged 25 and 55 is

$$\text{Salary} = 0.877 \times 25 - 1.929 = 21.925 - 1.929 = £19.996k = £19,996 \text{ at age } 25$$
$$\text{Salary} = 0.877 \times 55 - 1.929 = 48.235 - 1.929 = £46.306k = £46,306 \text{ at age } 55$$

Intuitively, you might think (I did) that if you were to swap the axes in this example you would swap the results as well. For example, since the most likely salary for an employee aged 55 years is £46,306 then the most likely age for someone earning £46,306 will be 55 years. However, that is not actually true. If you swap the axes the line is still fitted by minimising the squared difference between the points and the line on the vertical axis, but now that procedure uses the age values rather than the salary values. The line that is produced when the axes are swapped is different (Figure 11.7), and so the best estimate of the age of someone earning £46,306 is not 55 years it is 54.3 years.

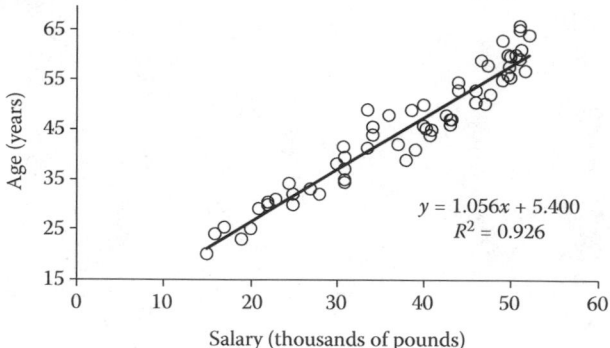

FIGURE 11.7

The same data as Figure 11.6 but plotted with the axes reversed. Note that in this case the equation translates salary into age.

Applying the new equation to translate the two salaries that were estimated for ages 25 and 55 produces two different age estimates

$$\text{Age} = 1.056 \times \text{salary} + 5.4$$
$$\text{Age} = 1.056 \times 19.996 + 5.40 = 21.116 + 5.40 = 26.5 \text{ years}$$
$$\text{Age} = 1.056 \times 46.306 + 5.40 = 48.9 + 5.4 = 54.3 \text{ years}$$

The important point here is that you must be very careful to ensure that you plot your data on the appropriate axes. If there is no inferred causal relationship between your two parameters the best option might be to use reduced major axis regression, which is described in Section 11.10. However, if there is really no causal relationship inferred you should think carefully about why you are using regression. Students often use regression when what they really need is correlation or a paired-sample test for differences between two parameters (Section 8.3). In the large majority of cases where regression is being used in geography there is a clear logic for deciding which parameter is the independent variable and which is the dependent and then you can use the standard method of simple linear regression.

11.4 Assumptions of Simple Linear Regression

If you go to a university library you will find huge tomes dedicated just to regression and there will be whole chapters on the assumptions. Personally I think that too much emphasis is placed on the assumptions, when in reality plotting the data on a scatter-graph and checking that the best-fit line looks sensible is all that is required for most studies.

Regression is a parametric method, so one of the underlying assumptions involves 'normality'. This is often taken to mean that the two samples should be normally distributed, but that is not actually true; it is the residuals from the regression that should be normally distributed. However, this is not easy to check, and also a bit of deviation from normality is not actually too problematic, as long as the line really does fit the data. My advice is to plot the data on a scatter-graph in a spreadsheet, fit a linear regression line (just right click and choose add trendline), and then check that the assumptions for parametric correlation and regression are not strongly violated. If that is true your regression line will not be misleading, even if it is not perfect.

The important assumptions for both parametric correlation and for regression are

1. The data are continuous
2. Most of the data are near the middle, or are evenly distributed
3. The relationship forms a straight line rather than a curve
4. There are no outliers or extreme values
5. Sample size is large enough

These assumptions and common violations are discussed in Section 10.2. The fifth assumption, that sample size is large enough, is more stringent for regression than for correlation, but it is still difficult to find a straight answer. There are various 'rules of thumb floating around, 15 is a common one for simple linear regression, but I have not seen any clear justification. You just have to plot your data on a scatter-graph and see if it looks sensible. The most common violations are described in Section 10.2, where there are lots of graphs for comparison. If your sample is smaller than 20 think carefully about whether you really need to define the shape of the relationship using regression or whether quantifying the strength of the relationship using correlation is enough.

For regression there are two more important assumptions

6. The data are homoscedastic
7. The residuals are independent

11.4.1 Homoscedasticity

This rather unfamiliar term tends to alarm students, but it simply means that the variability, or scatter, should not change along the horizontal axis. The issue should be clear from Figures 11.8 to 11.10.

If your data are not homoscedastic (which means they are heteroscedastic) it does not change the best-fit line, but it compromises the way that the uncertainty is calculated. For most applications of regression in geography it is not a huge problem, and if your data

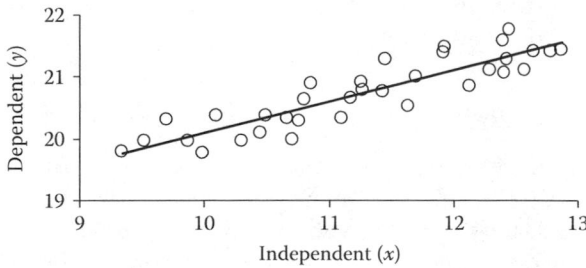

FIGURE 11.8
These data are homoscedastic. The variability does not change along the horizontal axis, so the scatter around the best-fit line is reasonably consistent. The standard error of the prediction will be a good estimate of the uncertainty of predictions for all of the values.

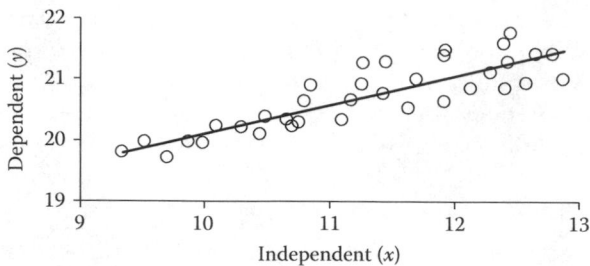

FIGURE 11.9

These data are heteroscedastic because the variability of low values on the *x* axis is less than the variability of high values on the *x* axis. If the standard error of the prediction is used as a measure of uncertainty it will exaggerate the uncertainty at the left side of the graph and underestimate uncertainty on the right side.

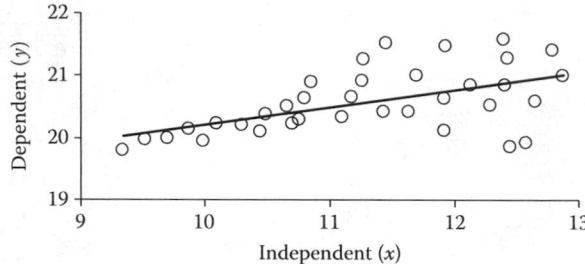

FIGURE 11.10

These data are very heteroscedastic because the variability of low values on the *x* axis is much less than the variability of high values on the *x* axis. If the standard error of the prediction is used as a measure of uncertainty it will grossly exaggerate the uncertainty at the left side of the graph and grossly underestimate uncertainty on the right side.

are a bit heteroscedastic you can go ahead and just mention it. If the data are strongly heteroscedastic it may be possible to correct the problem by 'transforming' the data in some way, such as using logarithms. Those methods are beyond the scope of this book. If your data are really not suitable for regression without having to employ some complicated transformation it is worth just pausing to consider why you are using regression and whether all you really want to do is measure the strength of the correlation. If that is the case then just use a non-parametric form of correlation, such as Spearman's rank correlation, and the heteroscedasticity does not matter. You can still plot the best-fit line just to illustrate the shape of the relationship.

11.4.2 Independence of Residuals

An important assumption of regression is that every pair of points is independent of every other pair. You can think of this as meaning that the order in which you collected the data should make no difference to the values. However, this is not always the case, particularly when you are dealing with data that is collected over time (time series) or across space. In such cases things that change over time, such as the seasons, or that change over space, along environmental or cultural gradients, can cause the values that you collect to be related to each other. If the value that you measure at one point in time or space is controlled to some extent by the value or values that precede it then you violate the assumptions of regression. You can test this assumption by looking at the residuals from the regression and seeing if they show a trend over time or across space. If your data are measured along a gradient, such as time, there is a statistical test that you can use to test for the independence of residuals (Durbin–Watson test) which is described in Section 11.11.

11.4.3 Outliers and Extreme Values

Simple linear regression is very sensitive to the presence of extreme values and outliers and it is very important to plot the data on a scatter-graph and check for them and for their effect on the regression line. If there are some outliers you should not just remove them so that you get a better regression line and stronger correlation, you need to try to understand why they are outliers and make a rational decision that you can justify in your report. The outliers are often the most interesting elements in a regression analysis, so take some time to look carefully at them.

The most common reason for marked outliers is human error. It is very easy to make a mistake when you type a lot of numbers into a spreadsheet and even if you download the data from some official source there is no guarantee that there will be no errors. Consider the data in Figure 11.11, for example. The relationship between age and salary is very strong, but there is one clear outlier; a person aged 52 but only earning the equivalent of a 25-year old. The likely explanation is that the two digits have been transposed. If I had collected these data I would check the original notebooks. If I had downloaded the data I would argue in the report that this is likely to be an error and then remove it.

Sometimes there are marked outliers that are not errors but which should, nonetheless, be removed. In Figure 11.12, for example, salary is plotted against age, but in this case there are two very high earners and their ages are separated by 30 years. These are not errors, this is the owner of the firm and his daughter. If your aim in plotting the graph is just to show the relationship between age and salary then they can be left in and explained. However, if your aim is to use the regression equation to calculate the most likely salary for employees of different age then leaving those two extremes in the analysis will cause a bias in the results. They are pulling the regression line up and away from the centre of the main set of data, so your salary estimates will be too high. When they are included only 60% of the variation in salary is accounted for by age ($R^2 = 0.6$), but when they are removed the line fits the data very well and 92% of the variance is explained.

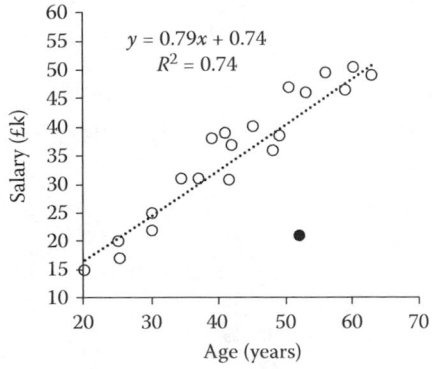

FIGURE 11.11
Where there are clear outliers they often represent errors. In this case the digits for age have probably been transposed, so that 25 has been plotted as 52.

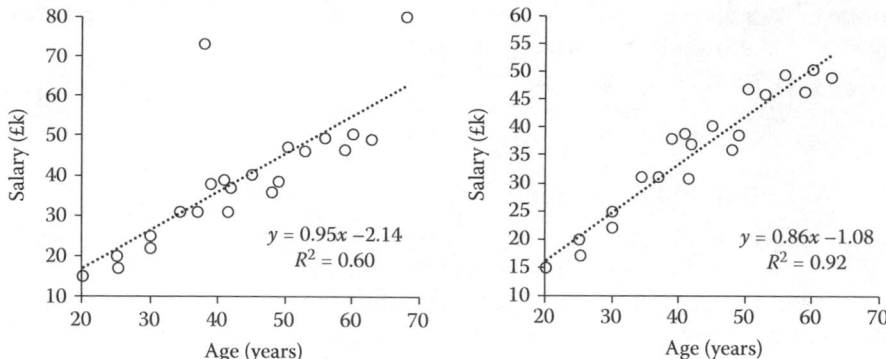

FIGURE 11.12
The relationship between age and salary plotted with and without two clear outliers.

11.5 Confounding Variables

One of the pitfalls of using regression, or correlation, is that you sometimes observe a strong relationship between two variables where there is no real cause and effect relationship. They appear related because they are both influenced by the same 'confounding' factor that is forcing the values to change in a particular direction.

A simple example might be the relationship between the number of visitors to a beach resort and the percentage of visitors that are eating ice cream. If you were to visit a beach many times over the year and record those two numbers I suspect there would be a close linear relationship and a very strong correlation. However, it would be rash to conclude that tendency to eat ice cream is in some way dependent on the number of people that surround you, making ice cream eating some sort of weird crowd behaviour that people tend to avoid when in small groups. The likely explanation is the change of the seasons. As summer approaches the weather improves, more people visit the beach, and also, and completely independently of how many people surround you, warm weather makes you more likely to eat ice cream. As autumn fades into winter the crowds thin and those who do visit are less likely to eat ice cream. In this case crowd numbers and proportion eating ice cream are correlated, but one does not cause the other, they both rise and fall with the seasons.

There is no simple solution to the problem of confounding variables. The key thing to remember is that the fact that two variables are strongly correlated, and that you can plot a regression line through them, does not mean that one causes the other. Correlation and regression only allow you to illustrate the relationship; to interpret it you need to understand the data and what they represent.

11.6 Interpreting Regression Results

11.6.1 Statistical Significance

Most statistical text books draw a clear distinction between testing the significance of the Pearson's correlation coefficient and testing the significance of a regression line. They generally argue that the significance of the regression line must be tested using the slope coefficient (b in the straight line equation). The appropriate test is whether the slope is significantly different from zero.

If you are using multiple regression with several independent variables, which is beyond the scope of this book, then the distinction between testing the significance of correlation and testing the significance of the slope is important. However, in the case of simple linear regression, where there is just one independent variable and one dependent variable, then it actually makes no difference. The slope of the line is only zero when there is no correlation between the two parameters. Testing the significance of the slope will give you exactly the same probability as calculating the significance of the Pearson's correlation coefficient (the r-value, or square root of R^2). The easiest way to test for significance is to take the square root of the R^2 value, which gives you Pearson's correlation coefficient, then check that value against tables of critical values for your sample size (Section 12.11).

No doubt some evil professors will not believe you if you tell them this and will force you to calculate the significance of the slope, in which case here is the method, which involves a t-test

$$t_b = \frac{\text{slope coefficient}}{\text{standard error of slope coefficient}} = \frac{b}{SE_b}$$

Unfortunately the standard error of the slope is a bit of a nuisance to calculate (Section 11.6.3), which is why it is easier to just test the r-value instead. The significance of the t-value can be checked using tables of critical values. The degrees of freedom are the number of pairs minus two. In a spreadsheet you can use the TDIST function but make sure the t-value is positive.

As an example take the regression equation in Figure 11.12, without the outliers, which is

$$y = 0.8604x - 1.08$$

The R^2 value is 0.9204 giving a Pearson's r-value of 0.9594 (the square root of R^2) and the number of pairs (N) is 20. The standard error of the slope (Section 11.6.3) is 0.0596 so the t value used to test the significance of the slope (b) coefficient is

$$t_b = \frac{0.8604}{0.0596} = 14.4$$

Just to make the point, the t value used to test the significance of the Pearson's r-value (t_r) is:

$$t_r = \frac{r\sqrt{N-2}}{\sqrt{1-r^2}} = \frac{0.9594\sqrt{20-2}}{\sqrt{1-0.9204}} = \frac{0.9594\sqrt{18}}{\sqrt{1-0.9204}} = \frac{0.9594\sqrt{18}}{\sqrt{0.0796}} = \frac{0.9594 \times 4.2426}{0.2821}$$

$$t_r = \frac{4.0704}{0.2821} = 14.4$$

To find the significance value either use tables of critical values (Section 12.7), with $N - 2$ degrees of freedom, or use the TDIST function in a spreadsheet (if t-value is negative change it to positive for this function). You can use a one-tail test if you predicted the direction of the slope of the regression line in advance. Otherwise, or if you are not sure, use a two-tail test. In this case the t-value is strongly significant (two-tail $p < 0.001$).

11.6.2 Effect Sizes in Regression: r, R^2 and Slope

If the assumptions of regression have been met, and the best-fit line looks sensible when plotted on a scatter-graph in a spreadsheet, then you can add not only the equation for the straight line but also an R^2 (R-squared) value. This metric, sometimes called the coefficient of determination, reflects the strength of the correlation between the two variables. It is well named because it is literally the square of the r-value (Pearson's product moment correlation coefficient). Where there is a clear cause and effect relationship between your two variables you can use the R^2 value to determine how much of the variability on the vertical axis is explained by the variability on the x axis. Put simply it tells you how much of y is explained by x. For example, if you were plotting salary (y) as a function of age (x), and obtained an R^2 value of 0.5 you could conclude that age explains about 50% of the variability in salaries (0.5 is a proportion 'out of one' so multiply by 100 to get a percentage). The other 50% remains unexplained.

When your main interest in a relationship is correlation rather than regression then you can just use the R^2 value to obtain the r-value by taking the square root and adding the sign. The r-value tells you the strength of the correlation. For correlation analysis the r-value is effectively the effect size. However, if you are really interested in the shape of a relationship, rather than just the strength of the correlation, then you need to be a bit more careful about how you use the r and R^2 values. Consider the two graphs showing the relationship between age and salary for two firms (Figure 11.13) that are offering similar starting salaries for graduates. Both firms show an equally strong correlation between salary and age, and the R^2 value of 0.9 tells us that 90% of the variability in salary is explained by age. However, this should not lead you to conclude that the relationship between age and salary is the same for these two firms.

In these examples r and R^2 are measures of the amount of variability in salary that is explained by age, but they do not tell you the magnitude of the change in salary with age. In many cases where regression is used you may be interested not just in the amount of variability that is explained but also in the magnitude of change in y as a function of x. In

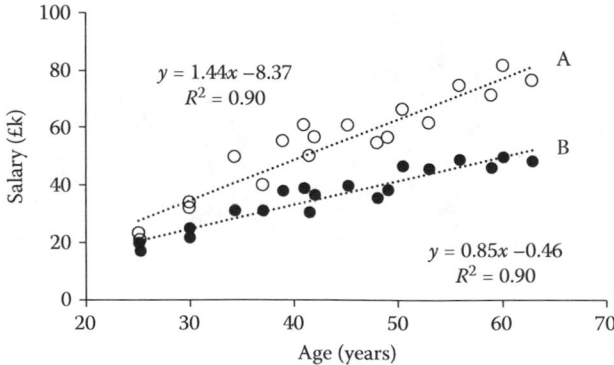

FIGURE 11.13

In this case although the R^2 values are equivalent, the slope parameters are very different. In many cases the slope parameter is a more useful effect size than the correlation coefficient.

such cases the r-value is not the appropriate effect size. The appropriate effect size is the 'slope' parameter (b) in the equation for the straight line. This is the number that, in this example, tells you how much extra salary you are likely to get each year. If you decide to work for firm B, the most likely annual increment in salary is £0.85k, or £850, but if you work for firm A it is £1440. That is a big difference and although the starting salaries are similar you can expect to earn more than £40k by the age of 40 at firm A whereas you would have to work to age 60 to achieve the same salary in firm B.

11.6.3 Uncertainty of the Slope Parameter

Where defining the slope of the best-fit line is the main reason for using regression it can also be useful to calculate an uncertainty for the slope. Where all of the points are close to the line the uncertainty is small and where there is a lot of scatter it will be large. The 95% confidence intervals for the slope (b) can be calculated based on the standard error of the slope (SE_b), which is a measure of the variability. The equation is

$$95\%CI = b \pm t \times SE_b$$

The value 't' is taken from the t-distribution and depends on the size of the sample. If your sample size is between 20 and 60 t is close to 2, so you may come across cases where authors just quote two standard errors. The correct values of t for different probabilities are presented in Section 12.7. If you can access the Excel 'analysis tools' in 'analysis' you can choose the regression option and the output includes the slope coefficient, which in this case is called the 'X coefficient' with a standard error next to it. The output also gives the results of a t-test for significance of the slope and the upper and lower 95% confidence limits.

If you cannot access the 'analysis tools' you can derive the standard error of the slope coefficient (SE_b) from two other functions: the standard error of the estimate (STEYX) and the sum of the squared differences between the values on the x axis and their mean (DEVSQ(x)). The equation is

$$SE_b = \frac{STEYX}{\sqrt{DEVSQ(x)}}$$

11.6.4 Estimating Goodness of Fit and Uncertainty

When you fit a regression line in a spreadsheet the easiest measure of how well the line fits the data is the R^2 value that is provided at the click of a mouse. However, there is a disadvantage in using just the R^2 value because it is calculated as a ratio between two numbers (Section 11.12) and so the original units are lost and it falls on a scale of zero to one. Sometimes when we use regression to calculate the 'most likely' values on the y axis for known values on the x axis, we would also like to give some indication of how uncertain those estimates are, and that requires a number that is in the same units as our original measurements. For example, when considering the relationship between age and salary, and predicting the salary you are most likely to earn at age 40, it would be nice to be able to put some uncertainty bounds around that number. If the salary is in pounds then we need the uncertainty in pounds as well.

The problem faced here is very similar to that of estimating a mean value from a sample. We often want to calculate not just the mean value of the sample but also to give some indication of how uncertain we are about the mean value in the population. The simplest solution for the mean is to use two standard errors of the best estimate of the mean, because in a reasonably large sample we can assume that if we were to take many samples of a given size from a population, then 95% of them will produce a mean value within about two standard errors of the mean calculated using an unbiased sample of the same size.

The same logic can be applied to producing uncertainty estimates for a regression line. The regression line is also based on a sample, and different samples from the same population would result in slightly different lines. In calculating the uncertainty of a mean we use the average squared difference between the individual measurements and the overall mean (that is the variance). In regression we use the average squared difference between the points and the line, measured as usual on the vertical axis. We can think of these distances as 'errors', since they are the difference between our best estimate for each point (the position on the regression line) and the real measurements. The average, or mean of these squared differences is called the 'mean squared error' (MSE). If a regression line has a small MSE the points are all very close to the line, so that estimates based on the regression line equation will be very good. If a regression line has a very large MSE then the points are distant from the line and the best estimates will not be very good at all.

The units of the MSE, as for variance, are the original units squared, which is not very convenient. The simple solution is to take the square root of the MSE to give the 'root mean squared error' (RMSE), which is now in the original units. This provides a very convenient metric for describing the amount of variability or spread of values around the best-fit regression line, and can be viewed as equivalent to the standard deviation as a measure of the spread around the mean in a single sample.

Although it is a common practice to use the RMSE as the measure of variability in regression, with just a slight modification it becomes much more useful. The RMSE is the square root of the squared differences between the points and the line, and so it describes the variability of the points around the line in the sample. If we want to use the regression line to estimate new values, however, we are effectively using a sample to make an estimate of a larger population and so we need to add a bit of uncertainty. This is achieved by replacing the sample size by the number of degrees of freedom, which is the sample size minus the number of parameters that were used to make the estimates. For simple linear regression using a straight line there are two parameters to estimate (intercept and slope) and so the degrees of freedom are $n - 2$. When the RMSE is calculated using the degrees of freedom rather than the full sample size the word 'mean' becomes misleading, and the metric is called the 'standard error of the prediction'. If you are working in a spreadsheet you do not need to calculate the standard error of the prediction by hand using the residuals, there is a function that will give you it automatically (SEYX). If you look on the help for this function it provides a very complicated equation, that makes no sense to me, but I have checked that it gives the same result as the much simpler method explained here.

The standard error of the prediction, like the standard error of the mean, is extremely useful because we can assume that, given a reasonably large sample, 95% of the time the true value in the population will lie within about 2SE of the estimate. Strictly speaking, as with the confidence limits applied to the calculation of the mean, the multiplier should only be 2 when the sample size lies between about 20 and 60, and for very large samples 1.96 is better. The correct multipliers for different sample sizes are based on the t-distribution (Section 12.7).

There are several other ways to estimate the uncertainty associated with predictions based on regression lines, all more complicated than that used here and I think they are beyond the scope of this book. There are plenty of other books that go into much more detail. As long as you are estimating values that fall within the range of the data that makes up your sample then I think the estimate plus or minus two standard errors of the prediction is good enough. If you want to predict values that lie well beyond your range of measurements then this measure of uncertainty will certainly be an underestimate, and you really need to be extremely careful. In particular you have to think carefully about the system that you are modelling and reflect on whether it is realistic to extend the regression line far beyond the range of the data that it is based on. In Figure 11.13, for example, we can easily use the regression equation to calculate the salaries of employees aged 10 or 100 years old, but neither estimate would be realistic. In many natural systems linear relationships only hold true over a limited range, and beyond that range it is not safe to extrapolate the relationship as a straight line.

11.7 Performing Simple Linear Regression Analysis

If you are intending to use regression analysis regularly, during a PhD for example, then you should learn to use one of the powerful tools such as SPSS or R and there are other books that go into much more detail. My favourites are Andy Field's book for SPSS (2013) and John Fox (2002) for R, though the huge tome by Field et al. (2012) is also very good. However, if you just want to use regression occasionally, or your main interest is in correlation and fitting a regression line is just a way to demonstrate the shape of the relationship, check the assumptions and calculate the strength of the correlation, then the most convenient option is to use a spreadsheet.

11.7.1 In a Spreadsheet

Put your data in two columns side by side, so that the pairs line up. Put the one you want to fall on the x axis (the independent [causal] variable) on the left. Pick up both series with the mouse and click 'insert' and choose a scatter-graph. The graph will pop-up and the series should be plotted on the correct axes. Now click on the points, right click and choose 'add trendline'. In the options you can choose to show the R^2 value and the equation. Do not choose 'set intercept'. Now the equation and R^2 value will appear on the graph. To check the significance take the square root of the R^2 value and add a sign: positive if the line is rising and negative if it is falling. Now you have the Pearson's correlation coefficient and you can check the significance using tables (Section 12.11) or a t-test. Instructions are given in Chapter 10. Do not use the default graph that Excel produces in your report, that is like writing 'I am lazy' on your work; tidy it up and make it fit your data properly. Adjust the axes to fit, add axis labels (click on the graph and go to the 'design' tab and 'add chart element'), make sure the fonts are big enough, etc. If a graph is well designed, with proper axis labels, you do not need a title, it is a waste of space.

If you want to take things a little further you can use the 'analysis tools' under the 'data' tab. These are sometimes hidden but instructions for finding them are given in Section 4.4. One of the options is regression. Be careful to enter the data sets into the correct boxes marked x (independent) and y. The output includes an analysis of variance, which you do not really need, and some statistical tests. The most useful is the probability listed for the 'X variable'. That is the two-tail statistical significance of the correlation coefficient and of the slope of the regression line. They are effectively the same thing for simple linear regression. Also listed is the standard error for the 'X-coefficient' which is actually the standard error of the slope of the regression.

If you tick some of the options, one of the outputs includes a table of 'predicted y' values together with the residuals and the standardised residuals. The residuals are the difference between the measured *y* values and the regression line measured on the vertical axis. They are in the original *y*-axis units. The standardised residuals are the same values translated into z-scores so that they have a mean of zero and a standard deviation of one. The logic of z-scoring is explained in Section 11.9. The residuals are useful because they are listed in the order in which the original data were entered. That means that if you suspect that the order in which you collected the data might influence the results you can apply the Durbin–Watson test.

11.7.2 Companion Site Calculators

Performing simple linear regression in a spreadsheet is not difficult, so you do not really need to use a pre-designed calculator. However, there are a few things that are a bit tricky to derive so I have designed some calculators that will help you out. They include calculators for checking that a linear fit is appropriate, using the number of runs test, applying reduced major axis regression, the Durbin–Watson test and a few other things that may be useful.

11.7.3 In SPSS

If you want to do fancy things with regression analysis then SPSS gives you many options. If you just want to perform simple linear regression there are so many options and so much potential output that it is all a bit bewildering. Enter your two series, side by side, in the 'data view' spreadsheet and on the 'variable view' page give them a name and assign the 'measure' as 'scale'. Now go to 'analyze', 'regression' and choose 'linear'. Put the dependent and independent data in the correct boxes and press 'ok'. The output is actually less useful than that given by Excel because there is no automatic scatter-graph to show you how well the line fits. To be honest I think it is more trouble that it is worth to use SPSS for simple linear regression analysis and I would stick to a spreadsheet.

11.7.4 *R* Commander

Enter your data as two columns with simple headers. The simplest option is to put the data in the top left corner of a spreadsheet and save it as a tab-delimited text file. In *R* Commander open that file using 'data' and 'import from text file'. Go to 'statistics', 'fit models' and choose 'linear regression'. Be careful to put the data in the right boxes; the independent variable, which goes on the *x* axis, is called the 'explanatory variable'. You can also give your regression model a name, which helps if you want to use it later.

The output is no more useful than that produced by Excel and there is no automatic scatter-graph with a fitted line so that you can check that it is a reasonable fit. The output is also a bit confusing if you are not used to using *R*. The two parts of the straight line equation are there; the slope of the regression line is '*A* estimate' and the intercept is labelled as such. The standard error of '*A*' is the standard error of the slope (SE_b in the terminology I have used). The 'multiple *R*-squared' is the R^2 value you need and final *p*-value is the two-tail probability.

When you have completed the analysis in *R* Commander you can then use your model to conduct other analyses. Most of the options are beyond the scope of this book and they apply mainly to more complicated models with several independent variables, but some of them are useful. Go to 'models' and choose 'confidence intervals' to obtain the 95% (or any other %) confidence intervals for the slope and the intercept. Or go to 'models' and choose 'numerical diagnostics' where you will find some tests for heteroscedasticity, non-linearity, autocorrelation and outliers. If you have very large data sets be careful with these tests, because small deviations from perfection can produce a significant result when the deviations are actually too small to make any difference. Looking at a scatter-graph and applying a bit of common sense is good enough for most purposes.

11.7.5 Examples

11.7.5.1 Example: Predicting the Weather

You take a spring break in northern Scotland and spend the whole week in the pub sheltering from the horrible weather. One of the whisky-soaked locals tells you that a cold May is good news, because it guarantees a warm summer. You suspect this is nonsense, so download the climate data for Northern Scotland for the last 90 years from the UK Meteorological Office. If the old soak is right then there should be a strong correlation between the temperatures of May and the summer (*JJA*), and if the relationship is linear you could use simple linear regression to predict the summer temperature for this year based on the May temperature. Since you intend to predict summer temperature from May temperature you place May on the horizontal axis, so that the equation translates May temperature into summer temperature (Figure 11.14).

As you suspected, the whisky has addled the old man's brains. There is virtually no correlation between the average temperatures of May and of the summer (*JJA*). In fact May temperature only explains about 1% of the variability in summer temperature. Because the correlation is close to zero the regression line is almost flat and the slope is not significantly different from zero. Trying to use June temperature to predict summer temperature would be completely pointless.

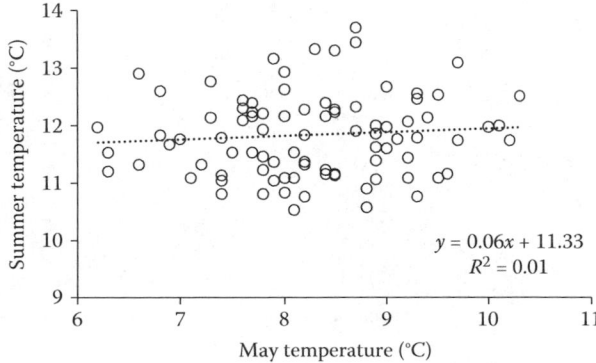

FIGURE 11.14
Relationship between summer temperature and the temperature of May. In this case summer temperature is placed on the vertical axis because the intention is to use the temperature of the current May to predict the most likely temperature of the coming summer.

11.7.5.2 Example: Predicting Degree Outcomes

In most universities the marks from the second year of study count towards the final degree mark, so we would expect a strong correlation between year 2 marks and final marks. Having received your year 2 mark can you then predict your final mark using simple linear regression? Is there a chance that you might improve or have you already blown it? In the example below (Figure 11.15) there are pairs of marks for 146 students and it is clear that there is a strong positive correlation ($r = 0.91$) and the year 2 marks explain about 83% of the variability in final year marks.

Note that the slope of the line is not very far from one (0.9), so it is quite close to a 1:1 line. The overall mean score (±95% confidence intervals) for year 2 is 59.63 ± 1.09 and for the final score it is 61.04 ± 1.08. Without looking more closely you might conclude that students are likely to get a final score that is almost exactly the same as their year 2 mark. However, although the two average scores are very similar, and would not be significantly different using an independent samples test, such as Student's t-test, that ignores the paired nature of the data. In fact more than two-thirds of the students increased their marks, and given a large sample of 146 the chances of such a large split occurring just by luck, if the chances of increasing and decreasing were equal, is less than a thousand to one (sign test, $p < 0.001$). A paired-sample t-test returns a two-tail probability of 9.0E-09, which means there are 8 zeros after the decimal place.

Even accepting that there is a reasonable chance of increasing your grades very slightly, you might look at the high R^2 value and conclude that the die is cast and there is not much you can do to change your grade. You will inevitably get a final mark that is, at best, just a tiny bit higher than your mark for year 2. Again that is not really true. There is indeed a very strong correlation, which is not surprising considering that the year 2 mark is included within the calculation of the final mark. However, even with an R^2 value of 0.83 there is still considerable uncertainty. The standard error of the prediction for these data (STEYX in Excel) is 2.72, so to estimate the 95% errors of the prediction we multiply that by a number taken from the t-distribution for a two-tail probability of 0.05 (Section 12.7) which is about 1.97, giving 95% confidence limits for the prediction of ±5.36. This is

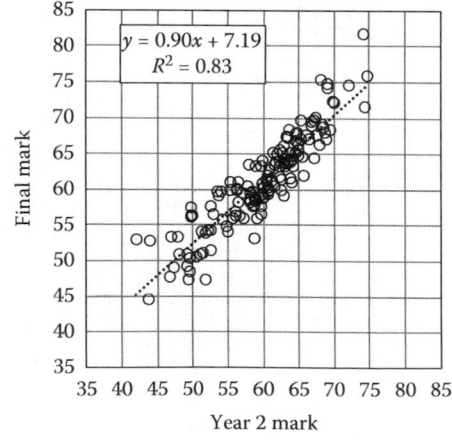

FIGURE 11.15

Relationship between examination marks from year 2 and the final year mark for a large sample of students. Final mark is placed on the vertical axis because the intention is to use the year 2 marks to predict the final year marks.

a rather rough estimate of the uncertainty, but it tells us that there is a 5% chance of your final mark lying within a range covering 10.72%, which in most universities is more than a whole degree class.

11.8 Tests for Comparing Two Regression Analyses

It is often useful to compare two regression analyses to see if they are significantly different. Common examples in geography dissertations involve comparing relationships in one place with those in another place, or comparing relationships measured in a system (e.g. a river) that has been disturbed in some way with the same relationships measured in a 'control' system. I have seen many dissertations that try to do this but they rarely use the data that they have collected to best effect. This kind of comparison is quite complicated and you need to be very clear about the differences you are testing for. You can test whether there is a significant difference between the two correlation coefficients, between the two slope values, or for a difference in the two sets of measured values. An example is perhaps the easiest way to explain.

Imagine that you have measured the average size of the stones forming the bedload at 15 points down a river. The local bedrock comprises mudstone and limestone and you measure the size of the two rock types separately (Figure 11.16). You wish to know whether the change in size with distance is the same for the two rock types.

The first question you might ask is whether the relationship between distance and size is equally strong for both rock types. The appropriate test here is for the significance of the difference between two correlation coefficients (Section 10.6). In this case the two correlation coefficients are -0.966 and -0.925 (the square root of the two R^2 values with the sign added) and they are not significantly different ($z = 0.75$, two-tail $p > 0.05$). We can conclude that the relationship between size and distance is equally strong for the two rock types.

Both rock types show a decline in size with distance, but note that the rate of change is not the same. The slope coefficient of the equation describes the decline in size per unit distance and is 81 mm per km for mudstone whereas for limestone it is only 17.6 mm per

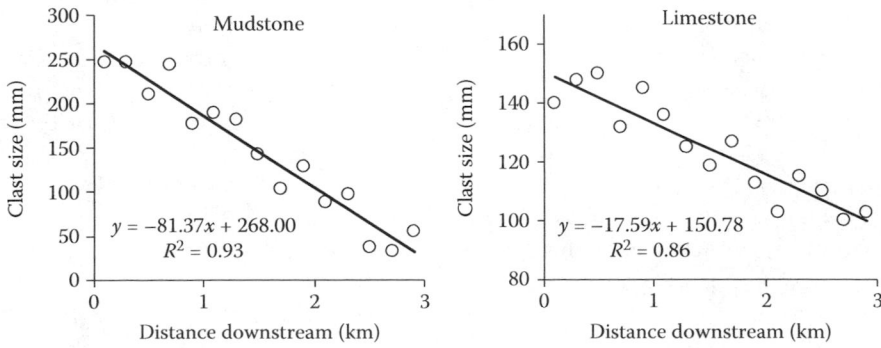

FIGURE 11.16

Two scatter-graphs, with simple linear regression lines and equations, comparing bedload size with distance down a river. Note that the vertical scale for mudstone is not the same as that for limestone.

km. The two slope coefficients can be compared using a *t*-test. If we call the two regression equations *A* and *B* then the parameters we need are

Slope for $A = b_A$

Slope for $B = b_B$

Standard error of $b_A = SE\ b_A$

Standard error of $b_B = SE\ b_B$

The equation is

$$t = \frac{b_A - b_B}{\sqrt{SEb_A{}^2 + SEb_B{}^2}}$$

Different ways to obtain the standard errors for the slopes are explained in Section 11.6.3. In this case the two values are 6.071 and 1.997 giving

$$t = \frac{(-81.37) - (-17.59)}{\sqrt{6.071^2 + 1.997^2}} = \frac{-63.78}{\sqrt{38.85 + 3.99}} = \frac{-63.78}{6.55} = -9.74$$

To find the probability associated with the difference we need the degrees of freedom, which is based on the number of pairs of samples used in each of the regression calculations:

$$\text{degrees of freedom} = (n_A - 2) + (n_B - 2)$$

In this case both regression equations are based on samples of 15 pairs, so the degrees of freedom are $13 + 13 = 26$. We are interested in the difference between the two slopes, and did not predict the direction of difference in advance, so we should apply a two-tail test. The two-tail probability associated with an absolute (ignore the sign) *t*-value of greater than 9 is millions to one (use tables or the TDIST function in Excel), so we can safely conclude that the slopes of the two regression lines are different. In the context of this study that tells us that mudstone breaks up, or is size sorted, much more quickly than limestone.

Another question that can be asked is whether the clasts of limestone tend to be larger or smaller than those of mudstone. This is not as easy as it sounds. You could ignore the distance down-river and just compare the 15 sizes for mudstone with the 15 sizes for limestone. Sample sizes are a bit small for an independent samples *t*-test so you could use the Mann–Whitney *U*-test, but entering the data into the companion site calculator for comparing two independent samples will give the results of those and several other tests. Neither test shows a significant difference ($p > 0.05$). The major difference between the two samples is not in the position of the middle (mean or median) but in variability. The *F*-test for equality of variance is strongly significant ($F = 19.7$, $p < 0.001$). However, treating the samples as independent ignores all of the information on change with distance, so a paired-sample approach, comparing the mean size of the rock types measured at the same locations, uses more information. Again, neither the parametric paired *t*-test nor the non-parametric Wilcoxon's matched-pairs signed-ranks test shows a significant difference. You might be tempted to conclude that the limestone and mudstone clasts in the river tend to be about the same size, however that would actually be quite wrong. The reality of the situation is clearer if the two sets of data are plotted on the same graph (Figure 11.17). Now it is clear that there is only a very short section of river where the clasts tend to be about the

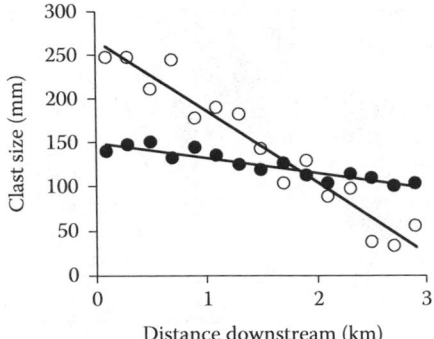

FIGURE 11.17
When the two data sets and their regression lines are plotted on the same graph (closed circles are limestone) it is clear that concluding that there is no difference in clast size for the two rock types is quite wrong.

same size. In the upper reaches of the river the mudstone clasts are consistently larger than the limestone clasts and in the lower reaches the opposite is true.

The reason that tests for two independent samples show no significant difference is because they ignore the locations where the samples come from; they just compare the position of the middle (mean or median) and in this case the middles are not very different. The paired sample tests for difference do not work well because although they use the extra information that allows the two samples from each location to be linked, they ignore the relative order of the sampling. The fact that the differences are in opposite directions at either end of the river pushes the average difference towards zero, hence the non-significant results.

A helpful solution here is to treat the samples as paired, but rather than looking at the individual values to calculate the difference in size at each measurement station. These differences can now be plotted on a scatter-graph and correlation and regression can be used to determine how the difference in size changes with distance downstream (Figure 11.18). If there was really no difference in size between limestone and mudstone, or if the difference was consistent along the river, there would be no correlation and the regression line would be flat, with a slope not significantly different from zero. If there was a consistent

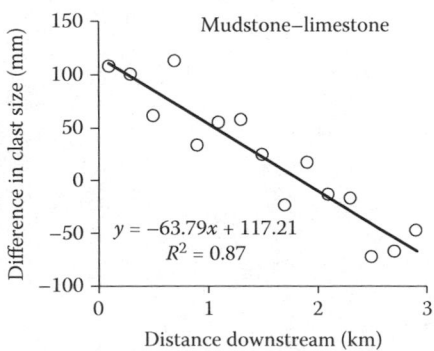

FIGURE 11.18
Plotting the difference in size as a function of distance downstream is an effective way of demonstrating the relationship between the two rock types.

difference between the sizes we would already have picked that up using the two-sample tests, so this extra step is an efficient way of testing for no difference. It is more powerful than the other tests because it uses more information. The other tests could not use the distance downstream as part of the analysis. It is abundantly clear from Figure 10.18 that the hypothesis that there is no difference in clast size is not true. The difference in size is very strongly correlated with distance downstream and both the correlation coefficient and slope coefficient are far from zero and strongly statistically significant ($p < 0.001$).

11.9 Standardising (z-Scoring) and Variance Scaling

Sometimes when we wish to compare two or more variables we are not interested so much in how they correlate with each other, but rather in how they co-vary either across time or across space. For example, you might want to plot changes in summer temperature, summer precipitation and annual tree growth increment on the same graph. That is very difficult to achieve because the scales of these three parameters are very different. With only two parameters you can solve the problem to some extent by placing a second vertical axis on the graph, but even then it is difficult to decide how to adjust the scale to get the most reasonable comparison. With three or more parameters it is often impossible.

Take the example of the three time series in Table 11.1. The three series have very different mean values and standard deviations, so if they are plotted on a graph with a single vertical axis it is very difficult to compare them (Figure 11.19). Plotting one of the series using a secondary vertical axis helps, but it makes it appear that series *B* and *C* are quite similar whereas series *A* is very different (Figure 11.19). However, this is a distorted view of the way that the three series evolve through time because neither of the vertical scales is suitable for plotting the values of series *A*.

TABLE 11.1

Three Times Series with Very Different Scales, Making It Difficult to Plot Them on the Same Graph

	Original Values				z-Scored Values		
Mean	13.3	160.7	3008	Mean	0	0	0
SD	5.87	67.25	1136.23	SD	1.00	1.00	1.00
	A	B	C		A	B	C
1920	4	51	2634	1920	−1.58	−1.63	−0.33
1930	6	70	3551	1930	−1.24	−1.35	0.48
1940	16	194	3975	1940	0.46	0.50	0.85
1950	9	123	1867	1950	−0.73	−0.56	−1.00
1960	18	211	4545	1960	0.80	0.75	1.35
1970	13	146	1647	1970	−0.05	−0.22	−1.20
1980	17	213	4215	1980	0.63	0.78	1.06
1990	10	132	1502	1990	−0.56	−0.43	−1.33
2000	18	211	2365	2000	0.80	0.75	−0.57
2010	22	256	3779	2010	1.48	1.42	0.68

Note: Standardising resolves the problem by giving them each the same mean and variance.

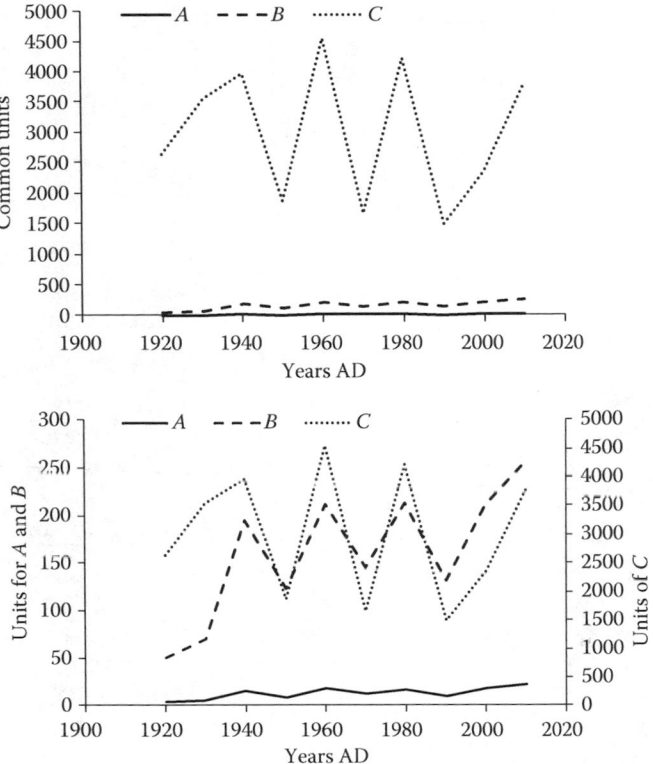

FIGURE 11.19

Three time series with very different mean and variance. Note that if a single vertical axis is used the scale is only appropriate for one of the series, so it is difficult to compare their evolution over time. If the series with the most extreme values is plotted using a secondary vertical axis it gives the impression that series *B* and *C* are quite similar but that series *A* is very different. However, this is only the case because neither axis is really appropriate for plotting the values of series *A*.

One solution, to make it easy to compare them, is to transform the data so that they all have the same mean value and also the same variance. We usually choose to do this by giving them all a mean of zero and a standard deviation of one. This procedure does not change the way that the series evolve across the time or space gradient of interest, it just re-scales them to make them easy to compare. The original units are replaced by units of standard deviation, but they are usually just referred to as *z*-scores.

To *z*-score a series you take the difference between each individual point and the overall mean of the whole series and then divide that by the standard deviation of the whole series

$$z\text{-score} = \frac{\text{Original value} - \text{Mean value}}{\text{Standard deviation}}$$

It is a straightforward procedure in a spreadsheet as long as you use the dollar sign to make sure that each equation refers to the overall mean and standard deviation. The three series in Table 11.1 are transformed into *z*-scores using this procedure, so that they all have a mean of zero and unit variance (standard deviation of one). When the three *z*-scored series are plotted (Figure 11.20) we see a rather different and more realistic picture of the

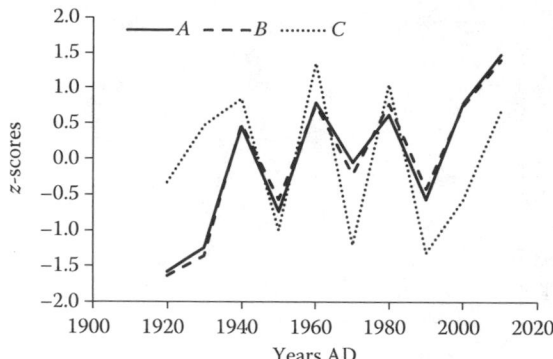

FIGURE 11.20
The same three series plotted as z-scores. Note that when the differences in mean values and variance are removed it is clear that it is series *A* and *B* that are most similar in terms of the way that they evolve over time.

similarities and differences in evolution over time. Now it is clear that series *A* and *B* evolve in almost exactly the same way and it is series *C* that is different.

When you have a series that has been z-scored, so that it has a mean of zero and unit variance, it is possible to give it whatever mean and variance you would like it to have. When the aim of your research is to translate the units of one set of data into the units of another, for example when we use 'proxies' such as tree-ring widths to reconstruct past summer temperature, then this procedure provides an alternative to regression. You first have to decide what mean and standard deviation you would like, and the obvious choice is to aim to give your tree-ring series the same mean and variance as the climate data that you are trying to replicate over the time period for which you have meteorological data. The first step is to z-score your tree ring or other proxy series, but rather than using the mean and standard deviation values for the whole record, use the values for the period that overlaps with your measured climate data. Now your z-scored series will have a mean of zero and unit variance over the calibration period (but not the whole series). You can now 'reverse' the z-scoring by adding back the mean and variance taken from the climate data using the formula:

$$\text{Variance scaled value} = (z\text{-score} \times \text{desired standard diviation}) + \text{desired mean}$$

This method has the advantage that there is no loss of variance in the reconstruction. This comes at a cost though, because the simple linear regression line is, by definition, the translation of the units on the *x* axis into the units on the *y* axis that minimises the mean squared error. Any other method of scaling the proxy to fit the climate data, including variance scaling as described here, must result in an increase in the mean squared error and therefore in the standard error of the prediction. Deciding whether to use regression, which minimises the error but underestimates the variability, or to match the variability but increase the error is quite a complicated problem. The lower the correlation between your proxy and climate the more variance you lose in regression, but if you use variance scaling on such data you end up with more noise than signal. The sensible cut-off for variance scaling seems to be an *r* value of 0.5; any lower and a horizontal line will give you a lower mean squared error (McCarroll et al. 2015). Using variance scaling to define the relationship between two variables is identical to using reduced major axis regression (Section 11.10).

It is not difficult to normalise data in a spreadsheet, but for convenience a simple calcula-tor is provided on the companion site. Simply enter up to 12 series, with sample sizes up to 300, and the z-scored series are provided.

11.10 Reduced Major Axis Regression

Simple linear regression assumes that one of the parameters is dependent on the other, and in that case the independent parameter is always placed on the horizontal (x) axis and the dependent variable on the vertical (y) axis. It is important that the axes are plotted cor-rectly otherwise the results of the regression analysis are wrong. However, there are cases where neither parameter can logically be considered to be dependent on the other, so the decision about which parameter should go on which axis is arbitrary and there are two ways to plot the data and two possible regression lines.

Where this is the case it is not appropriate to use simple linear regression and an alter-native called 'reduced major axis' (RMA) regression is used instead. It has the advantage that the regression line does not change when the axes are reversed. If we were to use this method to compare age and salary, for example, and estimated the most likely salary at age 30 years to be £40k, and then reversed the axes and calculated the most likely age for someone earning £40k, the best estimate would be 30 years. In Figure 11.5 the two possible regression lines based on simple linear regression are plotted together and the line that falls between them is based on RMA. This form of regression is equivalent to variance scaling, (Section 11.9) which is commonly used to reconstruct past climate using proxy evidence such as tree-ring measurements or indices based on historical documents. The disadvantage of RMA regression or variance scaling, when used for prediction, is that the errors are larger than those based on simple linear regression. The advantage is that there is no consistent underestimation of the variance in the reconstruction, which is inevitable with simple linear regression. The decision about which method to use hinges on how important it is to minimise the error in the reconstruction compared to how important it is to avoid underestimating variability in the reconstruction. It is a complicated problem that is still being debated (McCarroll et al. 2015).

Obtaining the RMA regression equation is not difficult. It is based on the mean and standard deviation of the two series. The tricky bit is plotting the line on a scatter-graph. As with any straight line, the general equation for an RMA regression line is

$$y = a + bx$$

Where a is the intercept and b is the slope.

For RMA regression we first estimate the magnitude of the slope based on the two stan-dard deviations

$$\text{Slope } (b) = \pm \frac{\text{Standard deviation of } y}{\text{Standard deviation of } x}$$

Next you have to decide whether the slope is positive or negative. You do this by look-ing at the data on a scatter-graph. The choice of which parameter to put on which axis is entirely arbitrary, it just determines where the standard deviations go in the equation used to calculate the slope. If it is not obvious whether the relationship is positive or negative then the relationship between the two parameters is almost certainly not significant so it doesn't matter what sign you give. However, a simple solution is to plot a simple linear regression line and see if that has a positive or negative slope.

Having determined the slope, and given it the appropriate sign, the intercept is obtained using this formula

$$\text{Intercept } (a) = \text{Mean of } y - b \times \text{Mean of } x$$

The data used to plot Figure 11.5, for example, has an x-axis mean of 44.80 and standard deviation of 12.17 and a y-axis mean of 155.95 and standard deviation of 13.19, The scatter-graph clearly shows a positive relationship (rising trend) so the slope and intercept of the RMA regression line are

$$\text{Slope } (b) = \pm \frac{13.19}{12.17} = 1.08$$

$$\text{Intercept } (a) = 155.95 - 1.08 \times 44.8 = 155.95 - 48.38 = 107.5$$

So the equation for the RMA regression line is

$$y = 107.5 + 1.08x$$

The easiest way to plot this regression line on a scatter-graph drawn in a spreadsheet is to take the highest and lowest values from the data plotted on the x axis and use the RMA equation to calculate the corresponding values for y. Now add those data to the scatter-graph as a separate series and replace the two points with a straight line. It is a rather fiddly procedure so a simple calculator is provided on the companion site. Enter the two data sets in the columns provided (n up to 500) and it will calculate the equation for the RMA regression line and plot it together with the data on a scatter-graph which you can then edit (Figure 11.21). The sign of the slope coefficient of a simple linear regression is used to determine the sign of the RMA slope coefficient.

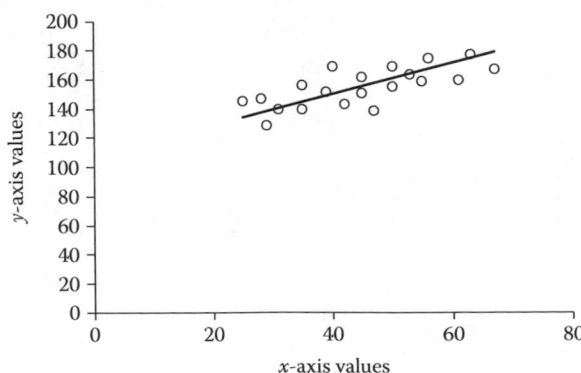

FIGURE 11.21
Example of the output graph from the companion site calculator for reduced major axis regression. Note that in this case you would have to edit the scale of the axes.

11.11 The Durbin–Watson Test for Autocorrelation of Residuals

As the name implies, this test was devised by Durbin and Watson (1950, 1951) to test for autocorrelation in the residuals from regression. It is usually applied to time series data. The residuals are the differences between the best-fit line and the measured values on the y axis. One of the assumptions of regression is that each error value is independent of the others, so that the value obtained in one year should not influence the value in the next year. In many systems that is not the case and there is a 'memory effect' that causes the values of one year to be influenced in some way by the values of the year before. Regression analysis using time series is an important tool in the finance industry and the confounding effect of time is an important source of bias, so people who work in 'econometrics' are very keen on this test. A correlation between the values of each year with the year before is known as first-order autocorrelation. Where there is a concern that there may be trends across space that could impart a bias the same methods can be used by substituting distance for time.

To apply the test by hand you first have to find the residuals from the regression. The regression option in the Analysis Toolpack in Excel has an option that will tabulate them for you. Otherwise, you do this by finding the best estimate of y for each value of x and then taking the difference between the estimate and the real measured value of y. An example will help here. Table 11.2 shows the measurements of a tree-ring index and of summer temperature for a region of northern Fennoscandia (part of the data used by McCarroll et al. 2013). In this case the intention is to use the tree-ring index, which extends for more than a thousand years, to reconstruct past summer temperature, so the tree-ring index is treated as the independent variable and the temperature as the dependent variable. The correlation between the two variables is very strong, and the relationship looks linear (Figure 11.22) so simple linear regression is appropriate. The estimates are obtained using the regression equation shown on the scatter-graph. For 2005, for example, the estimate of y (JJA_{temp}) is based on the tree index value of 1.42

$$\text{Estimated temperature} = 1.01 \times 1.42 + 10.49 = 1.43 + 10.49 = 11.9$$

Having calculated the residuals (e for error), the Durbin–Watson statistic is based on comparing the value in each year (e_t) to the value for the previous year (e_{t-1}). The differences are squared to get rid of the sign. The full equation is

$$DW = \frac{\sum (e_t - e_{t-1})^2}{\sum e_t^2}$$

The values for the two sides of the equation are shown in Table 11.2. Note that the values for the squared difference between each error and the error for the year before ($(e_t - e_{t-1})^2$) do not extend to the bottom of the table because there is no value for 1975 with which to compare. For this example the equation is solved by

$$DW = \frac{12.95}{8.23} = 1.57$$

TABLE 11.2

A Tree-Ring Growth Index and Mean Summer Temperature in Northern
Fennoscandia and the Steps Required to Calculate the Durbin–Watson Statistic

	Trees	JJA_{temp}	Estimate	Error	$(e_t - e_{t-1})^2$	e^2
2005	1.42	12.65	11.93	0.72	0.012	0.52
2004	1.07	12.41	11.57	0.83	0.438	0.70
2003	1.77	12.44	12.27	0.17	0.583	0.03
2002	1.43	12.87	11.94	0.94	0.025	0.87
2001	0.85	12.13	11.35	0.78	1.949	0.60
2000	1.73	11.62	12.24	−0.62	0.609	0.38
1999	0.77	11.43	11.27	0.16	0.041	0.03
1998	0.80	11.26	11.30	−0.04	0.112	0.00
1997	1.59	12.39	12.09	0.30	0.076	0.09
1996	0.66	11.17	11.15	0.02	0.531	0.00
1995	−0.39	10.84	10.09	0.75	1.569	0.56
1994	1.45	11.44	11.95	−0.50	0.053	0.25
1993	0.98	10.74	11.48	−0.73	0.861	0.54
1992	−0.02	10.66	10.47	0.19	0.171	0.04
1991	0.53	11.63	11.02	0.61	0.354	0.37
1990	1.40	11.91	11.90	0.01	0.000	0.00
1989	1.40	11.91	11.90	0.01	0.437	0.00
1988	1.12	12.29	11.62	0.67	2.557	0.45
1987	−0.03	9.53	10.46	−0.93	0.584	0.86
1986	0.92	11.26	11.42	−0.17	0.368	0.03
1985	1.95	11.69	12.46	−0.77	0.560	0.60
1984	0.29	10.76	10.79	−0.02	0.461	0.00
1983	0.90	10.69	11.40	−0.70	0.013	0.49
1982	−0.02	9.87	10.46	−0.59	0.093	0.35
1981	−0.21	9.99	10.27	−0.29	0.003	0.08
1980	1.87	12.14	12.37	−0.23	0.031	0.05
1979	1.48	11.93	11.98	−0.05	0.080	0.00
1978	−0.01	10.71	10.48	0.23	0.378	0.05
1977	0.38	10.49	10.88	−0.39	0.001	0.15
1976	0.65	10.79	11.15	−0.35		0.12
				Sum:	12.95	8.23

Testing the statistical significance of the *DW* statistic is a bit fiddly and involves tables
that have an unusual format. The method is often applied to multiple regression, where
there are several independent variables, so most of the tables of critical values that are
available are very complicated. A simplified version is provided here (12.13). To use the
tables choose your preferred level of significance ($p = 0.05$ or $p = 0.01$) and use the sample
size, which for simple linear regression is the number of pairs (N), to identify the upper
and lower critical values. If your value for *DW* is equal to or higher than the upper critical
value the result is not statistically significant and you can conclude that there is no problem
with positive autocorrelation of residuals. If your value is equal to or lower than the lower
critical value it indicates significant positive autocorrelation of residuals. If your value falls
between the two critical values then the test is inconclusive. In that case just mention it.

The tables of critical values used for the Durbin–Watson statistic are one tail because it
is usually used to test for positive autocorrelation. This means that if one value produces

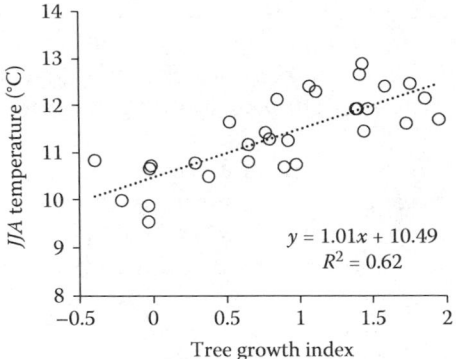

FIGURE 11.22

Scatter-graph showing how summer temperature might be estimated from a tree growth index. Biologically this makes no sense, because temperature does not depend on tree growth, but this is the usual way of using a natural 'proxy' such as tree growth, to reconstruct the climate of the past. The real reconstruction used a much bigger data set and reduced major axis regression to avoid underestimating the variance in the reconstruction.

a positive (or negative) residual the next value is also more likely than not to produce a positive (or negative) residual. Apparently it is also possible for time series to display negative autocorrelation, which would occur if a positive residual is likely to be followed by a negative residual, and vice versa. I cannot think of a geographical example where this is likely to be the case, but if you have a *DW* value that is much higher than two you can use the critical tables by subtracting your *DW* from four (the maximum possible value). Good luck with explaining those results.

11.11.1 How To Do It

If you run your regression using *R* Commander you can give the 'model' (your regression equation) a name and then go to 'models', Numerical diagnostics and choose the Durbin–Watson test. The output includes the *DW* statistic and a probability value. The default setting is a one-tail test for positive first-order autocorrelation, which is usually appropriate. When you run simple linear regression in SPSS you use 'analyze', 'regression' and 'linear' and the tab marked 'statistics' includes the Durbin–Watson option. Unfortunately the output only gives the *DW* value with no probability, so use the tables of critical values supplied here (Section 12.13). The easiest option for this test is to use the Real Statistics Toolpack for Excel. The Durbin–Watson test is one of the options under multiple regression (which also works for simple linear regression). The website includes a very clear explanation and also has tables of critical values. I have only found one free online calculator that works for this test (http://www.wessa.net/slr.wasp), but it does not give a probability. There is also a simple calculator on the companion site that will return the Durbin–Watson statistic.

11.12 Tests for Validation or Verification

When you are using regression on time series data, or across an environmental or cultural gradient, there is a danger that the relationship that you are trying to define may not be stable over time or over the gradient of interest. If your aim in using regression is to use the relationship defined by the straight line to extrapolate back or forward in time, for example, it is useful to check that the relationship is not just strong but also stable over time. A common example in geography is using regression to reconstruct the climate of the past using some proxy that we assume is strongly related to climate. Measures of tree growth, for example, are the most important methods available to reconstruct the summer temperature of the past and that is achieved by defining the relationship between temperature and growth over the recent period, when we have thermometer readings, and then extrapolating that relationship back in time. The reconstruction is only valid if the relationship between temperature and tree growth that we see today is the same as the relationship in the past.

Of course we can never be certain that the relationship between a proxy and climate has always been exactly the same, but we can add considerable confidence in that assumption by using what is known as 'verification'. This means that you define the relationship using a subsample rather than all of your data, and then test how good your predictions are by using the data that were held back. If your data have not been collected in any particular order you can use various random sampling approaches to divide your sample into a portion that is used to define the relationship (calibration) and a portion that is used to test it (verification). This is the essence of what are known as 'bootstrapping' approaches to estimating uncertainty. With time series or spatial gradients, however, random sampling is less powerful than simply splitting the data into two parts, because this allows you to see whether the relationship remains stable over the two parts, and therefore over time or space.

When you use a split-period design to test the stability of your relationship the easiest approach is to divide the data in half, and use one half to build the model (i.e. define the relationship using regression) and the other half to see how well it fits. If, for example, you have 100 years of temperature measurements, you can use the most recent 50 years to define the relationship (calibration) and the earlier 50 years to test it (verification). Then you can swap the two halves and do it again, just to be sure. The clearest way to define how well the relationship defined over the calibration period fits the verification period is to use the MSE, or the RMSE as defined in Section 11.4. A small MSE or RMSE means that the model fits very well and a large MSE or RMSE means that it does not fit very well. The question remains, however, how good is good enough?

There is no really easy answer to that question, but we can at least say that if we go to all of the effort of measuring a lot of proxy data and using it to give a best estimate of the temperature in the past, those estimates should be better than putting a horizontal line through the data and giving every year exactly the same value. That would clearly be a 'nonsense' estimate of the climate of the past. If our real estimate based on regression is no better than the 'nonsense' estimate then it is clearly not worth the effort. There are two useful measures of verification that can be used in this context, and the only difference is that they use a different horizontal line. The *reduction of error* (RE) statistic uses the average measured temperature (average of the *y*-axis values) over the calibration period, and the *coefficient of efficiency* (CE) statistic uses the average measured temperature over the verification period. The measure of 'goodness of fit' or error that they use is the MSE, or the average difference between the line (horizontal or regression) and the measured points.

The verification statistics RE and CE are very closely related to the much more familiar *R*-squared value. When you use all of the available data to produce a regression line, the MSE gives a measure of how well the line fits the data. Clearly the regression line should be better, and therefore give a lower MSE, than a horizontal line that just goes through the average of the measured temperature (y axis) values. The R^2 value is based on the ratio of the MSE based on the regression line and the MSE based on just using the mean for every year. The bigger the difference the better, so we use one minus that ratio, so that if the MSE values are the same, and the MSE is not better than the horizontal line, then you get a value of zero. One (unusual) way of writing the formula for R^2 is thus

$$R^2 = 1 - \frac{\text{MSE based on regression}}{\text{MSE based on the mean of } y}$$

The formulae of the RE and CE statistics are based on the same logic, the only difference is how you define the horizontal line. In RE it is the mean of the y-axis values over the calibration period and in CE it is the mean of the y-axis values in the verification period. Both metrics are calculated using the goodness of fit over the verification period only

$$RE = 1 - \frac{\text{MSE based on regression}}{\text{MSE based on calibration mean of } y}$$

$$CE = 1 - \frac{\text{MSE based on regression}}{\text{MSE based on verification mean of } y}$$

The CE statistic is more stringent than the RE statistic because the mean of the verification period is completely independent of the data used to build the regression model. There is no clear rule about how high the RE and CE statistics need to be to conclude that the regression model is stable over time (or space), but the RE definitely has to be more than zero, otherwise collecting all of that proxy data was a total waste of time, you might as well have just given every year the average of the period over which you have thermometer readings. Personally I think that with time series data the CE statistic should be more than zero as well, but not everyone agrees with that. As a rule of thumb, RE and CE values that approach one are fantastic, values close to zero are terrible and anything less than zero is disastrous.

Calculating the MSE in a spreadsheet, so that you can calculate the RE and CE statistics, is not very convenient, because you have to list all of the measured y values and compare them with the best estimates based on the regression equation. There is a short-cut though, because the spreadsheet will give you the standard error of the prediction based on the regression line, and working backwards we can convert that into the MSE.

The standard error of the prediction is given by the function SEYX. It is the square root of the sum of the squared errors divided by the degrees of freedom, so to obtain the sum of squared errors we can square it and multiply it by the degrees of freedom ($n - 2$). To translate the sum of squared errors into the MSE we divide by the sample size. The full equation is

$$MSE = \frac{SEYX^2 \times (n - 2)}{n}$$

For example, given a sample of 20 pairs of data if Excel returns the standard error of the prediction (SEYX) as 3.0 then

$$MSE = \frac{3^2 \times (20-2)}{20} = \frac{9 \times 18}{20} = \frac{162}{20} = 8.1$$

I am not aware of any online calculators that will perform the RE and CE statistics, and they are not supported by *R* Commander. If they are in SPSS somewhere they are well hidden. On the companion site there is a simple calculator that will perform both tests simultaneously.

11.13 More Complicated Regression Models

11.13.1 Non-Linear Regression

In simple linear regression, the aim is to fit a straight line through the data, however, it is possible to fit many other kinds of line to a set of points in a scatter-graph. Figure 11.23 displays the options that are available in Excel, together with the equation that describes the line. It is very tempting to try fitting one of these more complicated lines to your data, not least because they nearly always result in an increase in the R^2 value, giving the impression that they are 'better' than a linear fit. However, I would caution against this approach and avoid using non-linear regression unless you have a very clear theoretical reason for assuming that a particular type of curve is more logical than a straight line.

If you do use a curve, then use a simple one and check first to see if the relationship really is non-linear by using the number of runs test (Box 5.2). There is a calculator on the companion site that will do this for you. If you are dealing with relationships that are really non-linear you should reflect on why you are using regression analysis. If all you really want to do is show that the relationship is not linear, but still monotonic (not rising then falling) then you can fit the line but do not rely on the R^2 as a measure of the strength of the correlation; use Spearman's rho (section 10.5). The results are much easier to interpret and to defend in your discussion.

11.13.2 Multiple Linear Regression

In this chapter I have focussed on using regression to fit a line, which has one dimension, through a set of points that represents pairs of data plotted on two orthogonal axes (at right angles to each other). Mathematically there is no reason why the procedure cannot be scaled-up and applied to problems with more than two dimensions. For example, where there are three axes rather than two then the points do not plot in a single plane, they form a three-dimensional cloud, and regression can be used to find the two-dimensional surface that minimises the distance between the points and the surface. I am reliably informed that, in terms of the mathematics involved, there is no reason to stop at just two or three dimensions, but the very thought of it makes my brain hurt. I can use multiple regression, because it is quite easy in both *R* Commander and in SPSS, but I cannot honestly claim to fully understand it, placing multiple regression, along with many other things, beyond the scope of this book.

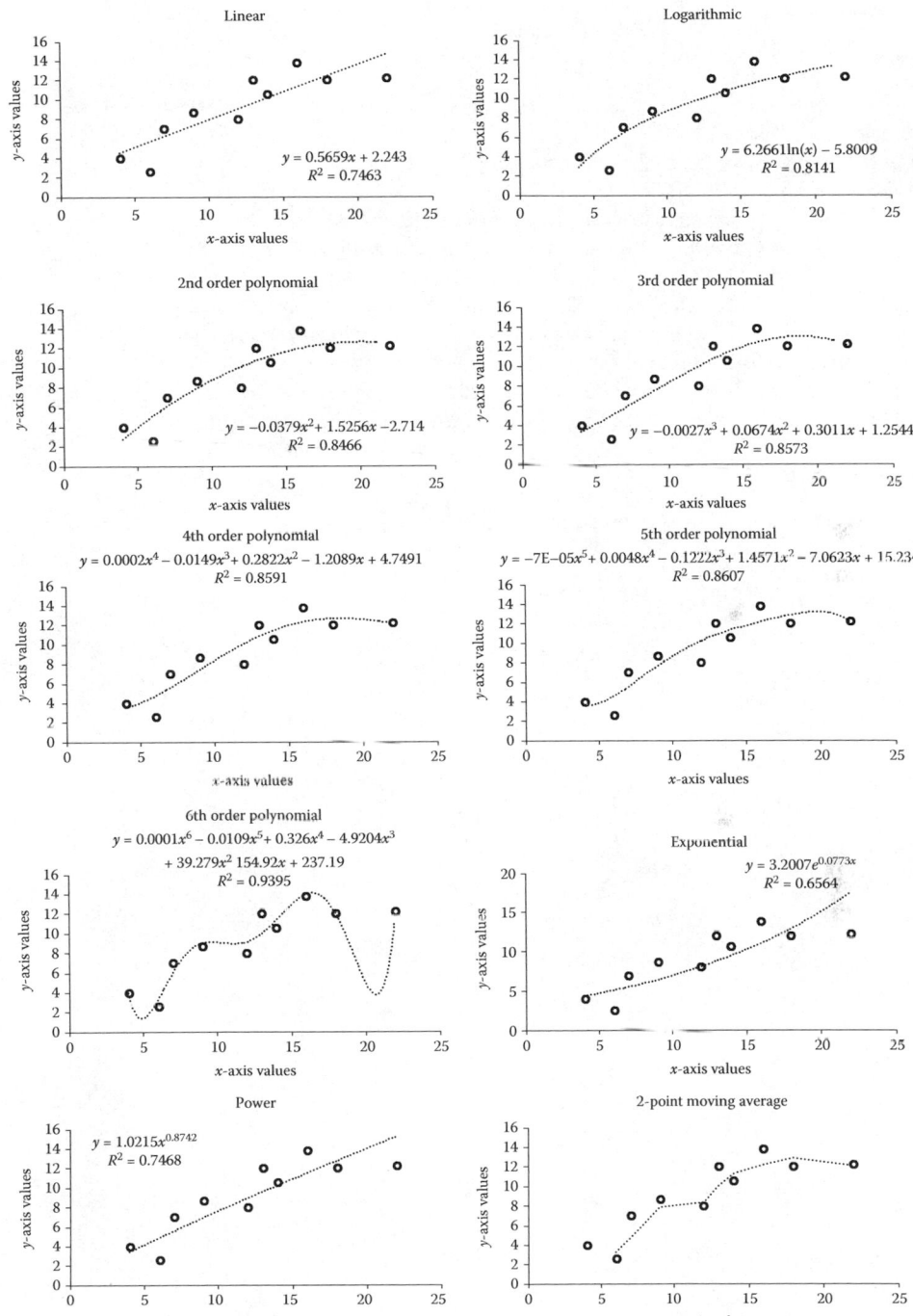

FIGURE 11.23
The options for fitting a 'trendline' in Excel. In this case there is no justification for using anything other than a straight line.

References

Durbin, J. and Watson, G. S. 1950. Testing for serial correlation in least squares regression, I. *Biometrika* 37: 409–428.

Durbin, J. and Watson, G. S. 1951. Testing for serial correlation in least squares regression, II. *Biometrika* 38(1–2): 159–179.

Field, A. 2013. *Discovering statistics using IBM SPSS statistics* (4th ed.). London: Sage, 916pp.

Field, A., Miles, J. and Field, Z. 2012. *Discovering statistics using R*. London: Sage, 958pp.

Fox, J. 2002. *An R and S-PLUS companion to applied regression*. London: Sage, 312pp.

McCarroll, D., Loader, N. J., Jalkanen, R., Gagen, M. H., Grudd, H., Gunnarson, B. E., Kirchhefer, A. J. et al. 2013. A 1200-year multiproxy record of tree growth and summer temperature at the northern pine forest limit of Europe. *Holocene* 23: 471–484.

McCarroll, D., Young, G. H. F., and Loader, N. J. 2015. Measuring the skill of variance-scaled climate reconstructions and a test for the capture of extremes. *Holocene* 25: 618–626.

12

Tables of Critical Values

![chapter rule]

12.1 Sign Test

	Two-Tail			Two-Tail			One-Tail			One-Tail	
$n \backslash p$	0.05	0.01	$n \backslash p$	0.05	0.01	$n \backslash p$	0.05	0.01	$n \backslash p$	0.05	0.01
5	–	–	28	8	6	5	0	–1	28	9	7
6	0	–	29	8	7	6	0	–1	29	9	7
7	0	–	30	9	7	7	0	0	30	10	8
8	0	0	31	9	7	8	1	0	31	10	8
9	1	0	32	9	8	9	1	0	32	10	8
10	1	0	33	10	8	10	1	0	33	11	9
11	1	0	34	10	9	11	2	1	34	11	9
12	2	1	35	11	9	12	2	1	35	12	10
13	2	1	36	11	9	13	3	1	36	12	10
14	2	1	37	12	10	14	3	2	37	13	10
15	3	2	38	12	10	15	3	2	38	13	11
16	3	2	39	12	11	16	4	2	39	13	11
17	4	2	40	13	11	17	4	3	40	14	12
18	4	3	41	13	11	18	5	3	41	14	12
19	4	3	42	14	12	19	5	4	42	15	13
20	5	3	43	14	12	20	5	4	43	15	13
21	5	4	44	15	13	21	6	4	44	16	13
22	5	4	45	15	13	22	6	5	45	16	14
23	6	4	46	15	13	23	7	5	46	16	14
24	6	5	47	16	14	24	7	5	47	17	15
25	7	5	48	16	14	25	7	6	48	17	15
26	7	6	49	17	15	26	8	6	49	18	15
27	7	6	50	17	15	27	8	7	50	18	16

Note: Reject the null hypothesis if the smaller of your two values is less than or equal to the tabulated value for your sample size (n). If in doubt use the two-tail table.

12.2 Chi-Square Distribution

12.2.1 Critical Values of χ^2 (Two-Tail)

$df \backslash p$	0.05	0.01	0.001	$df \backslash p$	0.05	0.01	0.001
1	3.841	6.635	10.828	26	38.885	45.642	54.052
2	5.991	9.210	13.816	27	40.113	46.963	55.476
3	7.815	11.345	16.266	28	41.337	48.278	56.892
4	9.488	13.277	18.467	29	42.557	49.588	58.301
5	11.070	15.086	20.515	30	43.773	50.892	59.703
6	12.592	16.812	22.458	31	44.985	52.191	61.098
7	14.067	18.475	24.322	32	46.194	53.486	62.487
8	15.507	20.090	26.124	33	47.400	54.776	63.870
9	16.919	21.666	27.877	34	48.602	56.061	65.247
10	18.307	23.209	29.588	35	49.802	57.342	66.619
11	19.675	24.725	31.264	36	50.998	58.619	67.985
12	21.026	26.217	32.909	37	52.192	59.893	69.346
13	22.362	27.688	34.528	38	53.384	61.162	70.703
14	23.685	29.141	36.123	39	54.572	62.428	72.055
15	24.996	30.578	37.697	40	55.758	63.691	73.402
16	26.296	32.000	39.252	41	56.942	64.950	74.745
17	27.587	33.409	40.790	42	58.124	66.206	76.084
18	28.869	34.805	42.312	43	59.304	67.459	77.419
19	30.144	36.191	43.820	44	60.481	68.710	78.750
20	31.410	37.566	45.315	45	61.656	69.957	80.077
21	32.671	38.932	46.797	46	62.830	71.201	81.400
22	33.924	40.289	48.268	47	64.001	72.443	82.720
23	35.172	41.638	49.728	48	65.171	73.683	84.037
24	36.415	42.980	51.179	49	66.339	74.919	85.351
25	37.652	44.314	52.620	50	67.505	76.154	86.661

Note: Reject the null hypothesis if χ^2 is greater than or equal to the tabulated value.

12.2.2 Critical Values of χ^2 for Two-Sample Chi-Square Tests

	One-Tail Test			Two-Tail Test		
$p =$	0.05	0.01	0.001	0.05	0.01	0.001
χ^2	2.71	5.42	9.55	3.84	6.64	10.83

Note: Reject the null hypothesis if χ^2 is greater than or equal to the tabulated value.

12.3 Kolmogorov–Smirnov One-Sample Test

	Two-Tail Tests						One-Tail Tests				
$n\backslash p$	0.05	0.01	$n\backslash p$	0.05	0.01	$n\backslash p$	0.05	0.01	$n\backslash p$	0.05	0.01
1	0.98	1.00	26	0.26	0.31	1	0.95	0.98	26	0.24	0.28
2	0.84	0.93	27	0.26	0.31	2	0.78	0.87	27	0.23	0.28
3	0.71	0.83	28	0.25	0.30	3	0.64	0.75	28	0.23	0.27
4	0.62	0.73	29	0.25	0.30	4	0.56	0.66	29	0.22	0.26
5	0.57	0.67	30	0.24	0.29	5	0.51	0.60	30	0.22	0.26
6	0.52	0.62	31	0.24	0.29	6	0.47	0.56	31	0.22	0.26
7	0.49	0.58	32	0.24	0.28	7	0.44	0.52	32	0.22	0.25
8	0.46	0.54	33	0.23	0.28	8	0.41	0.49	33	0.21	0.25
9	0.43	0.51	34	0.23	0.27	9	0.39	0.46	34	0.21	0.25
10	0.41	0.49	35	0.23	0.27	10	0.37	0.44	35	0.21	0.25
11	0.39	0.47	36	0.23	0.27	11	0.35	0.42	36	0.21	0.24
12	0.38	0.45	37	0.22	0.26	12	0.34	0.40	37	0.20	0.24
13	0.36	0.43	38	0.22	0.26	13	0.33	0.39	38	0.20	0.23
14	0.35	0.42	39	0.21	0.25	14	0.31	0.37	39	0.19	0.23
15	0.34	0.40	40	0.21	0.25	15	0.30	0.36	40	0.19	0.23
16	0.33	0.39	41	0.21	0.25	16	0.30	0.35	41	0.19	0.22
17	0.32	0.38	42	0.21	0.25	17	0.29	0.34	42	0.19	0.22
18	0.31	0.37	43	0.20	0.24	18	0.28	0.33	43	0.18	0.22
19	0.30	0.36	44	0.20	0.24	19	0.27	0.32	44	0.18	0.22
20	0.29	0.36	45	0.20	0.24	20	0.26	0.32	45	0.18	0.22
21	0.29	0.35	46	0.20	0.24	21	0.26	0.31	46	0.18	0.21
22	0.28	0.34	47	0.20	0.24	22	0.25	0.30	47	0.18	0.21
23	0.28	0.33	48	0.19	0.23	23	0.25	0.30	48	0.17	0.21
24	0.27	0.33	49	0.19	0.23	24	0.24	0.29	49	0.17	0.21
25	0.27	0.32	50	0.19	0.23	25	0.24	0.29	50	0.17	0.21

Note: Reject the null hypothesis if D is greater than or equal to the tabulated value for your sample size (n). If in doubt use the two-tail table.

12.4 One Sample Number of Runs Test for Randomness

12.4.1 One-Tail Test, $p = 0.05$ Small/Unequal Sample Sizes

n	4	5	6	7	8	9	10	11	12	13	14	15	16	17	18	19	20	n
	2	2	3	3	3	3	3	3	4	4	4	4	4	4	4	4	4	4
	3	3	3	3	4	4	4	4	4	5	5	5	5	5	5	5	5	5
4	8		3	4	4	4	5	5	5	5	5	6	6	6	6	6	6	6
5	9	9		4	4	5	5	5	6	6	6	6	6	7	7	7	7	7
6	9	10	11		5	5	6	6	6	6	7	7	7	7	8	8	8	8
7	9	10	11	12		6	6	6	7	7	7	8	8	8	8	8	9	9
8	–	11	12	13	13		6	7	7	8	8	8	8	9	9	9	9	10
9	–	11	12	13	14	14		7	8	8	8	9	9	9	10	10	10	11
10	–	11	12	13	14	15	16		8	9	9	9	10	10	10	10	11	12
11	–	–	13	14	15	15	16	17		9	9	10	10	10	11	11	11	13
12	–	–	13	14	15	16	17	17	18		10	10	11	11	11	12	12	14
13	–	–	13	14	15	16	17	18	18	19		11	11	11	12	12	12	15
14	–	–	13	14	16	17	17	18	19	20	20		11	12	12	13	13	16
15	–	–	–	15	16	17	18	19	19	20	21	21		12	13	13	13	17
16	–	–	–	15	16	17	18	19	20	21	21	22	23		13	14	14	18
17	–	–	–	15	16	17	18	19	20	21	22	22	23	24		14	14	19
18	–	–	–	15	16	18	19	20	21	21	22	23	24	24	25		15	20
19	–	–	–	15	16	18	19	20	21	22	23	23	24	25	25	26		
20	–	–	–	–	17	18	19	20	21	22	23	24	25	25	26	27	27	

12.4.2 One-Tail Test, $p = 0.01$ Small/Unequal Sample Sizes

n	4	5	6	7	8	9	10	11	12	13	14	15	16	17	18	19	20	n
	–	–	–	2	2	2	2	2	3	3	3	3	3	3	3	3	3	4
	–	2	2	2	2	3	3	3	3	3	3	4	4	4	4	4	4	5
4	9		2	3	3	3	3	4	4	4	4	4	4	5	5	5	5	6
5	9	10		3	3	4	4	4	4	5	5	5	5	5	5	6	6	7
6	–	11	12		4	4	4	5	5	5	5	5	6	6	6	6	6	8
7	–	11	12	13		4	5	5	5	6	6	6	6	7	7	7	7	9
8	–	–	13	14	14		5	5	6	6	6	7	7	7	7	8	8	10
9	–	–	13	14	15	16		6	6	6	7	7	7	8	8	8	8	11
10	–	–	–	15	15	16	17		7	7	7	8	8	8	8	9	9	12
11	–	–	–	15	16	17	18	18		7	8	8	8	9	9	9	10	13
12	–	–	–	15	16	17	18	19	19		8	8	9	9	9	10	10	14
13	–	–	–	–	17	18	19	19	20	21		9	9	10	10	10	11	15
14	–	–	–	–	17	18	19	20	21	21	22		10	10	10	11	11	16
15	–	–	–	–	17	18	19	20	21	22	23	23		10	11	11	11	17
16	–	–	–	–	17	18	20	21	22	22	23	24	24		11	12	12	18
17	–	–	–	–	–	19	20	21	22	23	24	24	25	26		12	12	19
18	–	–	–	–	–	19	20	21	22	23	24	25	26	26	27		13	20
19	–	–	–	–	–	19	20	22	23	24	24	25	26	27	27	28		
20	–	–	–	–	–	19	22	23	24	25	26	26	27	28	29	29		

Note: Reject the null hypothesis if the values are lower than or equal to the lower threshold (upper right side) or greater than or equal to the upper threshold (lower left side).

12.4.3 Two-Tail Test, $p = 0.05$ Small/Unequal Sample Sizes

n	4	5	6	7	8	9	10	11	12	13	14	15	16	17	18	19	20	n
	−	2	2	2	3	3	3	3	3	3	3	3	4	4	4	4	4	4
		2	3	3	3	3	3	4	4	4	4	4	4	4	5	5	5	5
4	9		3	3	3	4	4	4	4	5	5	5	5	5	5	6	6	6
5	9	10		3	4	4	5	5	5	5	5	6	6	6	6	6	6	7
6	9	10	11		4	5	5	5	6	6	6	6	6	7	7	7	7	8
7	−	11	12	13		5	5	6	6	6	7	7	7	7	8	8	8	9
8	−	11	12	13	14		6	6	7	7	7	7	8	8	8	8	9	10
9	−	−	13	14	14	15		7	7	7	8	8	8	9	9	9	9	11
10	−	−	13	14	15	16	16		7	8	8	8	9	9	9	10	10	12
11	−	−	13	14	15	16	17	17		8	9	9	9	10	10	10	10	13
12	−	−	13	14	16	16	17	18	19		9	9	10	10	10	11	11	14
13	−	−	−	15	16	17	18	19	19	20		10	10	11	11	11	12	15
14	−	−	−	15	16	17	18	19	20	20	21		11	11	11	12	12	16
15	−	−	−	15	16	18	18	19	20	21	22	22		11	12	12	13	17
16	−	−	−	−	17	18	19	20	21	21	22	23	23		12	13	13	18
17	−	−	−	−	17	18	19	20	21	22	23	23	24	25		13	13	19
18	−	−	−	−	17	18	19	20	21	22	23	24	25	25	26		14	20
19	−	−	−	−	17	18	20	21	22	23	23	24	25	26	26	27		
20	−	−	−	−	17	18	20	21	22	23	24	25	25	26	27	27	28	

12.4.4 Two-Tail Test, $p = 0.01$ Small/Unequal Sample Sizes

n	4	5	6	7	8	9	10	11	12	13	14	15	16	17	18	19	20	n
	−	−	−	−	2	2	2	2	2	2	2	3	3	3	3	3	3	4
		−	2	2	2	2	3	3	3	3	3	3	3	3	4	4	4	5
4	9		2	2	3	3	3	3	3	3	4	4	4	4	4	4	4	6
5	−	11		3	3	3	3	4	4	4	4	4	5	5	5	5	5	7
6		11	12		3	3	4	4	4	5	5	5	5	5	6	6	6	8
7	−	−	13	13		4	4	5	5	5	5	6	6	6	6	6	7	9
8	−	−	13	14	15		5	5	5	5	6	6	6	7	7	7	7	10
9	−	−	−	15	15	16		5	6	6	6	7	7	7	7	8	8	11
10	−	−	−	15	16	17	17		6	6	7	7	7	8	8	8	8	12
11	−	−	−	15	16	17	18	19		7	7	7	8	8	8	9	9	13
12	−	−	−	−	17	18	19	19	20		7	8	8	8	9	9	9	14
13	−	−	−	−	17	18	19	20	21	21		8	9	9	9	10	10	15
14	−	−	−	−	17	18	19	20	21	22	23		9	9	10	10	10	16
15	−	−	−	−	−	19	20	21	22	22	23	24		10	10	10	11	17
16	−	−	−	−	−	19	20	21	22	23	24	24	25		11	11	11	18
17	−	−	−	−	−	19	20	22	22	23	24	25	26	26		11	12	19
18	−	−	−	−	−	−	21	22	23	24	25	25	26	27	27		12	20
19	−	−	−	−	−	−	21	22	23	24	25	26	27	27	28	29		
20	−	−	−	−	−	−	21	22	23	24	25	26	27	28	29	29	30	

Note: Reject the null hypothesis if the values are lower than or equal to the lower threshold (upper right side) or greater than or equal to the upper threshold (lower left side).

12.4.5 One-Tail Test, Equal Sample Sizes

	$p = 0.05$			$p = 0.01$	
n	Lower	Upper	n	Lower	Upper
21	16	28	21	14	30
22	17	29	22	14	32
23	17	31	23	15	33
24	18	32	24	16	34
25	19	33	25	17	35
26	20	34	26	18	36
27	21	35	27	19	37
28	22	36	28	19	39
29	23	37	29	20	40
30	24	38	30	21	41
31	25	39	31	22	42
32	25	41	32	23	43
33	26	42	33	24	44
34	27	43	34	25	45
35	28	44	35	25	47
36	29	45	36	26	48
37	30	46	37	27	49
38	31	47	38	28	50
39	32	48	39	29	51
40	33	49	40	30	52
41	34	50	41	31	53
42	35	51	42	31	55
43	35	53	43	32	56
44	36	54	44	33	57
45	37	55	45	34	58
46	38	56	46	35	59
47	39	57	47	36	60
48	40	58	48	37	61
49	41	59	49	38	62
50	42	60	50	38	64

Note: Reject the null hypothesis if the values are lower than or equal to the lower threshold or greater than or equal to the upper threshold.

12.4.6 Two-Tail Test, Equal Sample Sizes

	Two-Tail $p = 0.05$			Two-Tail $p = 0.01$	
n	Lower	Upper	n	Lower	Upper
21	15	29	21	13	31
22	16	30	22	14	32
23	16	32	23	14	34
24	17	33	24	15	35
25	18	34	25	16	36
26	19	35	26	17	37
27	20	36	27	18	38
28	21	37	28	19	39
29	22	38	29	19	41
30	23	39	30	20	42
31	23	41	31	21	43
32	24	42	32	22	44
33	25	43	33	23	45
34	26	44	34	24	46
35	27	45	35	24	48
36	28	46	36	25	49
37	29	47	37	26	50
38	30	48	38	27	51
39	30	50	39	28	52
40	31	51	40	29	53
41	32	52	41	29	55
42	33	53	42	30	56
43	34	54	43	31	57
44	35	55	44	32	58
45	36	56	45	33	59
46	37	57	46	34	60
47	38	58	47	35	61
48	38	60	48	36	62
49	39	61	49	36	64
50	40	62	50	37	65

Note: Reject the null hypothesis if the values are lower than or equal to the lower threshold or greater than or equal to the upper threshold.

12.5 Kolmogorov–Smirnov Two-Sample Test

12.5.1 Two-Tail Tests, Small Unequal Sizes n = 5–20

n	5	6	7	8	9	10	11	12	13	14	15	16	17	18	19	20
							$p = 0.05$ (critical values × 100)									
5		82	79	77	76	74	73	72	71	71	70	70	69	69	68	68
6	99		75	73	72	70	69	68	67	66	66	65	64	64	64	63
7	95	91		70	68	67	66	65	64	63	62	61	61	60	60	60
8	93	88	84		66	64	63	62	61	60	59	59	58	58	57	57
9	91	86	82	79		62	61	60	59	58	57	57	56	55	55	54
10	89	84	80	77	75		59	58	57	56	55	55	54	54	53	53
11	88	83	79	76	73	71		57	56	55	54	53	53	52	51	51
12	87	81	77	74	72	70	68		54	53	53	52	51	51	50	50
13	86	80	76	73	71	68	67	65		52	51	51	50	49	49	48
14	85	79	75	72	70	67	66	64	63		50	50	49	48	48	47
15	84	79	74	71	69	66	65	63	62	60		49	48	47	47	46
16	83	78	74	70	68	66	64	62	61	60	58		47	47	46	46
17	83	77	73	70	67	65	63	61	60	59	58	57		46	45	45
18	82	77	72	69	66	64	62	61	59	58	57	56	55		45	44
19	82	76	72	69	66	64	62	60	59	57	56	55	54	54		43
20	81	76	71	68	65	63	61	59	58	57	56	55	54	53	52	
							$p = 0.01$ (critical values × 100)									

12.5.2 Two-Tail Tests, Small Unequal Sizes n = 15–30

n	15	16	17	18	19	20	21	22	23	24	25	26	27	28	29	30
							$p = 0.05$ (critical values × 100)									
15		49	48	47	47	46	46	45	45	45	44	44	44	43	43	43
16	58		47	47	46	46	45	45	44	44	43	43	43	43	42	42
17	58	57		46	45	45	44	44	43	43	43	42	42	42	41	41
18	57	56	55		45	44	44	43	43	42	42	42	41	41	41	40
19	56	55	54	54		43	43	42	42	42	41	41	41	40	40	40
20	56	55	54	53	52		42	42	41	41	41	40	40	40	39	39
21	55	54	53	52	52	51		41	41	41	40	40	39	39	39	39
22	54	53	53	52	51	50	50		40	40	40	39	39	39	38	38
23	54	53	52	51	50	50	49	49		40	39	39	39	38	38	38
24	54	53	52	51	50	49	49	48	47		39	38	38	38	37	37
25	53	52	51	50	50	49	48	48	47	47		38	38	37	37	37
26	53	52	51	50	49	48	48	47	47	46	46		37	37	37	36
27	52	51	50	50	49	48	47	47	46	46	45	45		37	36	36
28	52	51	50	49	48	48	47	46	46	45	45	44	44		36	36
29	52	51	50	49	48	47	47	46	45	45	44	44	44	43		35
30	51	50	49	49	48	47	46	46	45	45	44	44	43	43	42	
							$p = 0.01$ (critical values × 100)									

Note: Reject null hypothesis if D is greater than or equal to the tabulated value ÷ 100.

12.5.3 One-Tail Tests, Small Unequal Sizes $n = 5$–20

n	5	6	7	8	9	10	11	12	13	14	15	16	17	18	19	20
						$p = 0.05$ (critical values $\times$ 100)										
5		74	72	70	68	67	66	65	64	64	63	63	62	62	61	61
6	92		68	66	64	63	62	61	60	60	59	59	58	58	57	57
7	89	84		63	62	60	59	58	57	57	56	55	55	54	54	54
8	87	82	79		59	58	57	56	55	54	54	53	52	52	52	51
9	85	80	77	74		56	55	54	53	52	52	51	50	50	49	49
10	83	78	75	72	70		53	52	51	51	50	49	49	48	48	47
11	82	77	73	71	68	66		51	50	49	49	48	47	47	46	46
12	81	76	72	69	67	65	63		49	48	47	47	46	46	45	45
13	80	75	71	68	66	64	62	61		47	46	46	45	45	44	44
14	79	74	70	67	65	63	61	60	58		45	45	44	44	43	43
15	78	73	69	66	64	62	60	59	58	56		44	43	43	42	42
16	78	73	69	66	63	61	59	58	57	56	55		43	42	41	41
17	77	72	68	65	63	61	59	57	56	55	54	53		41	41	40
18	77	72	68	65	62	60	58	57	55	54	53	52	51		40	40
19	76	71	67	64	61	59	58	56	55	53	52	52	51	50		39
20	76	71	67	64	61	59	57	55	54	53	52	51	50	49	49	
						$p = 0.01$ (critical values $\times$ 100)										

12.5.4 One-Tail Tests, Small Unequal Sizes $n = 15$–30

	15	16	17	18	19	20	21	22	23	24	25	26	27	28	29	30
						$p = 0.05$ (critical values $\times$ 100)										
15		44	43	43	42	42	41	41	41	40	40	40	39	39	39	39
16	55		43	42	41	41	41	40	40	39	39	39	39	38	38	38
17	54	53		41	41	40	40	39	39	39	38	38	38	38	37	37
18	53	52	51		40	40	39	39	38	38	38	37	37	37	37	36
19	52	52	51	50		39	39	38	38	38	37	37	37	36	36	36
20	52	51	50	49	49		38	38	37	37	37	36	36	36	36	35
21	51	50	50	49	48	47		37	37	37	36	36	36	35	35	35
22	51	50	49	48	48	47	46		36	36	36	35	35	35	35	34
23	50	49	49	48	47	46	46	45		36	35	35	35	34	34	34
24	50	49	48	47	47	46	45	45	44		35	35	34	34	34	33
25	50	49	48	47	46	46	45	44	44	43		34	34	34	33	33
26	49	48	47	47	46	45	45	44	43	43	43		34	33	33	33
27	49	48	47	46	45	45	44	44	43	43	42	42		33	33	32
28	49	48	47	46	45	44	44	43	43	42	42	41	41		32	32
29	48	47	46	46	45	44	44	43	42	42	41	41	41	40		32
30	48	47	46	45	45	44	43	43	42	42	41	41	40	40	40	
						$p = 0.01$ (critical values $\times$ 100)										

Note: Reject null hypothesis if D is greater than or equal to the tabulated value $\div$ 100.

12.5.5 Two-Tail Tests, Equal Sample Sizes

$n\backslash p$	0.05	0.01	0.001	$n\backslash p$	0.05	0.01	0.001
5	5	5	–	40–42	13	15	18
6	5	6	–	43–45	13	16	19
7	6	6	–	46–47	14	16	19
8	6	7	8	48	14	16	20
9	6	7	9	49–52	14	17	20
10	7	8	9	53	14	17	21
11	7	8	10	54	15	17	21
12	7	8	10	55–58	15	18	21
13	7	9	10	59–61	15	18	22
14–15	8	9	11	62–63	16	19	22
16	8	10	11	64–68	16	19	23
17	8	10	12	69	16	20	23
18–19	9	10	12	70–75	17	20	24
20–22	9	11	13	76	17	21	24
23	10	11	14	77–78	17	21	25
24–25	10	12	14	79	18	21	25
26–27	10	12	15	80–82	18	21	25
28	11	13	15	83	18	21	26
29	11	13	15	84–88	18	22	26
30–32	11	13	16	89	19	22	26
33	12	14	16	90–91	19	22	27
34–37	12	14	17	92–96	19	23	27
38	12	15	17	97–98	19	23	28
39	12	15	18	99–100	20	23	28

Note: Reject null hypothesis if D is greater than or equal to the tabulated value. Do not use these tables for the one-sample test.

12.5.6 One-Tail Tests, Equal Sample Sizes

$n \backslash p$	0.05	0.01	0.001	$n \backslash p$	0.05	0.01	0.001
5	4	5	6				
6–7	5	6	7	43–47	12	15	18
8	5	7	8	48	12	15	19
9–10	6	7	8	49	13	15	19
11	6	8	9	50–52	13	16	19
12	6	8	10	53–55	13	16	20
13–14	7	8	10	56	13	17	20
15	7	9	11	57–58	14	17	20
16	7	9	11	59–63	14	17	21
17	8	9	11	64	14	18	21
18–21	8	10	12	65–66	14	18	22
22–24	9	11	13	67–70	15	18	22
25–26	9	11	14	71–75	15	19	23
27	9	12	14	76–77	16	19	23
28	10	12	14	78	16	19	24
29–31	10	12	15	79–83	16	20	24
32	10	13	15	84–87	16	20	25
33	10	13	16	88–90	17	21	25
34–36	11	13	16	91–96	17	21	26
37	11	14	16	97	17	22	26
38–40	11	14	17	98	18	22	26
41–42	12	14	17	99–105	18	22	27

Note: Reject null hypothesis if D is greater than or equal to the tabulated value. Do not use these tables for the one-sample test.

12.6 Wilcoxon's Matched-Pairs Signed-Ranks Test

	Two-Tail		One-Tail	
$n\backslash p$	0.05	0.01	0.05	0.01
5	–	–	0	–
6	0	–	2	–
7	2	–	3	0
8	3	0	5	1
9	5	1	8	3
10	8	3	10	5
11	10	5	13	7
12	13	7	17	9
13	17	9	21	12
14	21	12	25	15
15	25	15	30	19
16	29	19	35	23
17	34	23	41	27
18	40	27	47	32
19	46	32	53	37
20	52	37	60	43
21	58	42	67	49
22	65	48	75	55
23	73	54	83	62
24	81	61	91	69
25	89	68	100	76
26	98	75	110	84
27	107	83	119	92
28	116	91	130	101
29	126	100	140	110
30	137	109	151	120
31	147	118	163	130
32	159	128	175	140
33	170	138	187	151
34	182	148	200	162
35	195	159	213	173
36	208	171	227	185
37	221	182	241	198
38	235	194	256	211
39	249	207	271	224
40	264	220	286	238
45	343	291	371	312
50	434	373	466	397

Note: Reject the null hypothesis if T is less than or equal to the tabulated value.

12.7 Student's *t*-Tests

12.7.1 Two-Tail *t*-Tests

$df \backslash p$	0.05	0.01	0.001	$df \backslash p$	0.05	0.01	0.001
5	2.57	4.03	6.87	41	2.02	2.70	3.54
6	2.45	3.71	5.96	42	2.02	2.70	3.54
7	2.36	3.50	5.41	43	2.02	2.70	3.53
8	2.31	3.36	5.04	44	2.02	2.69	3.53
9	2.26	3.25	4.78	45	2.01	2.69	3.52
10	2.23	3.17	4.59	46	2.01	2.69	3.51
11	2.20	3.11	4.44	47	2.01	2.68	3.51
12	2.18	3.05	4.32	48	2.01	2.68	3.51
13	2.16	3.01	4.22	49	2.01	2.68	3.50
14	2.14	2.98	4.14	50	2.01	2.68	3.50
15	2.13	2.95	4.07	51	2.01	2.68	3.49
16	2.12	2.92	4.01	52	2.01	2.67	3.49
17	2.11	2.90	3.97	53	2.01	2.67	3.48
18	2.10	2.88	3.92	54	2.00	2.67	3.48
19	2.09	2.86	3.88	55	2.00	2.67	3.48
20	2.09	2.85	3.85	56	2.00	2.67	3.47
21	2.08	2.83	3.82	57	2.00	2.66	3.47
22	2.07	2.82	3.79	58	2.00	2.66	3.47
23	2.07	2.81	3.77	59	2.00	2.66	3.46
24	2.06	2.80	3.75	60	2.00	2.66	3.46
25	2.06	2.79	3.73	61	2.00	2.66	3.46
26	2.06	2.78	3.71	62	2.00	2.66	3.45
27	2.05	2.77	3.69	63	2.00	2.66	3.45
28	2.05	2.76	3.67	64	2.00	2.65	3.45
29	2.05	2.76	3.66	65	2.00	2.65	3.45
30	2.04	2.75	3.65	70	1.99	2.65	3.44
31	2.04	2.74	3.63	75	1.99	2.64	3.43
32	2.04	2.74	3.62	80	1.99	2.64	3.42
33	2.03	2.73	3.61	85	1.99	2.63	3.41
34	2.03	2.73	3.60	90	1.99	2.63	3.40
35	2.03	2.72	3.59	95	1.99	2.63	3.40
36	2.03	2.72	3.58	100	1.98	2.63	3.39
37	2.03	2.72	3.57	150	1.98	2.61	3.36
38	2.02	2.71	3.57	200	1.97	2.60	3.34
39	2.02	2.71	3.56	250	1.97	2.60	3.33
40	2.02	2.70	3.55	300	1.97	2.59	3.32

Note: Reject null hypothesis if absolute *t* value is greater than or equal to the tabulated value. Use degrees of freedom (*df*) not sample size. For a paired test *df* is number of pairs minus one. For independent samples *df* is the combined sample size minus two ($n_1 + n_2 - 2$).

12.7.2 One-Tail *t*-Tests

$df \backslash p$	0.05	0.01	0.001	$df \backslash p$	0.05	0.01	0.001
5	2.02	3.36	5.89	41	1.68	2.42	3.30
6	1.94	3.14	5.21	42	1.68	2.42	3.30
7	1.89	3.00	4.79	43	1.68	2.42	3.29
8	1.86	2.90	4.50	44	1.68	2.41	3.29
9	1.83	2.82	4.30	45	1.68	2.41	3.28
10	1.81	2.76	4.14	46	1.68	2.41	3.28
11	1.80	2.72	4.02	47	1.68	2.41	3.27
12	1.78	2.68	3.93	48	1.68	2.41	3.27
13	1.77	2.65	3.85	49	1.68	2.40	3.27
14	1.76	2.62	3.79	50	1.68	2.40	3.26
15	1.75	2.60	3.73	51	1.68	2.40	3.26
16	1.75	2.58	3.69	52	1.67	2.40	3.25
17	1.74	2.57	3.65	53	1.67	2.40	3.25
18	1.73	2.55	3.61	54	1.67	2.40	3.25
19	1.73	2.54	3.58	55	1.67	2.40	3.25
20	1.72	2.53	3.55	56	1.67	2.39	3.24
21	1.72	2.52	3.53	57	1.67	2.39	3.24
22	1.72	2.51	3.50	58	1.67	2.39	3.24
23	1.71	2.50	3.48	59	1.67	2.39	3.23
24	1.71	2.49	3.47	60	1.67	2.39	3.23
25	1.71	2.49	3.45	61	1.67	2.39	3.23
26	1.71	2.48	3.43	62	1.67	2.39	3.23
27	1.70	2.47	3.42	63	1.67	2.39	3.22
28	1.70	2.47	3.41	64	1.67	2.39	3.22
29	1.70	2.46	3.40	65	1.67	2.39	3.22
30	1.70	2.46	3.39	70	1.67	2.38	3.21
31	1.70	2.45	3.37	75	1.67	2.38	3.20
32	1.69	2.45	3.37	80	1.66	2.37	3.20
33	1.69	2.44	3.36	85	1.66	2.37	3.19
34	1.69	2.44	3.35	90	1.66	2.37	3.18
35	1.69	2.44	3.34	95	1.66	2.37	3.18
36	1.69	2.43	3.33	100	1.66	2.36	3.17
37	1.69	2.43	3.33	150	1.66	2.35	3.15
38	1.69	2.43	3.32	200	1.65	2.35	3.13
39	1.68	2.43	3.31	250	1.65	2.34	3.12
40	1.68	2.42	3.31	300	1.65	2.34	3.12

Note: Reject null hypothesis if absolute *t* value is greater than or equal to the tabulated value. Use degrees of freedom (*df*) not sample size. For a paired test *df* is number of pairs minus one. For independent samples *df* is the combined sample size minus two ($n_1 + n_2 - 2$).

12.8 Mann–Whitney *U*-Test

12.8.1 Small Samples of Equal Size: Sum of Ranks

	Two-Tail		One-Tail	
p	0.05	0.01	0.05	0.01
n				
4	10	–	11	0
5	17	15	19	16
6	26	23	28	24
7	36	32	39	34
8	49	43	51	45
9	62	56	66	59
10	78	71	82	74
11	96	87	100	91
12	115	105	120	109
13	136	125	142	130
14	160	147	166	152
15	184	171	192	176
16	211	196	219	202
17	240	223	249	230
18	270	252	280	259
19	303	283	313	291
20	337	315	348	324

Note: Reject the null hypothesis if the smaller of the two sums of ranks is less than or equal to the tabulated value.

12.8.2 Two-Tail Tests, Small Unequal Samples

$n\backslash n$	2	3	4	5	6	7	8	9	10	11	12	13	14	15	16	17	18	19	20
										$p = 0.05$									
2		–	–	–	–	–	0	0	0	0	1	1	1	1	1	2	2	2	2
3	–		–	0	1	1	2	2	3	3	3	4	5	5	6	6	6	7	8
4	–	–		1	2	3	3	4	5	6	7	8	9	10	11	12	13	13	14
5	–	–	–		3	5	6	7	8	9	11	12	13	14	16	17	18	19	20
6	–	–	0	1		6	8	9	11	13	14	16	17	19	21	22	24	26	27
7	–	–	0	1	3		10	12	14	16	18	20	22	24	26	28	30	31	34
8	–	–	1	2	4	6		15	17	19	22	24	26	29	31	34	36	39	41
9	–	0	1	3	5	7	9		20	23	26	28	31	34	36	39	42	45	48
10	–	0	2	4	6	9	11	13		26	29	33	36	38	42	45	49	52	55
11	–	0	2	5	7	10	13	16	18		33	37	40	44	48	51	55	58	62
12	–	1	3	6	9	12	15	18	21	24		41	45	49	53	57	61	65	68
13	–	1	3	7	10	13	17	20	23	27	31		50	54	59	63	67	72	76
14	–	1	4	8	11	15	19	23	26	30	34	39		59	64	69	73	78	83
15	–	1	5	8	12	16	20	24	28	33	37	41	47		69	75	80	85	90
16	–	2	5	10	13	17	23	26	32	36	42	45	51	55		80	86	92	98
17	–	2	6	10	14	19	24	29	35	39	44	49	54	59	64		93	99	104
18	–	2	6	11	16	20	25	31	37	43	47	52	59	63	70	75		106	112
19	0	3	7	12	17	22	28	34	39	45	51	57	63	69	75	81	87		119
20	0	3	8	13	18	24	31	36	42	48	54	61	67	74	78	85	92	95	
										$p = 0.01$									

Note: Reject the null hypothesis if U is less than or equal to the tabulated value.

12.8.3 One-Tail Tests, Small Unequal Samples

n	2	3	4	5	6	7	8	9	10	11	12	13	14	15	16	17	18	19	20
										$p = 0.05$									
2		–	–	0	0	0	1	1	1	1	2	2	3	3	3	3	3	4	4
3	–		0	1	2	2	3	4	4	5	5	6	7	7	8	9	9	10	11
4	–	–		2	3	4	5	6	7	8	9	11	12	13	13	15	13	17	18
5	–	–	0		5	6	8	9	10	12	13	15	16	18	19	20	22	23	25
6	–	–	1	2		8	10	12	14	16	17	19	21	23	24	26	28	30	32
7	–	0	1	3	4		13	15	17	19	22	24	26	28	30	32	35	37	39
8	–	0	2	4	6	8		18	20	23	26	28	31	33	36	39	41	44	47
9	–	1	3	5	7	9	11		24	27	30	33	37	39	42	45	48	51	54
10	–	1	3	6	8	11	13	16		31	34	38	41	44	48	51	55	58	62
11	–	1	4	7	10	13	15	18	22		38	42	46	50	54	57	61	65	69
12	–	2	5	8	11	14	18	21	24	28		47	51	55	60	64	68	72	76
13	0	2	5	8	12	15	20	23	27	31	35		56	60	65	70	75	79	84
14	0	2	6	10	13	18	22	26	30	34	38	43		66	71	76	82	87	92
15	0	3	7	11	15	19	24	28	32	38	42	46	52		77	83	88	94	100
16	0	3	7	12	16	21	26	30	36	40	46	50	56	61		89	95	101	107
17	0	4	8	12	17	23	28	33	39	44	49	54	60	66	71		102	109	114
18	1	4	9	14	19	24	30	35	41	47	53	59	65	71	76	82		116	123
19	1	4	9	15	20	26	33	38	44	50	56	62	69	75	82	88	94		130
20	1	5	11	16	22	28	35	40	47	53	60	67	73	80	87	93	100	107	
										$p = 0.01$									

Note: Reject the null hypothesis if U is less than or equal to the tabulated value.

12.8.4 Two-Tail Tests, Equal Sample Sizes

$n\backslash p$	0.05	0.01	n	0.05	0.01
4	0	–	28	272	235
5	2	0	29	294	255
6	5	1	30	317	276
7	8	4	31	341	298
8	13	8	32	365	321
9	17	11	33	391	344
10	23	16	34	418	369
11	30	21	35	445	394
12	37	28	36	473	420
13	45	34	37	503	447
14	55	41	38	533	475
15	64	52	39	564	504
16	75	59	40	596	533
17	87	70	41	628	564
18	99	81	42	662	595
19	113	93	43	697	627
20	127	105	44	732	660
21	142	118	45	769	694
22	158	133	46	806	728
23	175	148	47	845	762
24	192	164	48	884	796
25	211	180	49	924	830
26	230	198	50	965	864
27	250	216			

Note: Reject the null hypothesis if U is less than or equal to the tabulated value.

12.8.5 One-Tail Tests, Equal Sample Sizes

n	0.05	0.01	n	0.05	0.01
4	1	–	28	291	250
5	4	1	29	314	271
6	7	3	30	338	293
7	11	6	31	363	315
8	15	10	32	388	339
9	21	14	33	415	363
10	27	19	34	443	388
11	34	25	35	471	414
12	42	31	36	501	441
13	51	38	37	531	469
14	61	47	38	563	498
15	72	57	39	595	528
16	83	65	40	628	558
17	95	76	41	662	590
18	109	88	42	697	622
19	123	101	43	733	655
20	138	114	44	771	689
21	154	128	45	808	724
22	171	143	46	846	760
23	189	158	47	886	797
24	207	175	48	926	835
25	227	192	49	968	873
26	247	211	50	1010	913
27	268	230			

Note: Reject the null hypothesis if U is less than or equal to the tabulated value.

12.9 *F*-Test for Equality of Variance

12.9.1 Two-Tail Tests, Equal Sample Sizes

$n\backslash p$	0.05	0.01	$n\backslash p$	0.05	0.01
10	4.03	6.54	41	1.88	2.3
11	3.72	5.85	42	1.86	2.27
12	3.47	5.32	43	1.85	2.25
13	3.28	4.91	44	1.83	2.23
14	3.12	4.57	45	1.82	2.21
15	2.98	4.30	46	1.81	2.19
16	2.86	4.07	47	1.8	2.17
17	2.76	3.87	48	1.78	2.15
18	2.67	3.71	49	1.77	2.13
19	2.60	3.56	50	1.76	2.11
20	2.53	3.43	55	1.71	2.04
21	2.46	3.32	60	1.67	1.97
22	2.41	3.22	65	1.64	1.92
23	2.36	3.12	70	1.61	1.87
24	2.31	3.04	75	1.58	1.83
25	2.27	2.97	80	1.56	1.80
26	2.23	2.90	85	1.54	1.76
27	2.19	2.84	90	1.52	1.74
28	2.16	2.78	95	1.50	1.71
29	2.13	2.72	100	1.49	1.69
30	2.10	2.67	150	1.38	1.53
31	2.07	2.63	200	1.32	1.44
32	2.05	2.58	250	1.28	1.39
33	2.02	2.54	300	1.25	1.35
34	2.00	2.51	400	1.22	1.29
35	1.98	2.47	500	1.19	1.26
36	1.96	2.44	600	1.17	1.23
37	1.94	2.41	700	1.16	1.22
38	1.92	2.38	800	1.15	1.20
39	1.91	2.35	900	1.14	1.19
40	1.89	2.32	1000	1.13	1.18

Note: Reject the null hypothesis if *F* ratio is greater than or equal to the tabulated value.

12.9.2 One-Tail Tests, Equal Sample Sizes

$n\backslash p$	0.05	0.01	Sig.	0.05	0.01
10	3.18	5.35	41	1.69	2.11
11	2.98	4.85	42	1.68	2.09
12	2.82	4.46	43	1.67	2.08
13	2.69	4.16	44	1.66	2.06
14	2.58	3.91	45	1.65	2.04
15	2.48	3.7	46	1.64	2.02
16	2.4	3.52	47	1.63	2.01
17	2.33	3.37	48	1.62	1.99
18	2.27	3.24	49	1.62	1.98
19	2.22	3.13	50	1.61	1.96
20	2.17	3.03	55	1.57	1.9
21	2.12	2.94	60	1.54	1.85
22	2.08	2.86	65	1.51	1.8
23	2.05	2.78	70	1.49	1.76
24	2.01	2.72	75	1.47	1.73
25	1.98	2.66	80	1.45	1.7
26	1.96	2.6	85	1.43	1.67
27	1.93	2.55	90	1.42	1.64
28	1.9	2.51	95	1.41	1.62
29	1.88	2.46	100	1.39	1.6
30	1.86	2.42	150	1.31	1.47
31	1.84	2.39	200	1.26	1.39
32	1.82	2.35	250	1.23	1.34
33	1.8	2.32	300	1.21	1.31
34	1.79	2.29	400	1.18	1.26
35	1.77	2.26	500	1.16	1.23
36	1.76	2.23	600	1.14	1.21
37	1.74	2.21	700	1.13	1.19
38	1.73	2.18	800	1.12	1.18
39	1.72	2.16	900	1.12	1.17
40	1.7	2.14	1000	1.11	1.16

Note: Reject the null hypothesis if F ratio is greater than or equal to the tabulated value.

12.10 Page's Trend-Test

12.10.1 Three to Six Categories

$n\backslash p$	K = 3		K = 4		K = 5		K = 6	
	0.05	0.01	0.05	0.01	0.05	0.01	0.05	0.01
2	28	–	58	60	103	106	166	173
3	41	42	84	87	150	155	244	252
4	54	55	111	114	197	204	321	331
5	66	68	137	141	244	251	397	409
6	79	81	163	167	291	299	474	486
7	91	93	189	193	338	346	550	563
8	104	106	214	220	384	393	640	625
9	116	119	240	246	431	441	718	701
10	128	131	266	272	477	487	793	777
11	141	144	292	298	523	534	869	852
12	153	156	317	324	570	581	946	928
13	165	169	343	350	616	628	1022	1003
14	178	181	369	376	662	674	1098	1078
15	190	194	394	402	708	721	1174	1153
16	202	206	420	428	754	767	1249	1229
17	215	218	446	453	800	814	1325	1304
18	227	231	471	479	846	860	1401	1379
19	239	243	497	505	892	906	1477	1454
20	251	256	522	531	937	953	1528	1552
21	263	268	547	556	983	999	1603	1628
22	275	280	573	582	1029	1045	1678	1703
23	288	292	598	608	1075	1091	1753	1778
24	300	305	624	633	1121	1138	1828	1854
25	312	317	649	659	1167	1184	1903	1929
26	324	329	675	685	1213	1230	1977	2004
27	337	342	700	710	1258	1276	2052	2080
28	349	354	726	736	1304	1322	2127	2155
29	361	366	751	762	1350	1368	2202	2230
30	373	379	777	787	1396	1414	2276	2305

Note: Reject the null hypothesis if L is equal to or greater than the tabulated value.

12.10.2 Seven to Ten Categories

$n\backslash p$	K = 7		K = 8		K = 9		K = 10	
	0.05	0.01	0.05	0.01	0.05	0.01	0.05	0.01
2	252	261	362	376	500	520	670	696
3	370	382	532	549	736	761	987	1019
4	487	501	701	722	971	999	1301	1339
5	603	620	869	893	1204	1236	1614	1656
6	719	737	1037	1063	1436	1472	1927	1972
7	835	855	1204	1232	1668	1706	2238	2288
8	950	972	1371	1401	1900	1940	2549	2602
9	1065	1088	1537	1569	2131	2174	2859	2915
10	1180	1205	1703	1736	2361	2407	3169	3228
11	1295	1321	1868	1905	2592	2639	3478	3541
12	1410	1437	2035	2072	2822	2872	3788	3852
13	1525	1553	2201	2240	3052	3104	4097	4164
14	1639	1668	2367	2407	3281	3335	4405	4475
15	1754	1784	2532	2574	3511	3567	4714	4786
16	1868	1899	2697	2740	3741	3740	5022	5089
17	1982	2014	2862	2907	3970	4029	5330	5407
18	2097	2130	3028	3073	4199	4260	5638	5717
19	2217	2245	3193	3240	4428	4491	5946	6027
20	2325	2360	3358	3406	4657	4722	6253	6337
21	2439	2475	3523	3572	4886	4952	6561	6647
22	2553	2589	3687	3738	5112	5182	6868	6956
23	2667	2704	3852	3904	5343	5413	7176	7265
24	2781	2819	4017	4070	5572	5643	7483	7574
25	2895	2934	4181	4235	5801	5873	7790	7883
26	3009	3048	4346	4401	6029	6103	8097	8192
27	3123	3163	4511	4567	6257	6332	8404	8501
28	3236	3277	4675	4732	6486	6562	8711	8810
29	3350	3392	4840	4898	6714	6792	9017	9118
30	3464	3506	5004	5063	6942	7021	9324	9426

Source: Extracted from Table 2 of Page, E.B. (1963) Ordered hypotheses for multiple comparisons: A significance test for linear ranks. *Journal of the American Statistical Association* 58, 216–330 by permission of the American Statistical Association, www.amstat.org

Note: Reject the null hypothesis if L is equal to or greater than the tabulated value.

12.11 Pearson's Correlation Coefficient (r-Value)

12.11.1 Two-Tail Probabilities

$n\backslash p$	0.05	0.01	0.001	$n\backslash p$	0.05	0.01	0.001
3	0.997	1.000	1.000	41	0.308	0.398	0.495
4	0.950	0.990	0.999	42	0.304	0.393	0.490
5	0.878	0.959	0.991	43	0.301	0.389	0.484
6	0.811	0.917	0.974	44	0.297	0.384	0.479
7	0.754	0.875	0.951	45	0.294	0.380	0.474
8	0.707	0.834	0.925	46	0.291	0.376	0.469
9	0.666	0.798	0.898	47	0.288	0.372	0.465
10	0.632	0.765	0.872	48	0.285	0.368	0.460
11	0.602	0.735	0.847	49	0.282	0.365	0.456
12	0.576	0.708	0.823	50	0.279	0.361	0.451
13	0.553	0.684	0.801	55	0.266	0.345	0.432
14	0.532	0.661	0.780	60	0.254	0.330	0.414
15	0.514	0.641	0.760	65	0.244	0.317	0.399
16	0.497	0.623	0.742	70	0.235	0.306	0.385
17	0.482	0.606	0.725	75	0.227	0.296	0.372
18	0.468	0.590	0.708	80	0.220	0.286	0.361
19	0.456	0.575	0.693	85	0.213	0.278	0.351
20	0.444	0.561	0.679	90	0.207	0.270	0.341
21	0.433	0.549	0.665	95	0.202	0.263	0.332
22	0.423	0.537	0.652	100	0.197	0.256	0.324
23	0.413	0.526	0.640	110	0.187	0.245	0.310
24	0.404	0.515	0.629	120	0.179	0.234	0.297
25	0.396	0.505	0.618	130	0.172	0.225	0.285
26	0.388	0.496	0.607	140	0.166	0.217	0.275
27	0.381	0.487	0.597	150	0.160	0.210	0.266
28	0.374	0.479	0.588	160	0.155	0.203	0.258
29	0.367	0.471	0.579	170	0.151	0.197	0.250
30	0.361	0.463	0.570	180	0.146	0.192	0.243
31	0.355	0.456	0.562	190	0.142	0.186	0.237
32	0.349	0.449	0.554	200	0.139	0.182	0.231
33	0.344	0.442	0.547	250	0.124	0.163	0.207
34	0.339	0.436	0.539	300	0.113	0.149	0.189
35	0.334	0.430	0.532	350	0.105	0.138	0.175
36	0.329	0.424	0.525	400	0.098	0.129	0.164
37	0.325	0.418	0.519	450	0.092	0.121	0.155
38	0.320	0.413	0.513	500	0.088	0.115	0.147
39	0.316	0.408	0.507	750	0.072	0.094	0.120
40	0.312	0.403	0.501	1000	0.062	0.081	0.104

Note: Reject the null hypothesis if r is greater than or equal to the tabulated value. Note n is the number of pairs, not degrees of freedom.

12.11.2 One-Tail Probabilities

$n\backslash p$	0.05	0.01	0.001	$n\backslash p$	0.05	0.01	0.001
3	0.988	1.000	1.000	41	0.260	0.362	0.469
4	0.900	0.980	0.998	42	0.257	0.358	0.463
5	0.805	0.934	0.986	43	0.254	0.354	0.458
6	0.729	0.882	0.963	44	0.251	0.350	0.453
7	0.669	0.833	0.935	45	0.248	0.346	0.449
8	0.621	0.789	0.905	46	0.246	0.342	0.444
9	0.582	0.750	0.875	47	0.243	0.338	0.439
10	0.549	0.715	0.847	48	0.240	0.335	0.435
11	0.521	0.685	0.820	49	0.238	0.331	0.431
12	0.497	0.658	0.795	50	0.235	0.328	0.427
13	0.476	0.634	0.772	55	0.224	0.313	0.408
14	0.458	0.612	0.750	60	0.214	0.300	0.391
15	0.441	0.592	0.730	65	0.206	0.288	0.376
16	0.426	0.574	0.711	70	0.198	0.278	0.363
17	0.412	0.558	0.694	75	0.191	0.268	0.351
18	0.400	0.543	0.678	80	0.185	0.260	0.340
19	0.389	0.529	0.662	85	0.180	0.252	0.331
20	0.378	0.516	0.648	90	0.174	0.245	0.322
21	0.369	0.503	0.635	95	0.170	0.238	0.313
22	0.360	0.492	0.622	100	0.165	0.232	0.305
23	0.352	0.482	0.610	110	0.158	0.222	0.292
24	0.344	0.472	0.599	120	0.151	0.212	0.279
25	0.337	0.462	0.588	130	0.145	0.204	0.269
26	0.330	0.453	0.578	140	0.140	0.196	0.259
27	0.323	0.445	0.568	150	0.135	0.190	0.250
28	0.317	0.437	0.559	160	0.131	0.184	0.243
29	0.311	0.430	0.550	170	0.127	0.178	0.235
30	0.306	0.423	0.541	180	0.123	0.173	0.229
31	0.301	0.416	0.533	190	0.120	0.169	0.223
32	0.296	0.409	0.526	200	0.117	0.164	0.217
33	0.291	0.403	0.518	250	0.104	0.147	0.195
34	0.287	0.397	0.511	300	0.095	0.134	0.178
35	0.283	0.392	0.504	350	0.088	0.124	0.165
36	0.279	0.386	0.498	400	0.082	0.116	0.154
37	0.275	0.381	0.492	450	0.078	0.110	0.145
38	0.271	0.376	0.486	500	0.074	0.104	0.138
39	0.267	0.371	0.480	750	0.060	0.085	0.113
40	0.264	0.367	0.474	1000	0.052	0.074	0.098

Note: Reject the null hypothesis if *r* is greater than or equal to the tabulated value. Note *n* is the number of pairs, not degrees of freedom.

12.12 Spearman's Rank Correlation Coefficient (Rho)

12.12.1 Two-Tail Probabilities

$n\backslash p$	0.05	0.01	0.001	$n\backslash p$	0.05	0.01	0.001
4	–	–	–	31	0.36	0.46	0.57
5	1.00	–	–	32	0.35	0.45	0.56
6	0.89	1.00	–	33	0.35	0.45	0.55
7	0.79	0.93	1.00	34	0.34	0.44	0.55
8	0.74	0.88	0.98	35	0.34	0.43	0.54
9	0.70	0.83	0.93	36	0.33	0.43	0.53
10	0.65	0.79	0.90	37	0.33	0.42	0.53
11	0.62	0.76	0.87	38	0.32	0.42	0.52
12	0.59	0.73	0.85	39	0.32	0.41	0.51
13	0.56	0.70	0.82	40	0.31	0.41	0.51
14	0.54	0.68	0.80	41	0.31	0.40	0.50
15	0.52	0.65	0.78	42	0.31	0.40	0.50
16	0.50	0.64	0.76	43	0.30	0.39	0.49
17	0.49	0.62	0.74	44	0.30	0.39	0.48
18	0.47	0.60	0.73	45	0.29	0.38	0.48
19	0.46	0.58	0.71	46	0.29	0.38	0.47
20	0.45	0.57	0.69	47	0.29	0.37	0.47
21	0.44	0.56	0.68	48	0.29	0.37	0.47
22	0.43	0.54	0.67	49	0.28	0.37	0.46
23	0.42	0.53	0.65	50	0.28	0.36	0.46
24	0.41	0.52	0.64	55	0.27	0.34	0.43
25	0.40	0.51	0.63	60	0.26	0.33	0.42
26	0.39	0.50	0.62	70	0.24	0.31	0.39
27	0.38	0.49	0.61	75	0.23	0.30	0.37
28	0.38	0.48	0.60	80	0.22	0.29	0.36
29	0.37	0.48	0.59	90	0.21	0.27	0.34
30	0.36	0.47	0.58	100	0.20	0.26	0.33

Note: Reject the null hypothesis if rho is greater than or equal to the tabulated value.

12.12.2 One-Tail Probabilities

$n\backslash p$	0.05	0.01	0.001	$n\backslash p$	0.05	0.01	0.001
4	1.00			31	0.30	0.42	0.54
5	0.90	1.00		32	0.30	0.41	0.53
6	0.83	0.94		33	0.29	0.41	0.53
7	0.71	0.89	1.00	34	0.29	0.40	0.52
8	0.64	0.83	0.95	35	0.28	0.39	0.51
9	0.60	0.78	0.92	36	0.28	0.39	0.50
10	0.56	0.75	0.88	37	0.28	0.38	0.50
11	0.54	0.71	0.85	38	0.27	0.38	0.49
12	0.50	0.68	0.82	39	0.27	0.37	0.49
13	0.48	0.65	0.79	40	0.26	0.37	0.48
14	0.46	0.63	0.77	41	0.26	0.36	0.47
15	0.45	0.60	0.75	42	0.26	0.36	0.47
16	0.43	0.58	0.73	43	0.25	0.36	0.46
17	0.41	0.57	0.71	44	0.25	0.35	0.46
18	0.40	0.55	0.69	45	0.25	0.35	0.45
19	0.39	0.54	0.68	46	0.25	0.34	0.45
20	0.38	0.52	0.66	47	0.24	0.34	0.44
21	0.37	0.51	0.65	48	0.24	0.34	0.44
22	0.36	0.50	0.63	49	0.24	0.33	0.43
23	0.35	0.49	0.62	50	0.24	0.33	0.43
24	0.34	0.48	0.61	55	0.22	0.31	0.41
25	0.34	0.47	0.60	60	0.21	0.30	0.39
26	0.33	0.46	0.59	70	0.20	0.28	0.37
27	0.32	0.45	0.58	75	0.22	0.31	0.41
28	0.32	0.44	0.57	80	0.19	0.26	0.34
29	0.31	0.43	0.56	90	0.17	0.25	0.32
30	0.31	0.43	0.55	100	0.17	0.23	0.31

Note: Reject the null hypothesis if rho is greater than or equal to the tabulated value.

12.13 Durbin–Watson Test

Values of DW equal to or higher than the upper threshold are not significant, values equal to or lower than the lower threshold are significant. For intermediate values the test is inconclusive.

12.13.1 One-Tail Critical Values $p = 0.05$

n	Lower	Upper	n	Lower	Upper
10	0.88	1.32	41	1.45	1.55
11	0.93	1.32	42	1.46	1.55
12	0.97	1.33	43	1.46	1.56
13	1.01	1.34	44	1.47	1.56
14	1.04	1.35	45	1.48	1.57
15	1.08	1.36	46	1.48	1.57
16	1.11	1.37	47	1.49	1.57
17	1.13	1.38	48	1.49	1.58
18	1.16	1.39	49	1.50	1.58
19	1.18	1.40	50	1.50	1.58
20	1.20	1.41	55	1.53	1.60
21	1.22	1.42	60	1.55	1.62
22	1.24	1.43	65	1.57	1.63
23	1.26	1.44	70	1.58	1.64
24	1.27	1.45	75	1.60	1.65
25	1.29	1.45	80	1.61	1.66
26	1.30	1.46	85	1.62	1.67
27	1.32	1.47	90	1.63	1.68
28	1.33	1.48	95	1.64	1.69
29	1.34	1.48	100	1.65	1.69
30	1.35	1.49	110	1.67	1.71
31	1.36	1.50	120	1.69	1.72
32	1.37	1.50	130	1.70	1.73
33	1.38	1.51	140	1.71	1.74
34	1.39	1.51	150	1.72	1.75
35	1.40	1.52	160	1.73	1.75
36	1.41	1.52	170	1.74	1.76
37	1.42	1.53	180	1.74	1.77
38	1.43	1.53	190	1.75	1.77
39	1.43	1.54	200	1.76	1.78
40	1.44	1.54	500	1.85	1.86

12.13.2 One-Tail Critical Values $p = 0.01$

n	Lower	Upper	n	Lower	Upper
10	0.60	1.00	41	1.26	1.35
11	0.65	1.01	42	1.26	1.36
12	0.70	1.02	43	1.27	1.36
13	0.74	1.04	44	1.28	1.37
14	0.78	1.05	45	1.29	1.38
15	0.81	1.07	46	1.30	1.38
16	0.84	1.09	47	1.30	1.39
17	0.87	1.10	48	1.31	1.39
18	0.90	1.12	49	1.32	1.40
19	0.93	1.13	50	1.32	1.40
20	0.95	1.15	55	1.36	1.43
21	0.98	1.16	60	1.38	1.45
22	1.00	1.17	65	1.41	1.47
23	1.02	1.19	70	1.43	1.49
24	1.04	1.20	75	1.45	1.50
25	1.05	1.21	80	1.47	1.52
26	1.07	1.22	85	1.48	1.53
27	1.09	1.23	90	1.50	1.54
28	1.10	1.24	95	1.51	1.55
29	1.12	1.25	100	1.52	1.56
30	1.13	1.26	110	1.54	1.58
31	1.15	1.27	120	1.56	1.60
32	1.16	1.28	130	1.58	1.61
33	1.17	1.29	140	1.60	1.63
34	1.18	1.30	150	1.61	1.64
35	1.20	1.31	160	1.62	1.65
36	1.21	1.31	170	1.63	1.66
37	1.22	1.32	180	1.65	1.67
38	1.23	1.33	190	1.65	1.68
39	1.24	1.34	200	1.66	1.68
40	1.25	1.34	500	1.79	1.80

Source: Extracted from comprehensive tables computed using the TSP4.5 Econometrics package and provided by Stanford University: http://web. stanford.edu/~clint/bench/dwcrit.htm

12.14 Critical Values from the Standard Normal Distribution (Significance of z-Scores)

12.14.1 Single Test Critical Values

	Two-Tail			One-Tail		
$p =$	0.05	0.01	0.001	0.05	0.01	0.001
z-score	1.960	2.576	3.291	1.645	2.326	3.090

Note: Absolute (ignore the sign) z-scores are significant if they are greater than or equal to the tabulated value.

12.14.2 Applying a Bonferroni Correction for Multiple Testing

	Two-Tail			One-Tail		
$n \backslash p$	**0.05**	**0.01**	**0.001**	**0.05**	**0.01**	**0.001**
1	1.960	2.576	3.291	1.645	2.326	3.090
2	2.241	2.807	3.481	1.960	2.576	3.291
3	2.394	2.935	3.588	2.128	2.713	3.403
4	2.498	3.023	3.662	2.241	2.807	3.481
5	2.576	3.090	3.719	2.326	2.878	3.540
6	2.638	3.144	3.765	2.394	2.935	3.588
7	2.690	3.189	3.803	2.450	2.983	3.628
8	2.734	3.227	3.836	2.498	3.023	3.662
9	2.773	3.261	3.865	2.539	3.059	3.692
10	2.807	3.291	3.891	2.576	3.090	3.719
11	2.838	3.317	3.914	2.609	3.118	3.743
12	2.865	3.341	3.935	2.638	3.144	3.765
13	2.891	3.364	3.954	2.665	3.167	3.785
14	2.914	3.384	3.971	2.690	3.189	3.803
15	2.935	3.403	3.988	2.713	3.209	3.820
16	2.955	3.421	4.003	2.734	3.227	3.836
17	2.974	3.437	4.017	2.754	3.245	3.851
18	2.991	3.452	4.031	2.773	3.261	3.865
19	3.008	3.467	4.044	2.790	3.276	3.878
20	3.023	3.481	4.056	2.807	3.291	3.891
25	3.090	3.540	4.107	2.878	3.353	3.944
50	3.291	3.719	4.265	3.090	3.540	4.107

Note: In this table n is the number of tests involved, not the sample size. Reject the null hypothesis if the z-score (ignore the sign) is greater than or equal to the tabulated value.

Index